D0853787

TIME-SAVING TECHNIQUES FOR ARCHITECTURAL CONSTRUCTION DRAWINGS

TIME-SAVING TECHNIQUES FOR ARCHITECTURAL CONSTRUCTION DRAWINGS

Fred Nashed, AIA

VNR VAN NOSTRAND REINHOLD
New York

The information in this book came from many sources, including industry standards, manufacturers' literature, product representatives, and the author's experience. It is represented in good faith, but in spite of the fact that the author and publisher have made every reasonable effort to present accurate and dependable information, they do not guarantee or assume any liability for its accuracy or its applicability to any particular project. It is the responsibility of the user to apply his or her professional judgment in the use of information or drawings contained in this book. It behooves them to consult original sources, seek expert advice as needed, and consult applicable codes and authorities having jurisdiction before using the aforementioned material.

Library of Congress Catalog Card Number 92-18922
ISBN 0-442-00951-8

Printed in the United States of America.

Van Nostrand Reinhold
115 Fifth Avenue
New York, New York 10003

Chapman and Hall
2–6 Boundary Row
London, SE1 8HN, England

Thomas Nelson Australia
102 Dodds Street
South Melbourne 3205
Victoria, Australia

Nelson Canada
1120 Birchmount Road
Scarborough, Ontario MIK 5G4, Canada

16 15 14 13 12 11 10 9 8 7 6 5 4 3 2 1

Library of Congress Cataloging-in-Publication Data

Nashed, Fred.
Time-saving techniques for architectural construction drawings / Fred Nashed.
p. cm.
Includes bibliographical references and index.
ISBN 0-442-00951-8
1. Structural drawing. 2. Architectural drawing. I. Title.
T355.N37 1992
720'.28'4—dc20 92-18922
CIP

To my beloved wife, Gayle, without whose faith in me when I had my doubts and without whose constant assistance, this book would never have come to pass.

CONTENTS

PREFACE

Like most architects, I started as a draftsman. Knowing very little, I was exposed to bits and pieces of construction drawings assigned to me by the Job Captain, who to my uninitiated perception "knew everything." Very rarely did the Job Captain take the time to explain why certain things were done in a certain way. I had to discover the reasons on my own by reading books, attending seminars, wading through the secret language of codes and specifications, asking dumb questions, making mistakes and trying to justify them, and observing construction in progress and trying to hazard a guess about whether shop drawings agreed or conflicted with the construction drawings.

My first major project was a $27-million Bus Maintenance Facility that was entrusted to me in 1978. At the time, I was unfamiliar with the building type (industrial), the building code (had recently moved to the city), the office hierarchy (a large engineering firm that I had joined only four months before), or with the team (a joint-venture with another office that supplied most of the staff). In addition, the project had been on-going for about a year of programming, site visits, meetings with the client, etc., before I joined the firm. I was assigned to finish it because the architect of record decided to start his own practice. To add spice to the situation, the project manager was an electrical engineer. In researching this book, I tried recalling this particular project for use as a model to equip a person to handle any situation he or she might confront.

The author has opted to use the masculine form "he, him" in referring to persons in the text, instead of: "he or she, him or her" to make the narrative easier to read. This does not in any way suggest any gender bias on his part. Women's contributions to the profession are widely acknowledged. This decision was based simply on a choice between an easy-flowing narrative or a halting, difficult to understand one. I feel sure that readers of both sexes will understand that this decision is not intended to offend anyone or deny the existence of distinguished women practitioners among us.

This book is targeted toward all segments of the profession. It addresses the concerns of the fellow professional who would like to shorten the time needed to produce a set of drawings through improved management techniques. It gives insights to interns who aspire to jump-start their careers and prepare for the day when they will be put in charge of a project. It also targets students, explaining in detail how things are done and, more importantly, why. It explains the process of putting a set of construction drawings together—how to start the project, how to monitor progress, how to plan and organize the sheets, and how to manage the team to let each member achieve his best potential. It provides an easy to understand summary of useful information for quick reference when one has to make a quick decision, while also directing the reader to more in-depth references. It also acquaints the reader with office standards, including some unusual schedules. Finally, it provides a set of ready-to-use standard details with explanations under each detail. Because of limited space, wood construction details are not included in this section.

The book identifies the time-consuming tasks that confront the team preparing the construction drawings. It is divided into four chapters. The first provides proven techniques of good management. A well-planned project should not proceed in jerks and lurches ending in a mad scramble to meet impossible deadlines. This chapter provides the means to avoid that situation by using a precise monitoring technique to pace the effort. It shows how to involve the team and gain their cooperation. It describes the sequence of activities and explains standard office procedures. It provides a plan to prevent the most time-consuming phenomenon—a disorganized team with low morale that is duplicating its efforts and leaving chunks of its work for the last minute. It contains an example of a project mock-up. While this chapter provides the tools for good management, the overall success in guiding the team depends to a large extent on the team leader's self-confidence, experience, comportment, and, most of all, on his empathy with the team members.

The second chapter deals with office drafting standards.

Having worked for many firms using different standards to produce a good set of construction drawings, I have adopted standards that work well. This chapter contains schedules and forms and explains methods of assembling details. Reproducible blank forms are included in the appendix. It explains overlay drafting and gives other related information. Drafting standards help produce uniform and organized construction documents. Some offices leave it up to each individual to organize the sheet he or she is working on. The inevitable result is a chaotic set of drawings, with duplicated information, that takes longer to produce because each individual spends a lot of time figuring out where to put each detail and which detail to put next to the last detail drawn. The material included here prevents this from happening.

The third chapter provides information designed to expedite time-consuming tasks such as building code analysis and the proper use of the U.L. book. In addition, it includes specialized information that requires lengthy research. It is written in a concise, abbreviated form and is based on recent developments and industry standards. The topics range from items that affect the design of the building envelope to basic information such as how to figure net and gross areas. Because most architects are too busy to do research, this chapter provides a synopsis, a kind of *Reader's Digest*–type of approach, of some of the more important subjects.

The final chapter contains ready-to-use standard details that can be easily modified and "sticky-backed" to the project sheet. Explanations and pertinent information are written under each detail. Each subsection also includes general guidelines and pertinent data. A well-stocked standard details library has the potential of saving anywhere from 7 percent to 15 percent of the time spent on construction drawings. Large offices have known this fact for a long time. This chapter may be used as a nucleus of such a library.

Because this book is targeted toward professionals and students with a wide range of experience, I ask the more experienced reader's indulgence when you encounter parts that go into detailed explanations of basic information addressed to the less experienced among us. It is in their hands that the future of our profession lies. I sincerely hope that this book will be a help in making life somewhat easier for those individuals involved in the production of construction drawings, especially those who work on fast-track projects. Saving time on repetitious and tedious tasks will enable the team to spend more time on things that really matter such as aesthetics and proper detailing.

ACKNOWLEDGMENTS

I wish to thank those companies and individuals who were gracious enough to grant my requests for permission promptly and unconditionally. I also wish to thank those colleagues who made it possible for me to acquire the experience that made this book possible, especially Ben Brewer, FAIA, Sikes, Jennings, Kelly and Brewer, Houston, TX, and former president of the national AIA; Joe Milton, AIA, of Henry, Milton, Roberts, Houston; Ronald J. Shaw, AIA, and Bill Grady, AIA, of F & S. Partners, Inc., Dallas, TX; Mermod Jaccard, AIA, and Ed Reichert, AIA, of CTJ&D, Architects, Houston; David Graeber, FAIA, of Graeber, Simmons and Cowan, Austin, TX; Bernard Johnson, Inc., for the opportunity to lead the team that prepared the construction drawings for the $27-million award-winning Houston Bus Maintenance Facility; and Lloyd Walter, FAIA, of WRC&P, Inc., Winston-Salem, NC, for the opportunity to work on two multimillion dollar projects

Special thanks are due Donald Stull, FAIA, and David Lee, FAIA, current president of the Boston Society of Architects, for giving me the chance to organize their office manual. This activity and the team's enthusiasm inspired me to write this book. During my employ at Stull and Lee, Inc., I refined some of the team leadership themes described in Chapter 1.

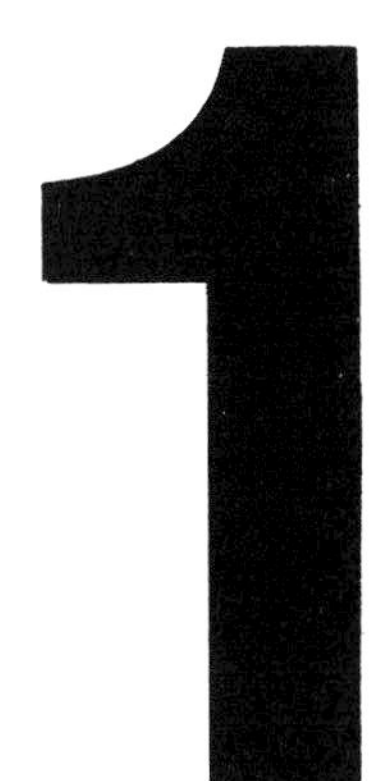

MANAGING THE PROJECT

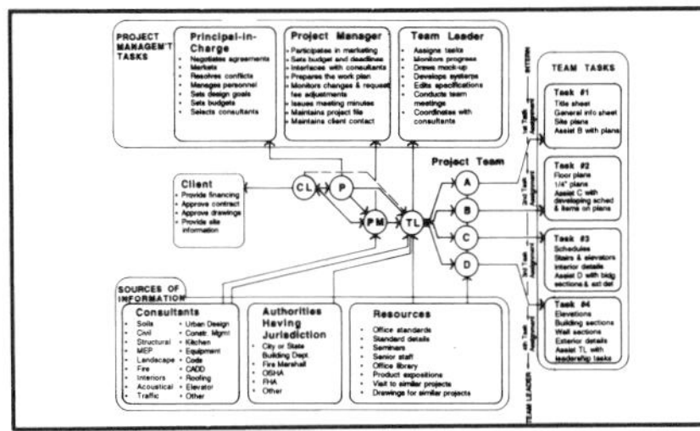

Efficient project management is essential to the time-saving approach to the production of construction drawings. Employing good management techniques ensures that the work proceeds smoothly, raises the morale of the team, and, in most cases, results in getting the project finished on time and within the budget.

In contrast, poorly managed projects proceed in an erratic way, the Project Manager (PM) seems to be fighting brushfires most of the time, deadlines are often missed, and drafting personnel alternate between idleness and overwork. Everybody is relaxed at the start of the project, because team members do not have a clear idea of what needs to be accomplished or what the deadlines are. As these deadlines approach, the team leader hits the panic button and everybody participates in a mad scramble—otherwise referred to by the French term *charrette*—to catch up. Members assigned to other teams are hijacked to do grunt work below their level of experience. This builds up resentments and throws the schedules of the other projects into disarray. Overtime is authorized, affecting the private lives of the staff and cutting into the profitability of the project. Some team leaders perceive this as a way to establish an esprit de corp and introduce excitement. I can think of better ways to instill team spirit and bring excitement into the lives of team members.

Good management employs proven methods to plan the project, staff it properly, monitor its progress, control its budget, and ensure the quality of the work at each step. This must be coupled with a yardstick to measure the performance of team members and a well-defined way for advancement in the office as well as in the profession.

The following are characteristics inherent in good management: a chain of command with well-defined areas of responsibility understood by all members of the staff (Section 1.1); a guideline for planning the project to define the budget, list the tasks, and set deadlines (Section 1.2); a definition of project activities to be performed (Section 1.3); and to provide a mechanism for monitoring progress (Section 1.4).

This chapter deals only with project management issues. It does not address the other prerequisites of successful practice such as design and marketing, which are beyond the scope of this book.

The following sections are based on the assumption that a project manager or a principal or senior architect handles the administrative functions such as the budget, staffing, client issues, record keeping, monitoring progress, etc. Because these individuals are usually experienced in these activities, this chapter touches very lightly on this area. There are many books written on the subject of project management, some of which are listed under "Sources of More Information" at the end of this chapter.

A second assumption in this field, is that the project is a medium or large project that requires an in-house Team Leader. Most of the material in the following sections focuses on the methodology to be used by this individual to manage the team and control the project. This is an area seldom addressed in depth in books devoted to the subject of project management.

The last assumption is that the office is structured around the team approach. This is not always the case. According to the AIA's *Architect's Handbook of Professional Practice*, Volume 1, there are six forms of organization for architectural firms (Fig. 1-1). These forms are shaped by many factors, including a decision by the principal(s), based on the size of the firm, whether it is organized along single or multi-disciplines, and the type of clients and projects it specializes in as well as other factors.

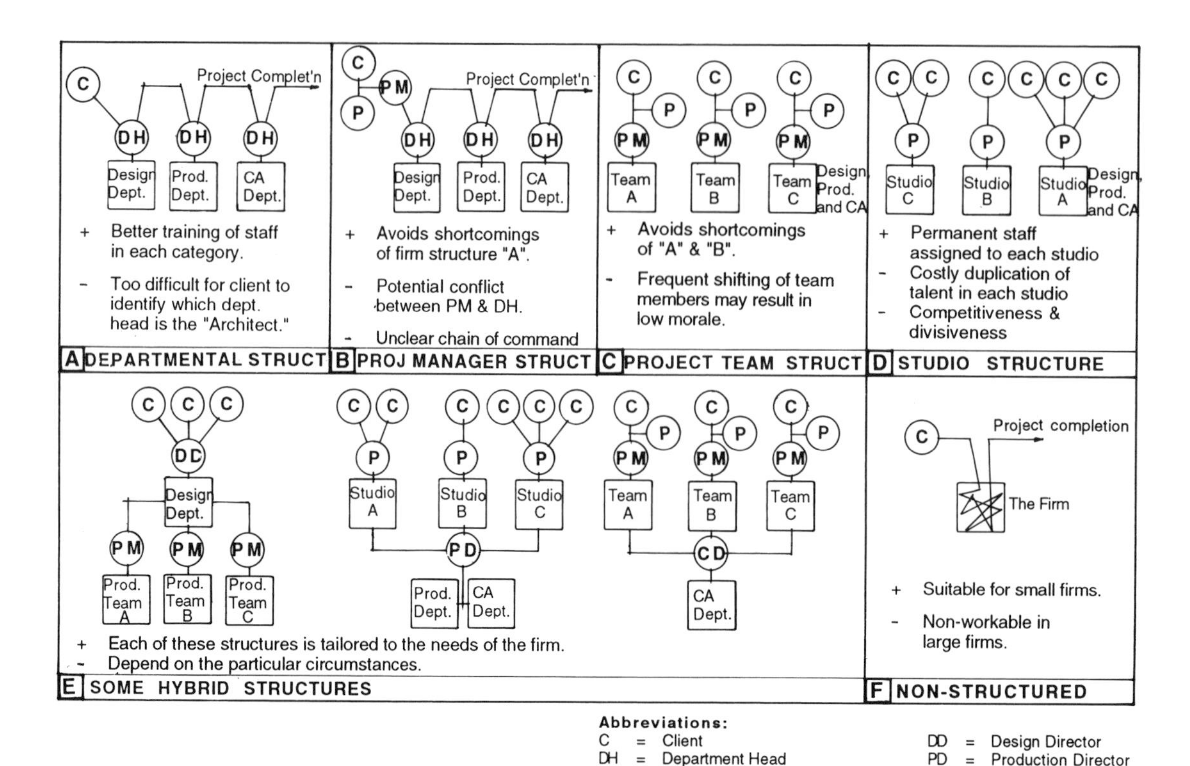

FIGURE 1-1. Forms of Architectural Firm Structures (Adapted from Chapter 1.9, *The Architect's Handbook of Professional Practice*. 1987. American Institute of Architects. Reproduced with permission under license #91116; further reproduction prohibited.)

1.1 THE CHAIN OF COMMAND

1.1.1 General

Most architectural firms are informal in their personal relationships. Everybody calls the principals by their first names and, in most cases, work is done in a pleasant atmosphere. This does not alter the fact that there is a distinct chain of command, with a distinct pecking order, that defines the work, assigns tasks, sets a budget, identifies deadlines, and maintains control over the proceedings.

The six forms of firm structure described in the introduction to this chapter have their strengths and weaknesses. I have worked in firms organized according to structures type A, C, and E3 (Fig. 1-1). The most fulfilling experiences were acquired in type C structures, because the staff was exposed to all facets of the profession the team was well supervised and developed a team spirit that promoted cooperation and boosted morale. Of course, the personalities involved may have had something to do with that, but I think this type of structure has more advantages than disadvantages. Structures D and E share many of the same advantages inherent in type C.

In all these structures, there is a shared feature: a team headed by a leader directing their efforts, answering their questions, and reporting to a Principal or to a Project Manager. For large projects, the assignment of a Project Manager in addition to the Team Leader is a definite plus. The PM maintains client relations for several projects, determines the budget for each, establishes a work plan, manages contract and administrative issues, and resolves disputes. These are time-consuming activities that require a certain skill and patience which may not necessarily be present in the character of a very technically or design-oriented person such as the Team Leader, who is responsible for the preparation of the drawings, adherence to codes, coordination with consultants, etc.

Figure 1-2 shows an ideal team structure, indicating a summary of the role assigned to each team member, the relationships between them, and the resources available to help them in preparing the contract documents. The following subsections elaborate on these roles and suggest a new method of dividing tasks among the team members preparing the drawings.

1.1.2 The Client

Although, technically speaking, the client is not a member of the team, his decisions concerning the budget, deadlines, and the quality and scope of the project have such an important impact on the architectural team that a brief description of the role played by this entity is necessary. A client may be an individual, a private organization, or a government agency.

The client provides information for programming the project, approves or modifies the design at every phase, arranges for financing, and provides site information.

The architect must retain control of one of the three parameters that define the project—cost, quality, or scope. For example, if the owner has a limited budget and wants a first-class building, the architect must convince the client to limit the size (scope) of the building to stay within the budget or to use lower-cost finishes (quality). Of course, if the budget is no object, which almost never happens, the architect's task becomes much easier.

1.1.3 The Principal-in-Charge

Principals bring to the project a wealth of knowledge and expertise based on their experience. They set the firm's goals, assign projects to teams, and maintain amicable relations with the client. A principal's duties vary from firm to firm depending on its structure. Where a firm is headed by several principals, a principal may specialize in one or two aspects of the practice, such as design or marketing, etc. As a general rule, a principal performs the following duties: negotiates or approves architect-owner and architectural consultant fees and signs all agreements for the project or delegates that task to a project manager; generates new work through marketing or personal contacts; resolves all conflicts; evaluates personnel performance and makes decisions concerning staff hiring, promotions, or terminations; critiques design, sets design goals, and approves the design before presentation to the client; and sets the budget for each project.

In small firms and in firms subdivided into studios, the Principal takes on the function of the project manager in addition to the duties listed above.

1.1.4 The Project Manager (PM)

The person chosen to perform the task of project manager must have certain qualifications suitable for this pivotal position. A project manager must be, above all, a diplomatic person. This does not imply that he must cave in to every demand to make the client happy. When the request is unreasonable, unethical, or presents a financial loss to the firm, the PM must find a more acceptable alternative or decline firmly and explain the reasons. This must be done in a way that does not alienate the client. A project manager must have the educational background and hands-on experience to answer most of the client's questions on the spot and gain his confidence. This does not mean providing inaccurate answers. Answers to questions that require obvious research may be deferred and provided as promptly as possible.

Depending on the organization of the firm, the Project Manager may be entrusted with most of the functions of a principal and is almost autonomous in his dealings with the client and the project team.

The more common Project Manager position, however, has more limited authority. The person entrusted with this responsibility manages the project within certain limitations, including a budget set by the Principal, agreements negotiated and signed by the Principal, team members assigned to the project by the Principal, and presentations to the client made either in the presence of the Principal or by the Principal.

In either position, the PM should be familiar with codes, relative costs of different methods of construction, technical issues, and last, but not least, he must have an assertive character. This does not mean an overbearing or abrasive character that generates resentment and alienation, but a kind of firmness coupled with resiliency that is necessary to this position. Not every PM lives up to this ideal, only the successful ones.

As stated above, not all PMs have the same duties and responsibilities. The following is a listing of the responsibilities of the average Project Manager:

1. Participates in the marketing effort and conducts presentations to the client.
2. Sets the budget, the project scope, and selects the team or is given a budget and participates in team selection.
3. Selects consultants and negotiates contracts (this may be handled by the Principal) and maintains a working relationship with them.
4. Identifies tasks and assigns them to appropriate staff or makes recommendations to the Team Leader if other than the PM.
5. Monitors changes and identifies extra services, if outside the scope of the project, to request client authorization for a fee adjustment.
6. Attends all meetings, takes minutes, and issues them promptly for distribution to all participants and other parties affected by the decisions and conclusions.
7. Establishes a project file, signs all correspondence, and routes all communications to the appropriate persons.
8. Monitors progress and ascertains conformance to deadlines and the budget.
9. Checks drawings for coordination, conformance to all applicable codes, and requirements mandated by the client-architect agreement before issuing for construction. A PM may be authorized to seal and sign the drawings.
10. Resolves contractor disputes and authorizes payments.
11. Prepares project data to be used on similar projects in the future and for marketing.
12. Maintains client contact.

At some offices, the PM may be referred to as Project Architect or Project Director.

1.1.5 The Team Leader (TL)

Like most positions in architectural offices, this position is called by many names. It is referred to as Project Architect,

Job Captain, Project Manager, etc. I prefer to call it Team Leader because that is the function of that individual—to lead the drafting team.

On major projects, the Team Leader should be registered and preferably have some experience in the project type. In many instances, the TL leads the same team through the Schematic, Design Development, and Construction Drawings phases. This is the best way to handle a project because the team will make every effort to carry the design intent from phase to phase.

The Team Leader's duties include the following:

1. Plans the team tasks based on the budget and target dates set by the PM.
2. Establishes a personal project file, making sure to place original correspondence of relevant information in the central file.
3. Starts mechanisms for monitoring progress such as the Project Sheet Status Report and the Progress Graph.
4. Draws a mock-up of every sheet to act as a road map for the team member entrusted with the sheet.
5. Does a thorough code analysis and selects U.L. designs to satisfy fireproofing requirements.
6. Defines materials to be specified and meets with product representatives to get relevant information on which to base the choices.
7. Develops exterior wall systems and selects standard details.
8. Conducts weekly team meetings according to an agenda.
9. Coordinates with consultants, supplies them with screened background plans to draw their systems on, and gets the feedback necessary for developing the architectural drawings. (PM may attend some of these meetings if the TL lacks that experience.)

Subsequent sections will address each item mentioned above.

1.1.6 The Project Team

Team members differ in educational background, experience, and aptitudes. It is important that junior staff be started on a training program to acquire the experience they need to advance in the profession. The method most often used is to leave these individuals to follow their own initiative in seeking the training they need to qualify for the professional registration exam. Some firms designate a senior staff member as a sponsor to supervise the intern's work while an adviser from another firm meets with him on a regular basis to discuss progress and offer advice. This is part of the Intern-Architect Development Program (IDP) to prepare for the exam.

While these training programs are good for overall training, they leave a lot to be desired as an in-depth training program for producing a good set of construction drawings. I believe in implementing a more comprehensive approach, which I will refer to as The Four Task Method (Fig. 1-2). This method utilizes the intern's need to learn by providing a clear four-step path that encourages the intern to proceed along a training program similar to a college curriculum. Each step is successively more difficult than the preceding one. After finishing the fourth step (or task), the intern is ready to take the Construction Documents and Services part of the registration exam. After passing the exam, this individual becomes eligible for team leadership. Offices that opt to use this method will find that it serves their own best interest. Because the firm provides an incentive for the intern to pursue his learning while working, the intern will do that work with enthusiasm and speed, a win-win situation. It also encourages the participants to continue working for the firm for the duration of the program and beyond.

For this method to succeed, the office must have an adequate library, a monthly seminar for all team members, and a knowledgeable technical adviser-sponsor to provide answers and guidance. While the Four Task Method serves new graduates, it works equally well for more experienced staff because it provides each team member autonomy in handling a well-defined task, it offers more seasoned staff a chance to guide a less experienced team member, and opens an opportunity for a team member to acquire skill in a more difficult task higher up the ladder as well as to participate in all adjunct activities required to finish each task. Conformance to the code and specifications as well as coordination with drawings from the other disciplines are examples of adjunct activities.

This contrasts with what many offices practice—a method that goes like this: "Frank is good at drawing plan details; let him handle this part on this project also." Poor Frank finds himself relegated to drawing stairs on project after project until he quits or protests strongly. The latter behavior may be held against him because he was not "cooperative."

Needless to say, the Team Leader must check the drawings periodically and involve each team member in correcting his mistakes while explaining the reason for each change.

The following is a description of each of the four tasks mentioned above and is meant to be an example only. Please refer to the sheet list for the sample project in the Project Sheet Status Report (Figs. 1-20 and 1-21) and the project mock-up. (Figs. 1-9 to 1-15)

Task #1

Assigned to team member A. This team member is new to the profession and is assigned to assist team member B and may be assigned to do miscellaneous tasks required by the Team Leader in addition to the following.

1. Title Sheet
   —Sketched by the TL and drafted by A.
2. General Information
   —Symbols and abbreviations from office standards file.
   —Code analysis by TL and team member D.
   —Area calculations done by A and checked by TL.
   —Other components as required.

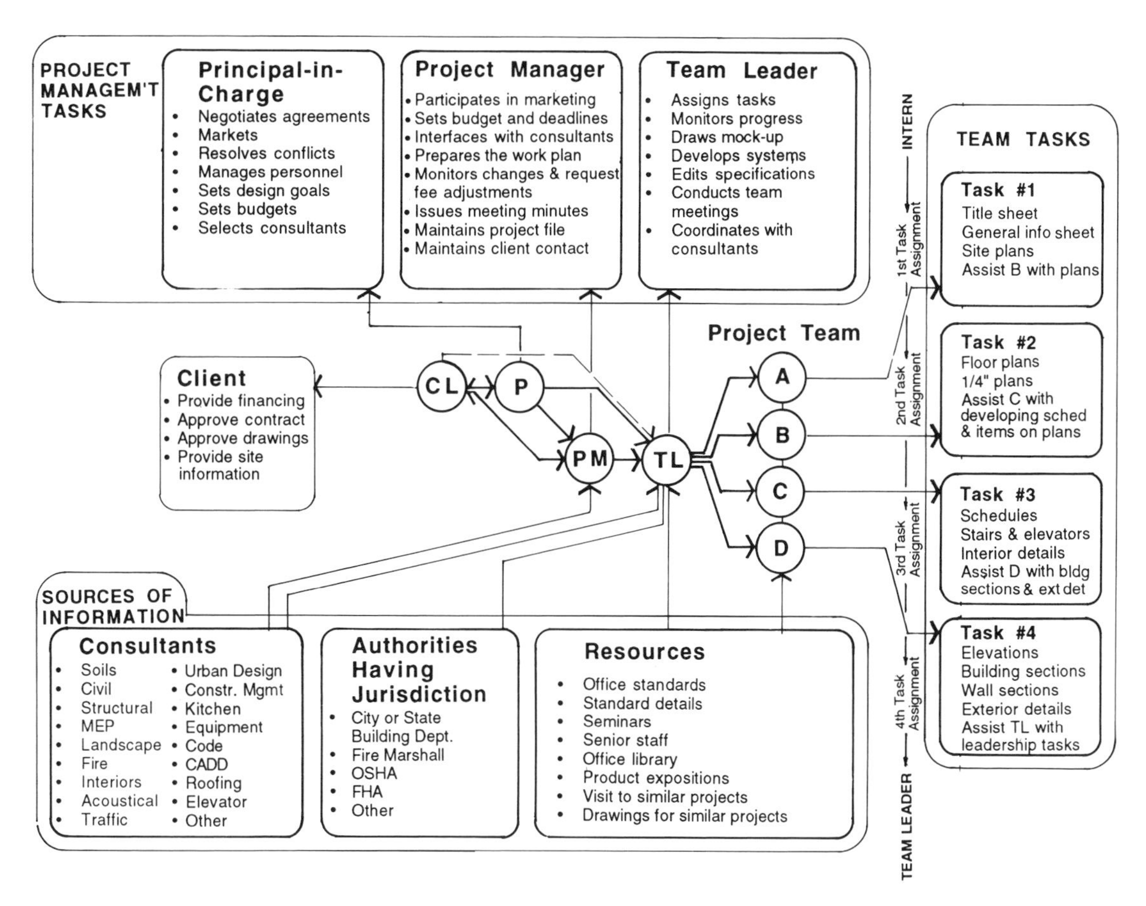

FIGURE 1-2. Team Organization Chart

3. Site Plan
   —First-floor plan is reduced to site plan scale (1/16″ = 1′-0″, 1:20, or 1:40); B guides A in this work.
   —Site plans from other disciplines may require splicing into the project sheet format. It is preferable that the PM give blank sheets (with title block) to these disciplines to draw their plans on and avoid that task.
4. Reflected Ceiling Plans
   —B guides A in developing the plans and details.

Task #2 Assigned to team member B.
This team member is responsible for all items related to the plans.

1. Plans (See Chapter 2 for guidelines).
2. Enlarged plans including core plans (if the project is designed around core areas that combine stairs, toilet rooms, etc.).
3. Supervises plan details and reflected ceiling plans done by A.
4. Equipment plans (if in the project).
5. Tasks done with input from C.
   —Room finish schedule and partition types.
   —Door schedule, door types, and frame details.

Task #3
Assigned to team member C. This team member is responsible for all items related to the interior of the building.

1. Stair and elevator details (Chapter 4). Escalator details (if in the project).
2. Interior elevations and miscellaneous interior details.
3. Explains the method of determining partition types to B and checks the plans for correctness.
4. Determines door types, explains what is required for the door sheet, assigns to B, and checks for correctness.
5. Tasks done by C with input from D.
   —Glazing details
   —Miscellaneous exterior details

Task #4 Assigned to team member D.
This team member is responsible for all items related to the exterior of the building. He may substitute for the TL during his absence if required to do so.

1. Elevations and building sections (Chapter 2).
2. Wall sections and section details (Chapter 2).
3. Supervises tasks done by C (see Task #3).

4. Assists the Team Leader in code and material research, specifications, coordination, and team management tasks.

This method promotes cooperation between members of the team. It provides exposure to each successive task before assuming responsibility for it in subsequent projects. In addition, it provides a clear path for advancement in the firm. After finishing the four tasks, the team member becomes eligible to assume team leadership on a small project if deemed capable of that assignment.

It is a good idea for the TL to inform the team at the outset that each team member will be assigned a task, will be assisted by another member, and will, in turn, assist a third member, and that, at the end of the project, each team member will fill out a confidential evaluation of the two team members that he collaborated with in regard to cooperation and performance. This makes the team members more considerate in their relationships.

The Four Task Method is analogous to the method used in the construction industry. The Team Leader's role is similar to the role of the General Contractor (GC). The latter does work using his own construction crew. The TL has a similar function. He does the mock-up and may develop a sample drawing for each task. The subcontractors perform finite specialized tasks, coordinate their work with other subcontractors, and are directed by the GC. Each team member is entrusted with a semi-independent defined task that he is responsible for and must coordinate with the other team members at the direction of the TL.

1.1.7 Team Management Guidelines

1. On large projects more than one team member may be assigned to each task. This is determined by the number of hours estimated to finish the sheets and the duration of the project. For example, if the total number of hours allocated to the construction drawing (CD) phase is 6,000 and the duration is 6 months (1,000 man hours), then the team size would be 6 (6,000 ÷ 1,000 = 6 persons). This enables the TL to assign two persons to each of the two most complicated tasks. If, on the other hand, the project is a small one requiring only 3,000 hours for the CD phase, the team size would be 3 (3,000 ÷ 1,000 = 3 persons), which requires that two tasks be combined and assigned to one team member.
2. The Team Leader should plan a shared activity outside the office each month. This may be handled by inviting suggestions from the team members. One Team Leader I know used to invite the team at 4 P.M. on Friday to have a beer at the neighboring tavern. This was the highlight of the week. I prefer having a monthly outing based on consensus. Examples of activities are bowling, skating, a water sport, a ball game, lunch out as a group, etc. This feature of team management is the best tool for creating a team spirit, helping the members to know one another better, and fostering friendly relations with one another and the TL.
3. The TL should involve team members in the coordination effort, which includes the architectural set (correct references between the sheets) as well as coordination with the drawings from consultants. The leader may challenge the team by offering a reward to the team member who can find the most inconsistencies. This works best if the team is youthful. This approach has tremendous educational value because the team members gain experience while having fun. This may also be applied to other "fun" jobs like checking shop drawings.

   Needless to say, this does not mean that the TL would depend solely on this method of checking. He must double-check everything before issuing the drawings. These activities make the task much easier, educate the team, reward the more alert among them, and lighten the TL's burden.
4. Give each team member a copy of the code analysis, making sure to highlight the parts that are applicable to the sheets each is entrusted with (Chapter 2).
5. Distribute to each the section of the specifications that relates to the work assigned to him, with instructions to read it carefully to discover and correct any inconsistencies present in the drawings.
6. Keep the team apprised of all developments related to the project. Distribute meeting minutes, news items, etc. Be upbeat; try not to be gloomy or express negative sentiments about the office or "the management." Doing so invariably alienates and demoralizes the team. Of course, being human, the TL will have some down days. During these periods, it is a good idea for him or her to do tasks that do not require contact with the team until the feeling passes.
7. The TL should visit each workstation daily. Engage each team member in conversation and be genuinely interested and encouraging. Ask questions and listen carefully to the answers. Answer any questions or guide the team member to the proper source for answers. If the question cannot be answered on the spot, the TL should make a record of it, promise to provide the answer, and do so as quickly as possible. Neglecting to do so will eventually lead to a loss of faith and a reputation that the TL cannot be depended upon to provide answers. Such a reputation is difficult to get rid of and may inspire team members to bypass the TL and seek answers from other team leaders or find ad hoc answers that may not be correct.

   I have observed team leaders whose main goal was to preserve what they perceived to be an image of infallibility. To avoid being confronted with a question they may not be able to answer on the spot, they either pretended to be too busy to talk to anybody, showed obvious displeasure at being interrupted, or avoided their workstations most of the day. Some team leaders started talking at team meetings and didn't let anyone else get a word in edgewise. The inevitable result was a resentful and unenthusiastic team that devised its own answers and produced a below-average set of drawings. An intensive redlining exercise by the Team Leader to correct the "stupid" mistakes followed. In this atmosphere, more than the average number of mistakes crop up in the final set of drawings, triggering a disproportionate number of addenda and change orders that cost the office time and money. In this particular case, several team members left the office for "personal" reasons.

8. The Team Leader should tailor the task to each member's ability. Estimate the number of hours required to finish it accordingly. Make sure that each team member is assigned an equal burden. I have known instances where a relatively skilled team member was yanked from project to project on short notice to handle the more difficult tasks. This unpredictable existence caused him to quit in disgust. This mismanagement of skilled people is unjustifiable and results in a waste of time, as the individual slowly finishes the task at hand and eases into the new task.

   Sure, there will be occasions when unforeseen pressing tasks and brushfires crop up, but they should be kept to a minimum. If this happens too often, the individual should be rewarded in some fashion.
9. The TL should spend time on the drafting board to draw parts of the project and thereby set the standard for difficult tasks such as a typical wall section, an enlarged plan of the toilet room, a glazing detail, etc. In addition to setting the standard, this activity demonstrates the Team Leader's ability to do good work and shows that he is part of the team. This may be unnecessary if the team is composed of experienced personnel capable of doing quality work because IT may imply that the TL does not trust them to do their part properly.
10. At the end of the project, celebrate with an office party. This recognizes a job well done and closes the project in preparation for gearing up for the next effort.

The team approach has a proven track record. In addition to motivating the team and raising their morale, it enables the work to proceed more smoothly and expeditiously. Team members stay longer at companies that use the team approach than do employees who work at firms that use other methods.

To recap, the four-task approach provides a step-by-step training method that points each member of the team toward an achievable goal, namely team leadership; it establishes a mechanism for collaboration that directs each member to work with two other team members (if one leaves the team or goes on vacation, there is a natural substitute familiar with the work to take his place); also, it exposes each team member to all facets of the practice by participation in coordination, specifications, code analysis, etc.; (this prepares the participants for becoming good team leaders), last, but not least, the shared activity ties the team together and brings enjoyment to the work.

1.2 PLANNING THE PROJECT

1.2.1 General

As mentioned at the beginning of this chapter, I will touch lightly on the part of project management handled by the Project Manager. The purpose of this section is to provide background information so the Team Leader will have a better understanding of the overall project management concept.

Understanding the part of the project covered by the PM helps to prevent duplication of effort and gives a preview of what the TL may someday do. It also lessens the perception held by some Team Leaders that the PM is having an easy time or the perception by some Project Managers that he or she only has to massage the egos of the client and the Principal while assigning everything to the Team Leader.

1.2.2 The Project Scope

After the owner selects the architect to design the project, the Project Manager (or partner) meets with the owner or his representatives to define the scope of services required to finish the project. One tool used for that purpose is the Phase/Service Matrix shown in figure 1-3. This matrix is based on documents B161 and B162 in the *Architect's Handbook of Professional Practice* issued by the AIA. As the project progresses, other services may be requested by the client. It is the duty of the PM to evaluate whether these additional services are sizable enough to warrant a request for a fee adjustment based on a change in the scope of services.

1.2.3 The Work Plan

After defining the scope of services, the PM fills out the appropriate owner/architect agreement defining the form of compensation, additional services, reimbursable expenses, and other conditions (see B series documents in the *Architect's Handbook*).

Based on this information, the PM develops a work plan. An example is shown in figure 1-4.

1.2.4 The Task Outline

A task is defined as an activity that requires an amount of effort that lasts for a period of time and has a start and a finish date. An example of a task outline is shown in figure 1-5.

After defining the tasks, the PM determines their sequence as shown in the example. The PM then assigns each task to a designated team member (for the PM, the "team" includes consultants, a specification writer, and an estimator as well as the in-house personnel or the Team Leader if other

PHASE/SERVICE MATRIX

Project FRANKLIN ELEMENTARY Project # 7701

	PHASE 1: PREDESIGN SERVICES 1	PHASE 2: SITE ANALYSIS SERVICES 2	PHASE 3: SCHEMATIC DESIGN SERVICES 3	PHASE 4: DESIGN DEVELOPMENT SERVICES 4	PHASE 5: CONSTRUCTION DOCUMENTS SERVICES 5	PHASE 6: BIDDING OR NEGOTIATIONS SERVICES 6	PHASE 7: CONSTRUCTION CONTRACT ADMINISTRATION SERVICES 7	PHASE 8: POST-CONSTRUCTION SERVICES 8	PHASE 9: SUPPLEMENTAL SERVICES 9a	PHASE 9: (CONT'D) SUPPLEMENTAL SERVICES 9b
A	.01 Project Administration	.01 Project Administration	.01 Project Administration	.01 Project Administration	● .01 Project Administration	.01 Project Administration	.01 Project Administration	.01 Project Administration	.61 Special Studies	.79 Materials and Systems Testing
B	.02 Disciplines Coordination/Document Checking	.02 Disciplines Coordination/Document Checking	.02 Disciplines Coordination/Document Checking	.02 Disciplines Coordination/Document Checking	● .02 Disciplines Coordination/Document Checking	.02 Disciplines Coordination/Document Checking	.02 Disciplines Coordination/Document Checking	.02 Disciplines Coordination/Document Checking	.62 Renderings	.80 Demolition Services
C	.03 Agency Consulting/Review/Approval	.03 Agency Consulting/Review/Approval	.03 Agency Consulting/Review/Approval	.03 Agency Consulting/Review/Approval	● .03 Agency Consulting/Review/Approval	.03 Agency Consulting/Review/Approval	.03 Agency Consulting/Review/Approval	.03 Agency Consulting/Review/Approval	.63 Model Construction	.81 Mock-up Services
D	.04 Owner-supplied Data Coordination	.04 Owner-supplied Data Coordination	.04 Owner-supplied Data Coordination	.04 Owner-supplied Data Coordination	● .04 Owner-supplied Data Coordination	.04 Owner-supplied Data Coordination	.04 Owner-supplied Data Coordination	.04 Owner-supplied Data Coordination	.64 Life Cycle Cost Analysis	.82 Still Photography
E	.05 Programming	.13 Site Analysis and Selection	.21 Architectural Design/Documentation	.21 Architectural Design/Documentation	● .21 Architectural Design/Documentation	.34 Bidding Materials	.41 Office Construction Administration	.50 Maintenance and Operational Programming	.65 Value Analysis	.83 Motion Pictures and Videotape
F	.06 Space Schematics/Flow Diagrams	.14 Site Development Planning	.22 Structural Design/Documentation	.22 Structural Design/Documentation	● .22 Structural Design/Documentation	.35 Addenda	.42 Construction Field Observation	.51 Start-up Assistance	.66 Quantity Surveys	.84 Coordination with Non-Design Professionals
G	.07 Existing Facilities Surveys	.15 Detailed Site Utilization Studies	.23 Mechanical Design/Documentation	.23 Mechanical Design/Documentation	◐ .23 Mechanical Design/Documentation	.36 Bidding/Negotiations	.43 Project Representation	.52 Record Drawings	.67 Detailed Construction Cost Estimates	.85 Special Disciplines Consultation
H	.08 Marketing Studies	.16 On-site Utility Studies	.24 Electrical Design/Documentation	.24 Electrical Design/Documentation	◐ .24 Electrical Design/Documentation	.37 Analysis of Alternates/Substitutions	.44 Inspection Coordination	.53 Warranty Review	.68 Energy Studies	.86 Special Building Type Consultation
I	.09 Economic Feasibility Studies	.17 Off-site Utility Studies	.25 Civil Design/Documentation	.25 Civil Design/Documentation	● .25 Civil Design/Documentation	.38 Special Bidding Services	.45 Supplemental Documents	.54 Postconstruction Evaluation	.69 Environmental Monitoring	
J	.10 Project Financing	.18 Environmental Studies and Reports	.26 Landscape Design/Documentation	.26 Landscape Design/Documentation	● .26 Landscape Design/Documentation	.39 Bid Evaluation	.46 Quotation Requests/Change Orders		.70 Tenant-related Services	
K		.19 Zoning Processing Assistance	.27 Interior Design/Documentation	.27 Interior Design/Documentation	● .27 Interior Design/Documentation	.40 Construction Contract Agreements	.47 Project Schedule Monitoring		.71 Graphics Design	
L			.28 Materials Research/Specifications	.28 Materials Research/Specifications	● .28 Materials Research/Specifications		.48 Construction Cost Accounting		.72 Fine Arts and Crafts Services	
M	.29 Project Development Scheduling	.29 Project Development Scheduling	.29 Project Development Scheduling	.29 Project Development Scheduling	.30 Special Bidding Documents/Scheduling		.49 Project Closeout		.73 Special Furnishings Design	
N	.31 Project Budgeting	.31 Project Budgeting	.32 Statement of Probable Construction Cost	.32 Statement of Probable Construction Cost	● .32 Statement of Probable Construction Cost				.74 Non-Building Equipment Selection	
O	.33 Presentations	.33 Presentations	.33 Presentations	.33 Presentations	● .33 Presentations				.75 Project Promotion/Public Relations	
P					● .71 GRAPHICS DESIGN				.76 Leasing Brochures	
Q									.77 Expert Witness	
R									.78 Computer Applications	

● = Time Estimates and Records
◐ = Time Estimates Only
○ = No Time Estimates or Records

AIA® FORM F860 · PHASE/SERVICE MATRIX · JANUARY 1978 EDITION · © 1978

THE AMERICAN INSTITUTE OF ARCHITECTS, 1735 NEW YORK AVE., N.W., WASHINGTON, DC 20006

AIA FMS THE AMERICAN INSTITUTE OF ARCHITECTS FINANCIAL MANAGEMENT SYSTEM

Source: Compensation Guidelines for Architectural and Engineering Services; Washington, DC: The American Institute of Architects, 1978, p. 30.

Because AIA Documents are revised from time to time, users should acertain from the AIA the current editions of the Documents reproduced herein.

FIGURE 1-3. Phase/Service Matrix Form (From *You and Your Architect.* 1987. The American Institute of Architects. Reproduced with permission under license #91116; further reproduction prohibited.)

than the PM. On some jobs where the contract is negotiated rather than put out for bids, the contractor is also part of the project team).

1.2.5 The Schedule

The schedule is based on the task outline. It graphically shows the start and finish dates for each task. In most cases it takes the form of a bar chart (Fig. 1-6). This becomes the work plan for every member of the team. Each participant in the project must sign off on these dates. To accomplish this, the PM invites them to a kick-off meeting, explains the project, distributes the schedule, and makes adjustments as necessary.

1.2.6 The Project Budget

Every major project must show profit to enable the firm to operate profitably, especially during lean times. To ensure this, the PM or the Principal negotiates a fee that covers the following: cost of personnel—the salaries required to execute the tasks described in Section 1.2.4; overhead—rent and utilities, holidays and vacations, insurance, etc.; expenses—consultant fees, printing and typing costs, travel and long-distance phone calls, mailing, supplies, etc.; contingencies—unforeseen expenses; and profit—a percentage of the total budget.

There are several ways to compute the budget. One example is shown in figure 1-7. Computing a reasonably accurate budget enables the PM to assign adequate staff to do the required work within the allocated time.

1.3 PROJECT ACTIVITIES

1.3.1 General

The Team Leader has responsibilities parallel to those of the Project Manager, especially on large projects. Based on the information provided by the PM, the TL plans the project, assigns tasks, constructs a Team Task Assignment Schedule, and sees to it that the project is done according to it and within the budget determined by the PM. To do this properly, the Team Leader must have a good working relationship with the Project Manager, one that is based on mutual respect. He must realize that the PM must be apprised of all developments of any significance and that all decisions which impact the budget must be communicated to the Project Manager as they occur.

Every project's Construction Drawings phase goes through three distinct activities before it is ready to be issued for bids. These activities are Project Start-up, the Main Effort, and Coordination and Conclusion.

The following is a description of each activity and the planning required to guide the team and inspire them to do a good job.

1.3.2 Project Start-Up

Before the team is chosen, the TL must prepare certain data and plan the assignment for each team member. The TL may choose to enlist the most experienced team member to participate in the start-up activity. This will prepare that individual for eventually becoming a Team Leader. Project start-up must be done in as short a time as possible, but it must be done carefully to ensure that the project proceeds in

COMPONENT	"BROADSCOPE" WORK PLAN	"NARROWSCOPE" WORK PLAN
DESCRIPTION, REQUIREMENTS, OBJECTIVES	Title and description Projected scope Projected schedule Construction cost limitation Any other owner requirements Level of amenity sought Other project objectives	Title, description, project number, PM responsibilities Approved scope Approved construction cost Anticipated delivery approach Any owner-mandated consultant contracts Key owner representatives and decision structure Reference to owner-architect agreement Level of amenity sought Other project objectives
TASKS	Major tasks or phases needed to accomplish project Sequence of major tasks or phases Responsibility for each major task or phase	Detailed list of tasks (and groupings into phases) Task sequence Responsibility for each task
SCHEDULE	Milestone dates for each major task or phase	For each task: —Anticipated duration —Assumptions and contingencies used in establishing duration —Earliest and latest start dates —Earliest and latest finish dates
BUDGET	Internal budget targets for each major task or phase	For each task: —Personnel hours needed, by each staff category —Total personnel cost —Total consultant costs —Other expenses required
ARCHITECT'S COMPENSATION	Estimated (or proposed) compensation	Approved compensation Billing approach or pattern Anticipated revenue, by month or other time period

FIGURE 1-4. Components of the Project Work Plan (From *Managing Architectural Projects: The Process*. 1981. The American Institute of Architects. Reproduced with permission under license #91116; further reproduction prohibited.)

TASK OUTLINE FOR DESIGN OF A SMALL OFFICE BUILDING

First Cut Outline	Final Outline
A. Obtain client data 1. Site location 2. Financing availability 3. Goals for project 4. Zoning	1. Obtain client data 2. Check zoning 3. Develop program 4. Site access plan
B. Develop program 1. Gross space required 2. Parking 3. Walking 4. Mass transit 5. Environmental 6. Utilities 7. Tenant uses	5. Schematic development 6. Identify outside consultants 7. Design development 8. Initiate specifications 9. Code review 10. Drawing production
C. Schematic development 1. Layout plans and elevations 2. Identify alternate structures 3. Investigate building materials 4. Identify outside consultants 5. Layout "miniset" of drawings 6. Preliminary cost estimate	11. Updated cost estimate 12. Final specifications 13. Bidding 14. Contract administration 15. Punch list 16. Project management
D. Design development 1. Finalize plans 2. Initiate specifications 3. Layout all drawings 4. Code review	
E. Drawing production 1. Updated cost estimate 2. Coordination of consultant drawings 3. Final details 4. Final specifications	
F. Bidding	
G. Contract administration 1. Site meetings 2. Change orders 3. Punch list	
H. Project management 1. Budgets and schedules 2. Team selection 3. Review meetings 4. Client relations 5. Financial management 6. Project records 7. Quality control 8. Managing changes	

First cut and final task outlines are shown for a small office building. Note that project management is included as a separate task in both outlines.

FIGURE 1-5. Example of a Task Outline (Reprinted, by permission, from Burstein and Stasiowski, *Project Management for the Design Professional*, 25)

an orderly and speedy fashion, with each member knowing exactly what to do and when to do it. Nothing wastes time more than team members milling around not knowing what is to be done next.

One of the first tasks is a thorough code analysis. This provides answers to many questions commonly raised by the team. Questions concerning stair design, fireproofing, type of roofing, glazing, etc., are all determined by the building code. A guide form is included in Section 3.1. Based on the fire-rating information found through this analysis, U.L. designs can be chosen (see Sec. 3.2).

The Team Leader then visualizes all the sheets required for the Construction Drawing phase. The TL starts with the design development drawing list and designates sheets to provide the rest of the information needed by the contractor to construct the building. Other sheets may be required as the project proceeds. This list is written on a form similar to the Project Sheet Status Report shown in figures 1-20 and 1-21. Each sheet is assigned a number based on the numbering system established by the office (Chapter 2). Based on the sheet titles and the hours allocated to finish each sheet, the TL assigns sheets to each team member.

Next, the TL constructs the Progress Graph. This graph as the name implies is used to monitor progress (see Sec. 1.4.). Based on the deadlines identified in this graph, a Team Task Assignment Schedule (Fig. 1-8) is started to determine the

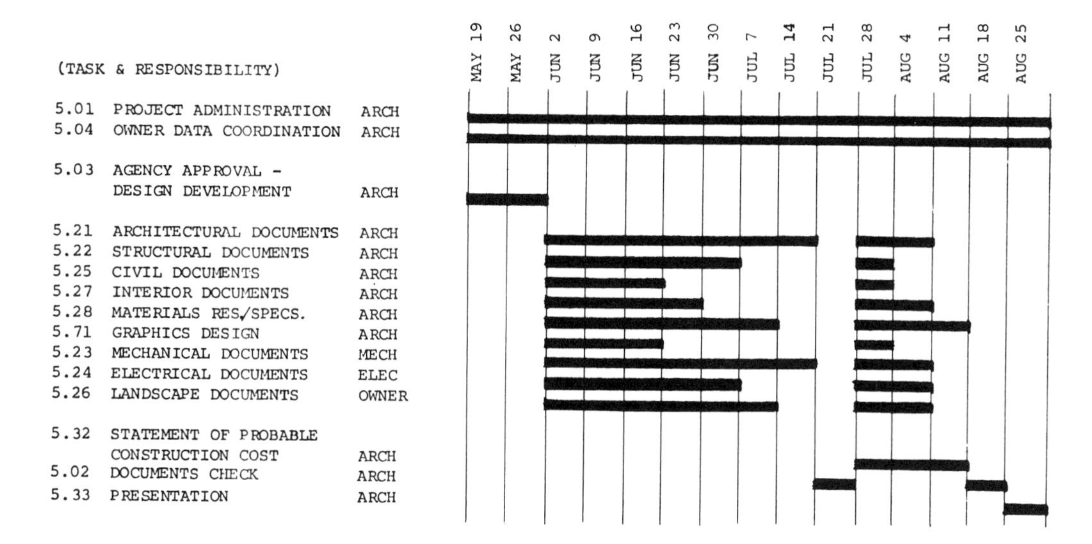

FIGURE 1-6. Example of a Bar Chart Schedule (From *Managing Architectural Projects: The Process*. 1981. The American Institute of Architects. Reproduced with permission under license #91116; further reproduction prohibited.)

start and finish dates of activities on each sheet. Please note that the tasks are identified by a letter rather than the initials of a team member. At the first team meeting, the TL would ask the team members to indicate their choice of the tasks each would like to undertake in order of preference. This is done with the understanding that not all preferences can be granted. Giving the team members the opportunity to express their wishes and making every effort to satisfy those choices gives them the incentive to do their best. Of course, the actual task assignment must be based on the level of experience of each participant.

Based on the sheet list, the TL draws a mock-up of all the sheets at ¼ inch = 1-inch scale (Fig. 1-9 through 1-15). This is the best guide for the team. It gives them an overview of the whole project and shows how each sheet is arranged.

A good mock-up takes from two days to plan and draw a small project to a week or more for a large one. It is time well spent. Mock-ups are real time-savers. Without them each team member would spend countless hours planning each sheet and would include extraneous, and sometimes repetitious, information. The result is that each sheet would not be consistent in its organization throughout the project.

The mock-up (sometimes referred to as a mini-set or project cartoon) minimizes the unplanned, unauthorized addition of sheets and details to the project. The Team Leader must make it quite clear at the outset that any additions must be authorized by him before implementation, making sure, however, not to deter the team members from making suggestions. I have seen instances where a project contained several sheets with two or three details each. These could have been combined into one sheet titled "Miscellaneous Details." Every time an extra sheet is added unnecessarily to the project, it adds expense for printing and wastes time because each sheet starts slow, ends slow, and the team member usually takes a break between sheets.

The mock-up gives the TL control of what is included in the project. It identifies standard details to be used on the project (Chapter 4), before team members start wasting time drawing them from scratch. It provides a graphic basis for estimating the time required to finish each sheet (Chapter 2). Finally, it prevents duplication of details. This can happen because some details belong in a gray area shared by two groups of sheets. For example, exterior guardrails may belong in the miscellaneous details sheet or the stair details sheet. Two team members, each working on a different group of sheets, may spend valuable time developing the same detail without realizing that they are doing so.

One of the start-up activities is for the TL to establish a personal project file. This puts all the information at the Team Leader's fingertips. This file is independent from the central project file maintained by the office. Originals of all correspondence, minutes, copies of telephone conversation records, and other relevant data must be sent to the central file. An example of suggested headings is shown in figure 1-16. Under item 3 in the file (Product Information) the TL identifies materials and systems to be used on the project. This must be done as early as possible to give enough time for the TL and members of the team to contact product representatives and gather information crucial to choosing the right products to include in the specifications and help define the details.

Some offices use a standard form, to be filled out during the meeting with the product rep. An example is shown in figure 1-17. This information should be kept in a general office file established to serve as a record of the project at

PROJECT ESTIMATING SHEET Project: *Example* Number: ______ Client: *DOE*

Prepared by: DB
Date: 11/26/79

DIRECT LABOR COSTS (Category of Personnel / Rate per Hour)

Phase/task	A: Principal $22/hr	A: Proj. Mgr. $16/hr	A: Arch/Eng $13/hr	A: Technician $10/hr	A: Drafting $8/hr	A: Secretary $6/hr	B: Direct Labor Costs
A	20 hrs. / $440	20 / 320	200 / 2600	40 / 400	40 / 320	20 / 120	340 hrs. / $ 4200
B1	8 / 176	16 / 256	40 / 520	0 / 0	0 / 0	8 / 48	72 / 1000
B2	0 / 0	20 / 320	40 / 520	0 / 0	20 / 160	20 / 120	100 / 1120
B3	12 / 264	40 / 640	120 / 1560	0 / 0	20 / 160	40 / 240	232 / 2864
B4	20 / 440	60 / 960	400 / 5200	0 / 0	0 / 0	0 / 0	480 / 6600
B5	0 / 0	20 / 320	40 / 520	400 / 4000	0 / 0	40 / 240	500 / 5080
C1	8 / 176	20 / 320	120 / 1560	60 / 600	20 / 160	40 / 240	268 / 3056
C2	20 / 440	40 / 640	160 / 2080	0 / 0	120 / 960	20 / 120	360 / 4240
C3	8 / 176	20 / 320	160 / 2080	80 / 800	0 / 0	12 / 72	280 / 3448
D	8 / 176	12 / 192	80 / 1040	0 / 0	20 / 160	20 / 120	140 / 1688
E	20 / 440	60 / 960	40 / 520	0 / 0	0 / 0	12 / 72	132 / 1992
F	20 / 440	40 / 640	80 / 1040	0 / 0	12 / 96	12 / 72	164 / 2288
G1a	8 / 176	20 / 320	80 / 1040	20 / 200	80 / 640	120 / 720	328 / 3096
G1b	8 / 176	20 / 320	80 / 1040	20 / 200	80 / 640	120 / 720	328 / 3096
G1c	8 / 176	20 / 320	80 / 1040	20 / 200	80 / 640	120 / 720	328 / 3096
G2	20 / 440	80 / 1280	160 / 2080	80 / 800	120 / 960	240 / 1440	700 / 7000
G3	8 / 176	40 / 640	40 / 520	20 / 200	20 / 160	60 / 360	188 / 2056
H	60 / 1320	160 / 2560	40 / 520	0 / 0	20 / 160	40 / 240	320 / 4800
TOTALS	256 hrs. / $ 5632	708 / 11,328	1960 / 25,480	740 / 7400	652 / 5216	944 / 5664	5260 hrs. / $ 60,720

FEE COMPUTATION

Phase/task	C: Overhead (B x 1.5)	D: Other Direct Costs	E: Est. Cost B + C + D	F: Contingencies	G: Total Budget (E + F)	H: Profit	I: Project Value (G + H)
A	6,300	2640	13,140	1,314	14,454	1,445	15,899
B1	1,500	420	2,920	292	3,212	321	3,533
B2	1,680	160	2,960	296	3,256	326	3,582
B3	4,296	490	7,650	765	8,415	842	9,257
B4	9,900	3,200	19,700	1,970	21,670	2,167	23,837
B5	7,620	4,800	17,500	1,750	19,250	1,925	21,175
C1	4,584	420	8,060	806	8,866	887	9,753
C2	6,360	260	10,860	1,086	11,946	1,195	13,141
C3	5,172	200	8,820	882	9,702	970	10,672
D	2,532	200	4,420	442	4,862	486	5,348
E	2,988	280	5,260	526	5,786	579	6,365
F	3,432	520	6,240	624	6,864	686	7,550
G1a	4,644	1,200	8,940	894	9,834	983	10,817
G1b	4,644	1,200	8,940	894	9,834	983	10,817
G1c	4,644	1,200	8,940	894	9,834	983	10,817
G2	10,500	600	18,100	1,810	19,910	1,991	21,901
G3	3,084	2,800	7,940	794	8,734	873	9,607
H	7,200	1,400	13,400	1,340	14,740	1,474	16,214
TOTALS	91,080	21,990	173,790	17,379	191,169	19,116	210,285

FIGURE 1-7. Example of a Project Estimating Sheet (Reprinted, by permission, from Burstein and Stasiowski, *Project Management for the Design Professional*, 54)

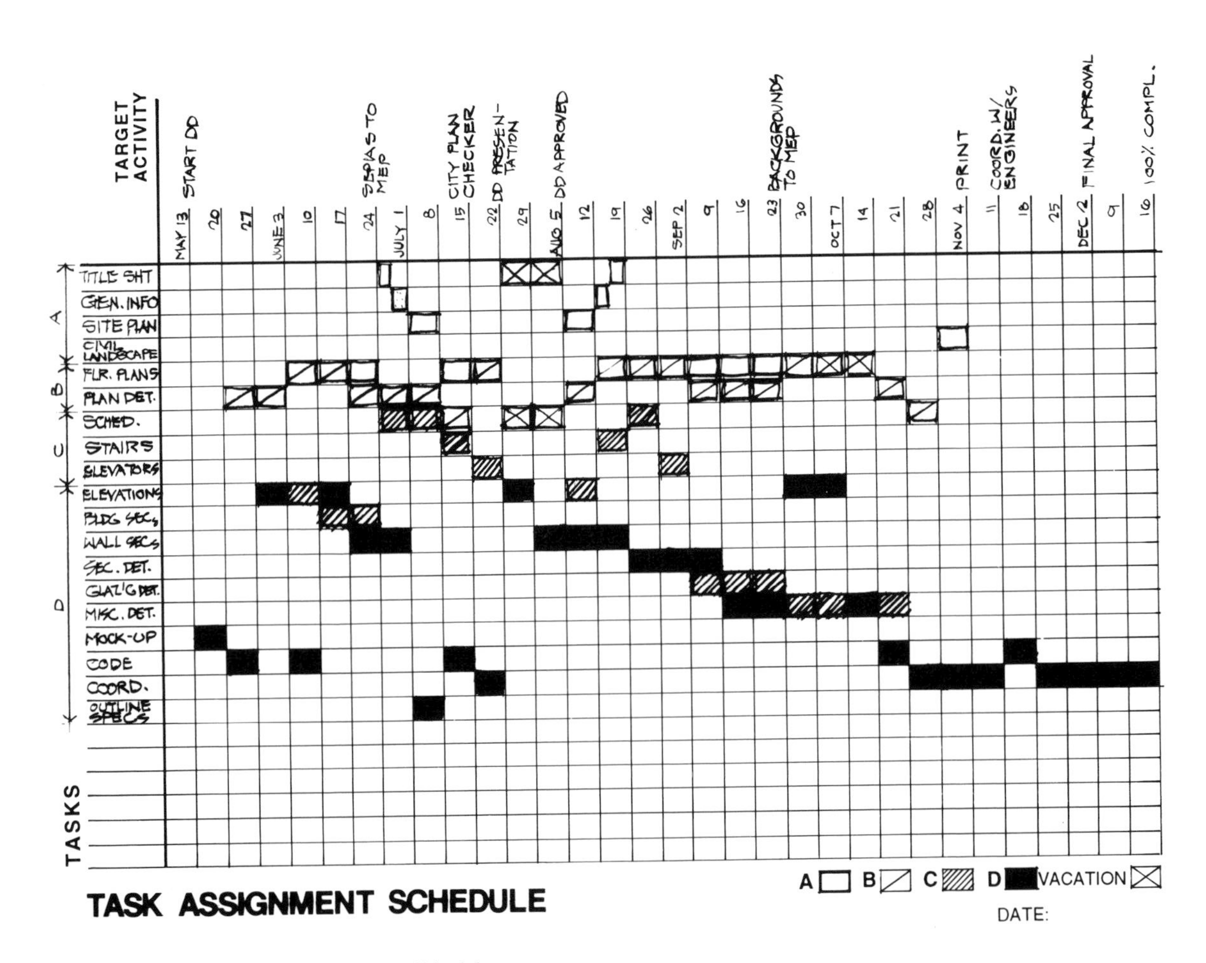

FIGURE 1-8. Team Task Assignment Schedule

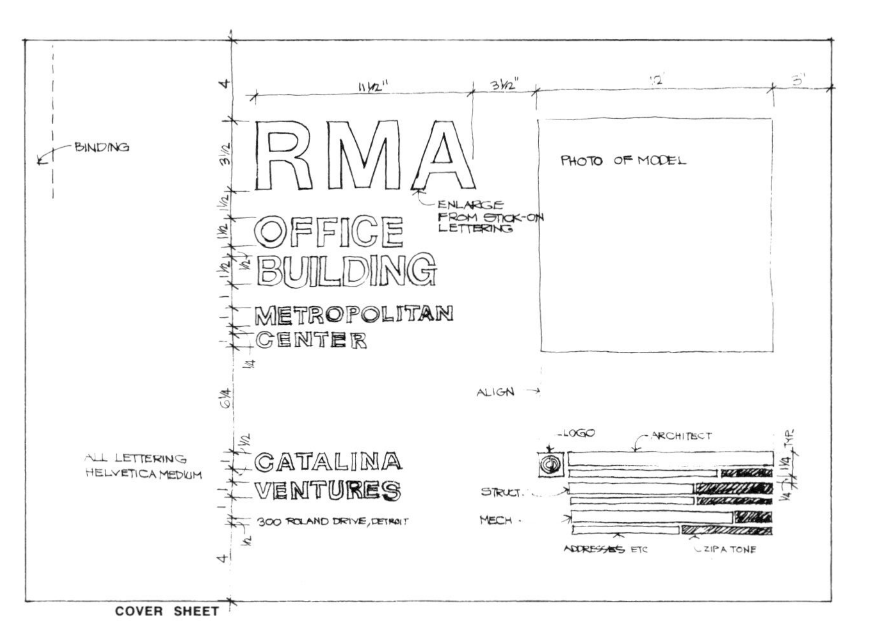

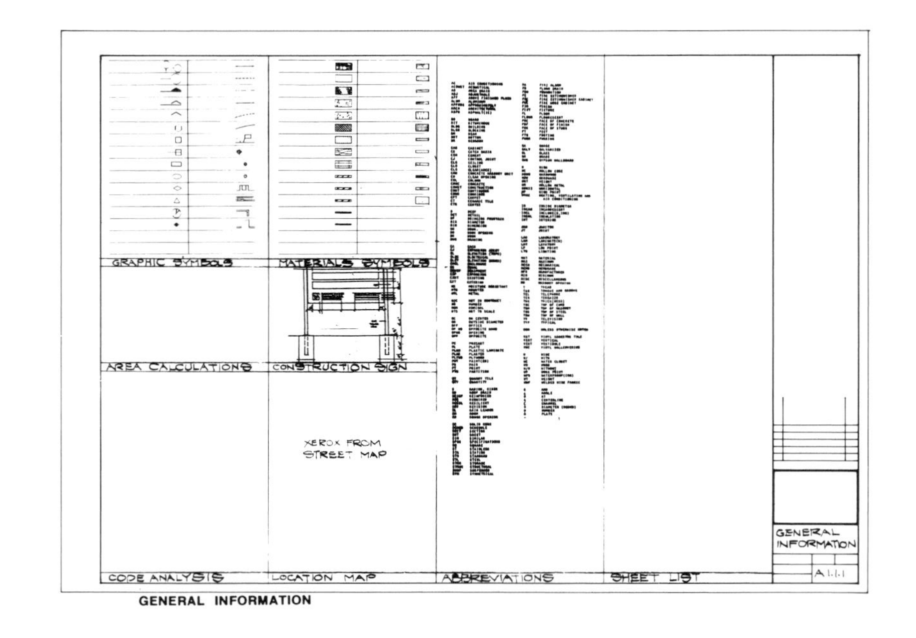

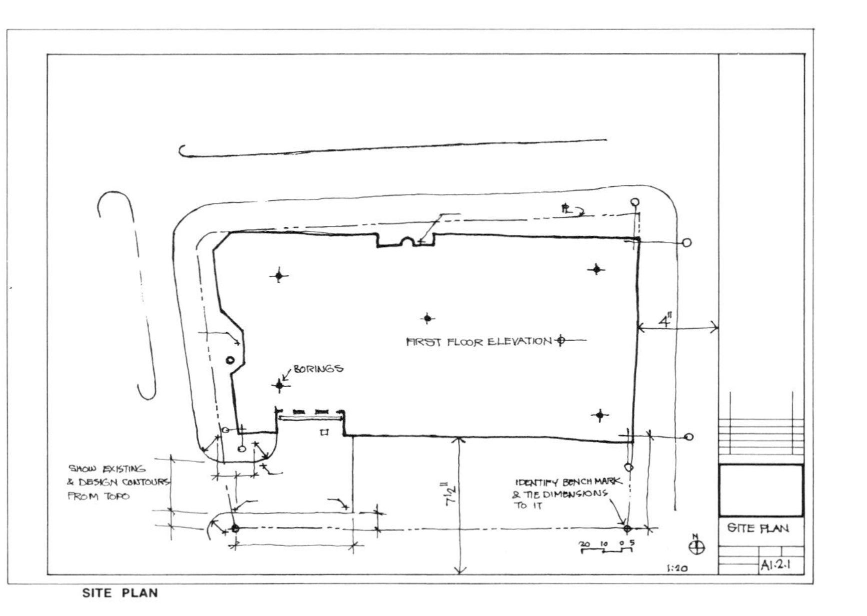

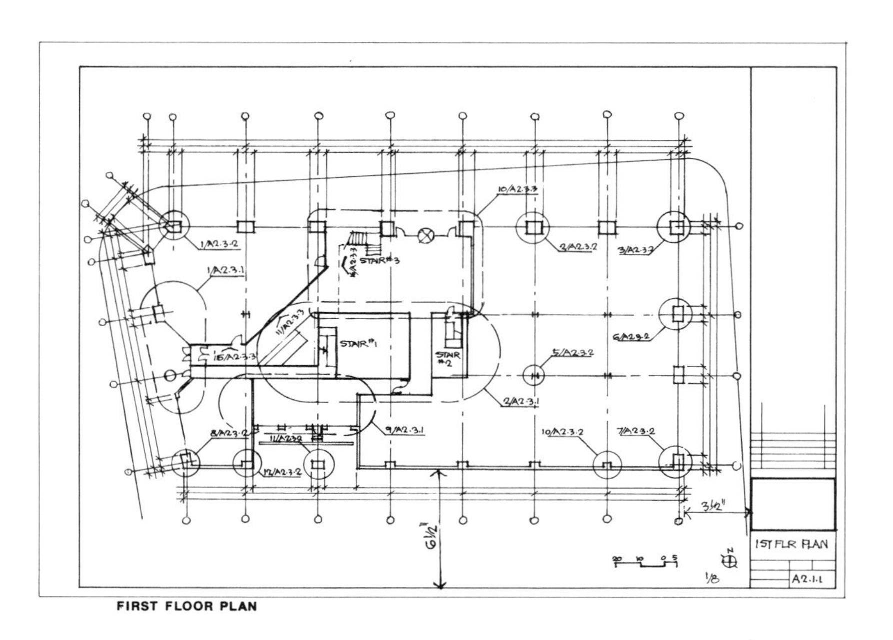

FIGURE 1-9. The Project Mock-Up

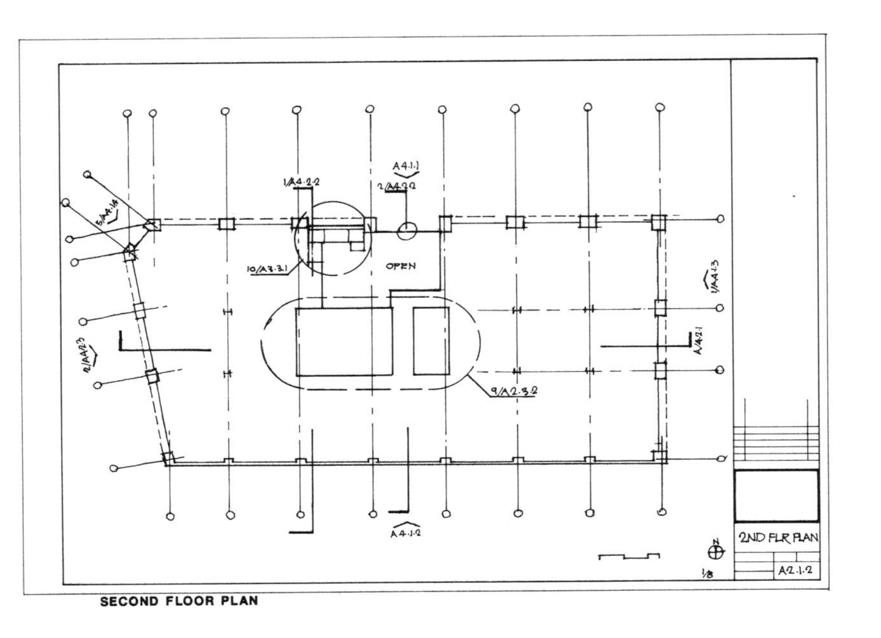

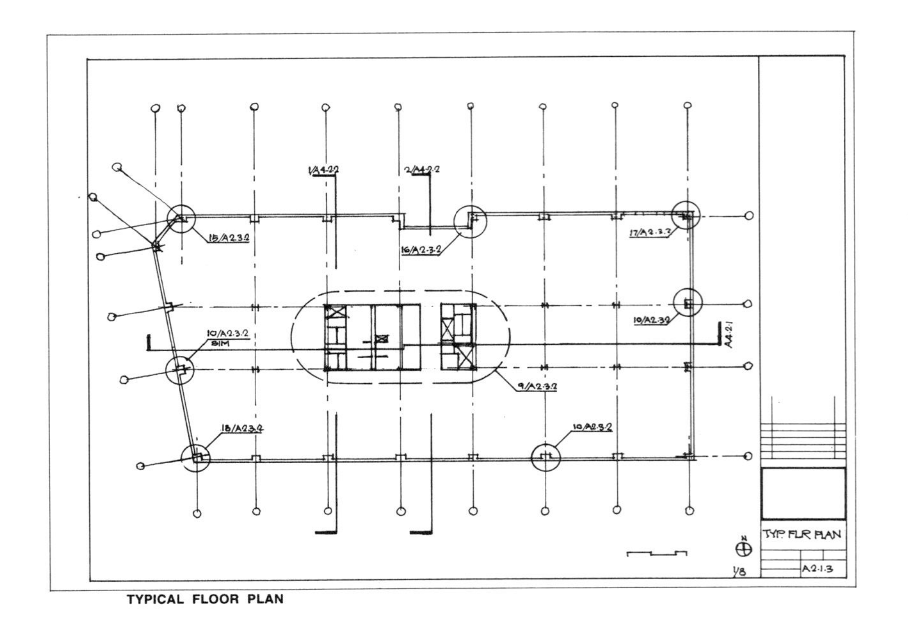

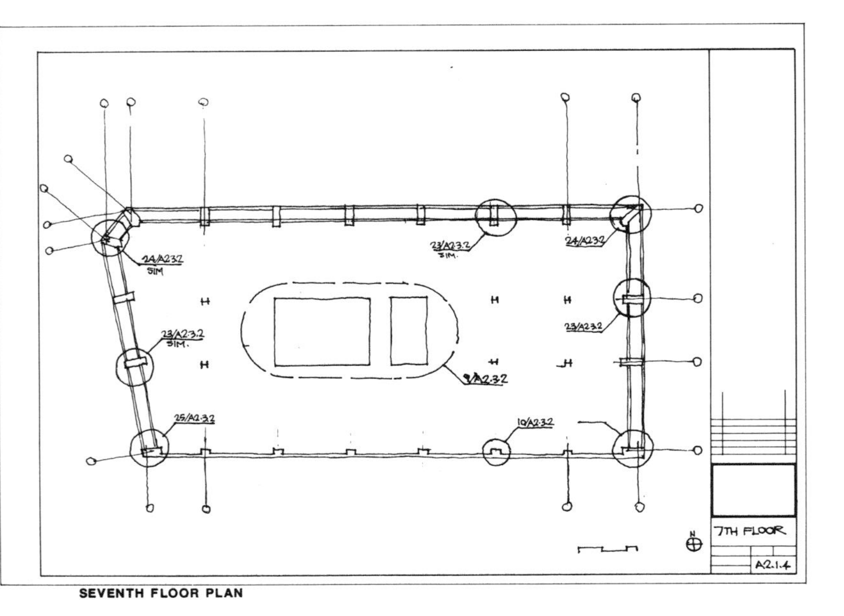

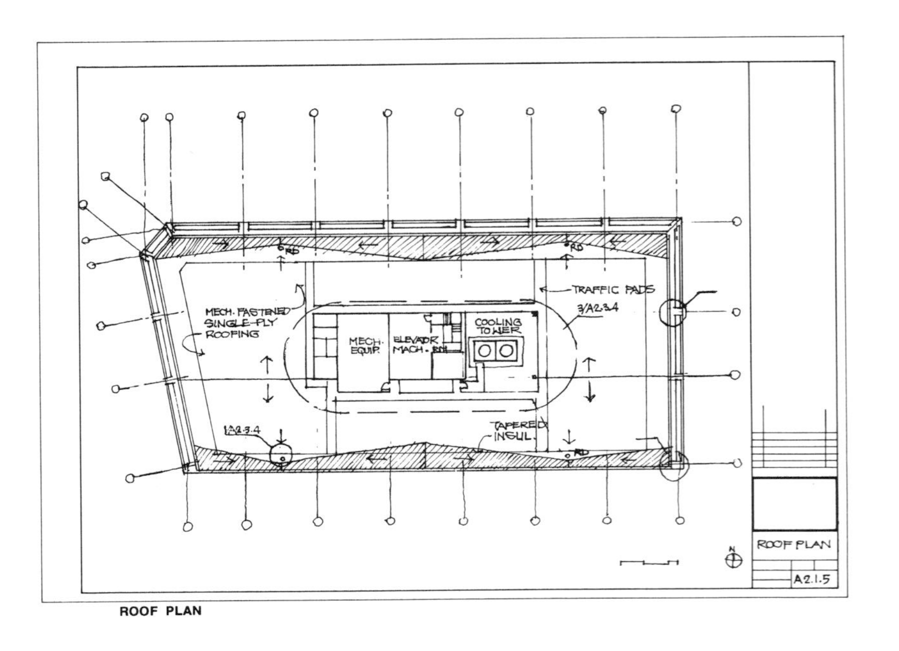

FIGURE 1-10. The Project Mock-Up (cont.)

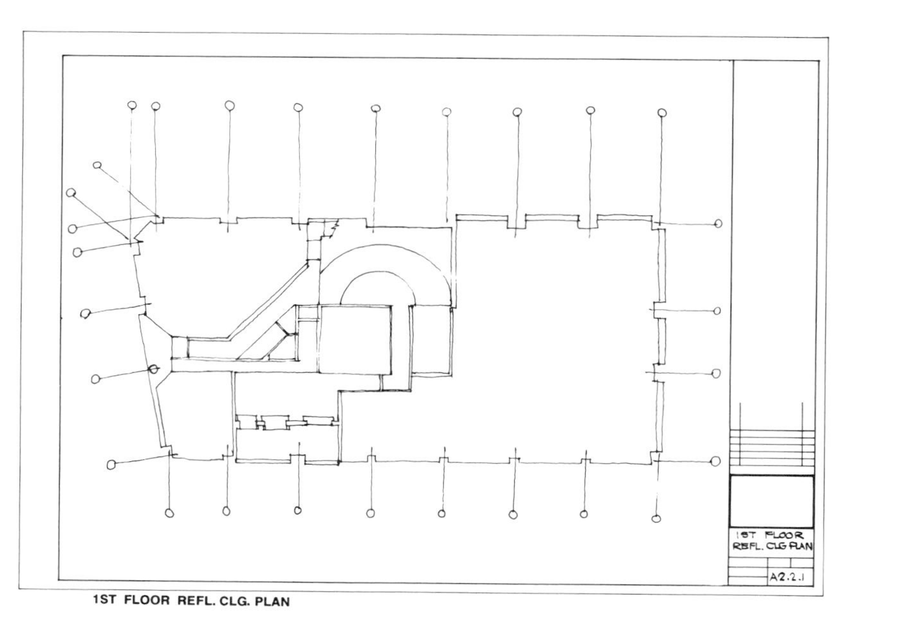

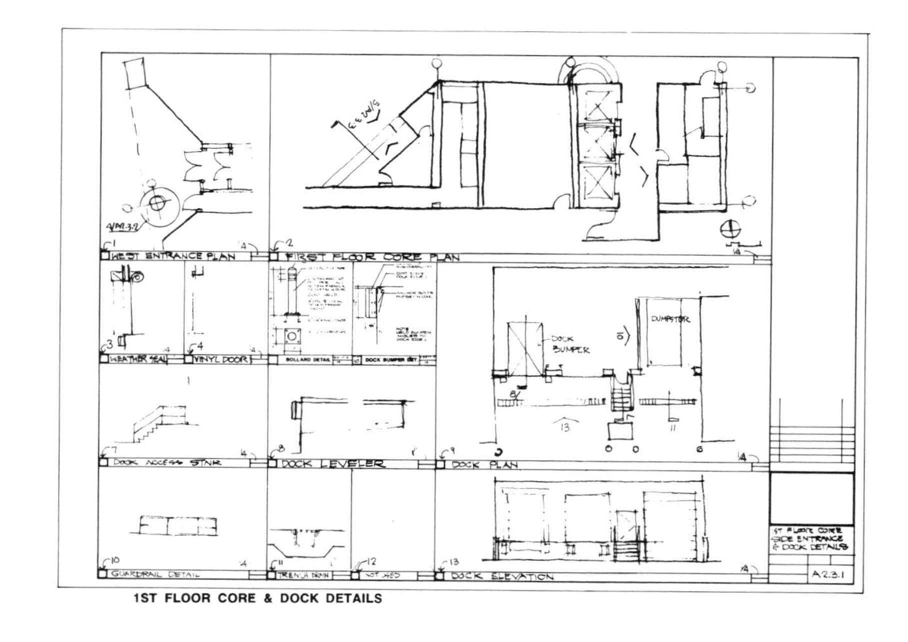

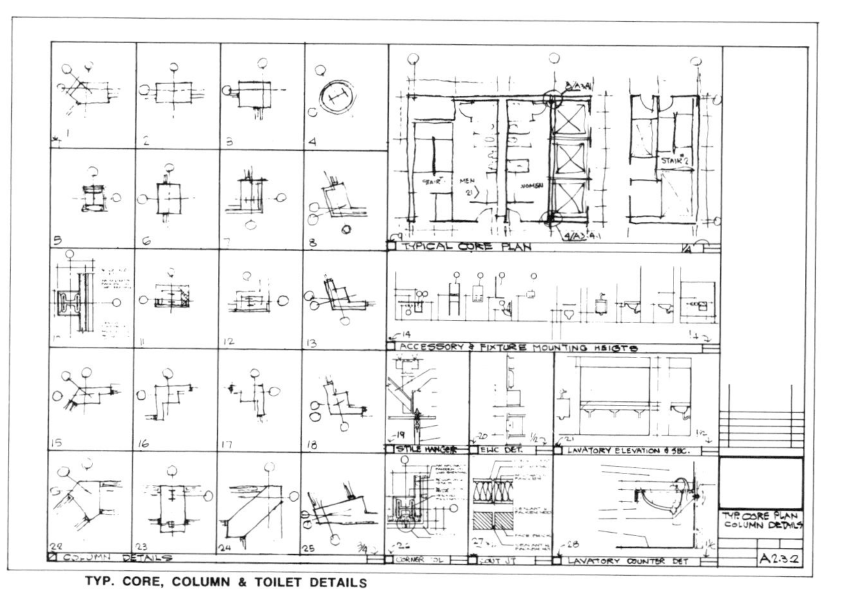

FIGURE 1-11. The Project Mock-Up (cont.)

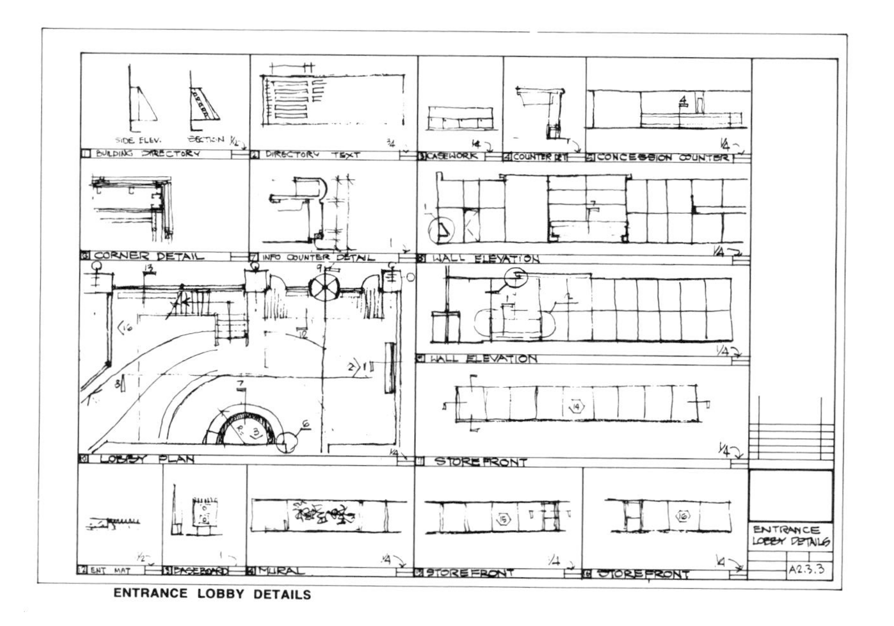

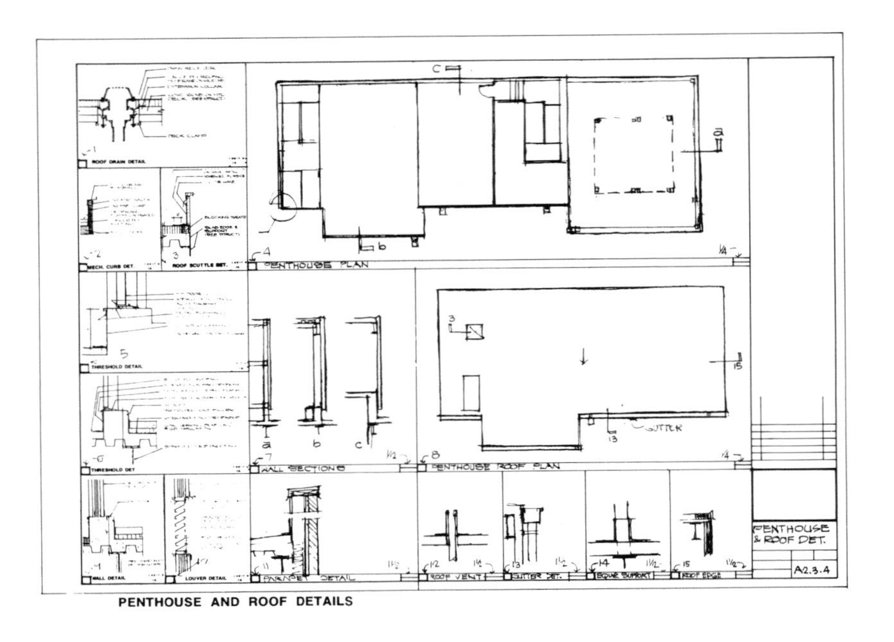

PENTHOUSE AND ROOF DETAILS

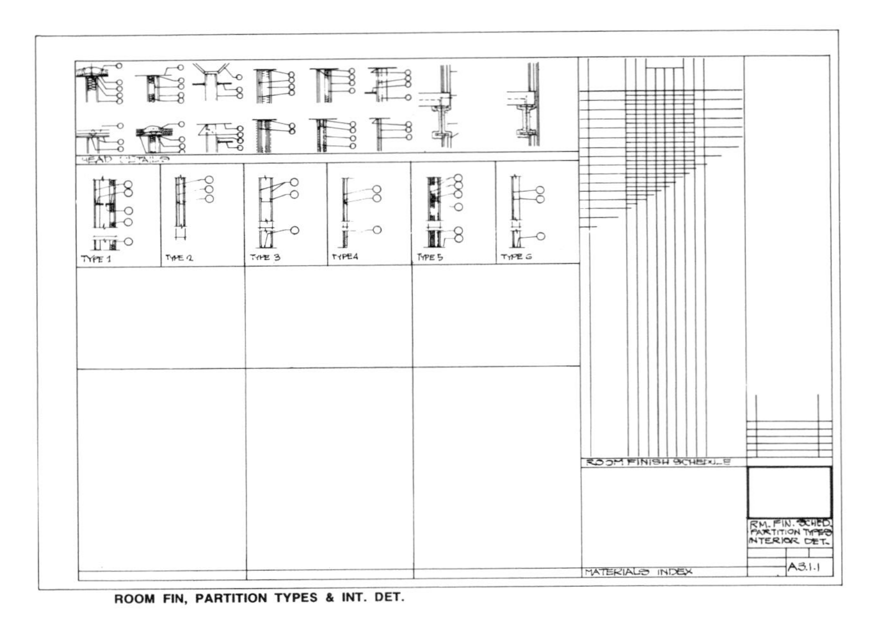

ROOM FIN, PARTITION TYPES & INT. DET.

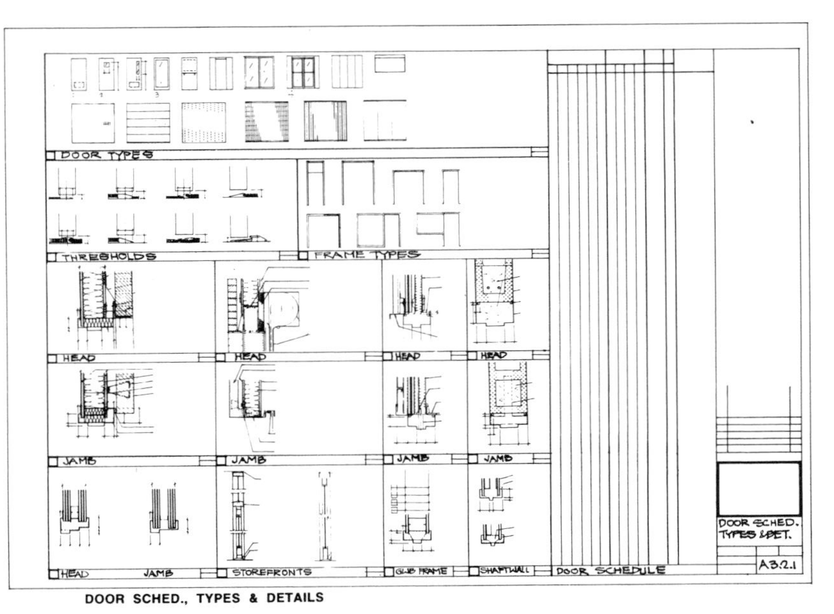

DOOR SCHED., TYPES & DETAILS

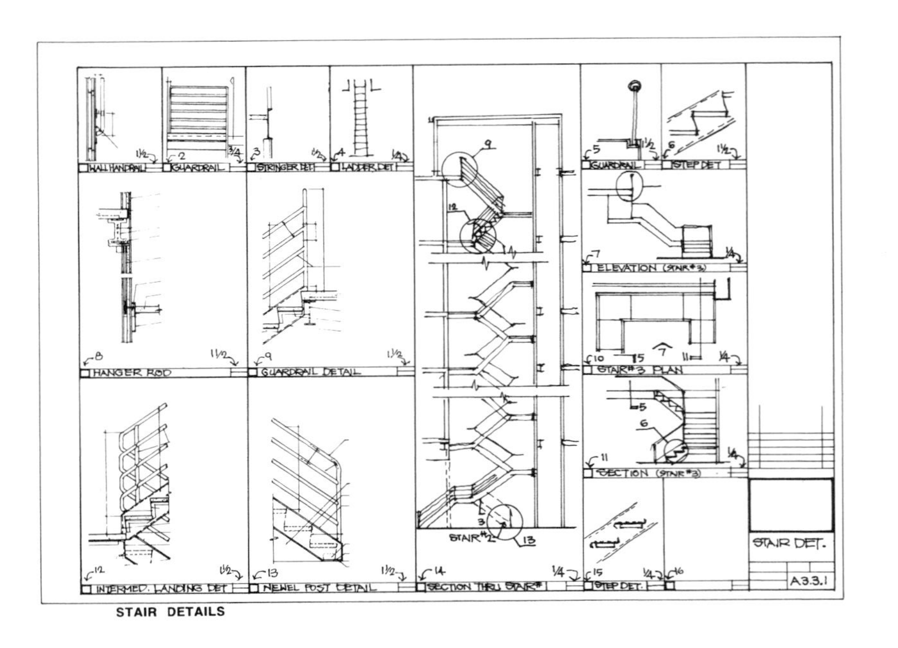

STAIR DETAILS

FIGURE 1-12. The Project Mock-Up (cont.)

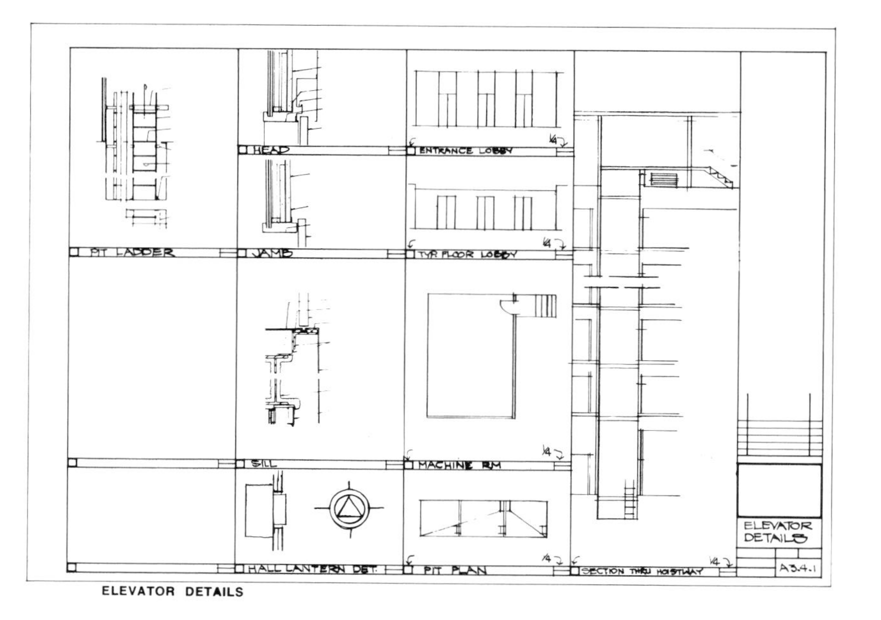

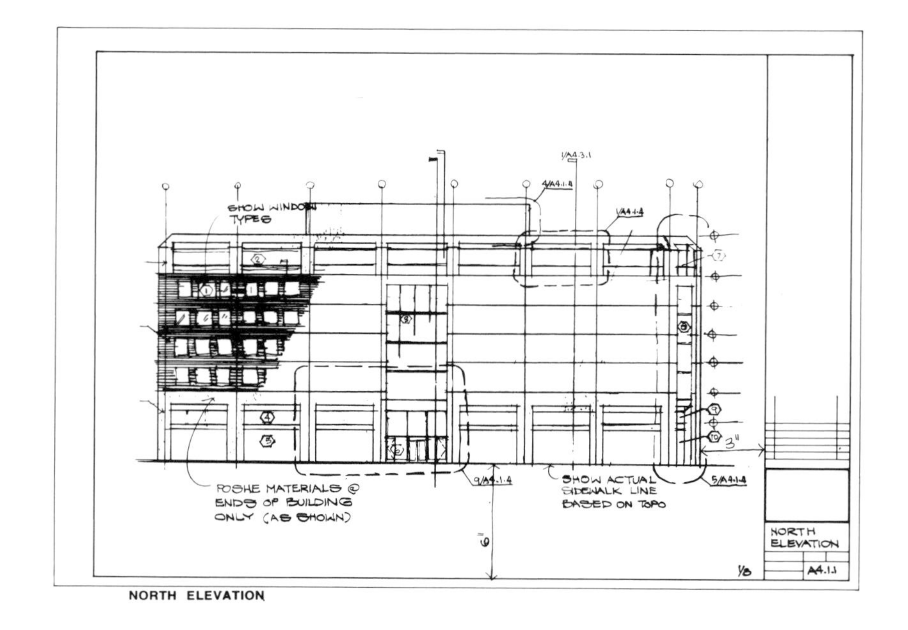

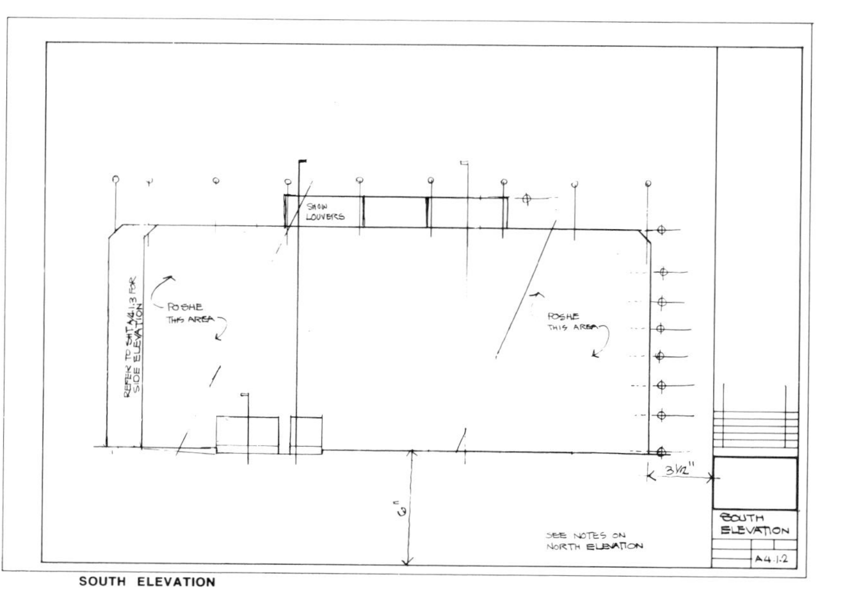

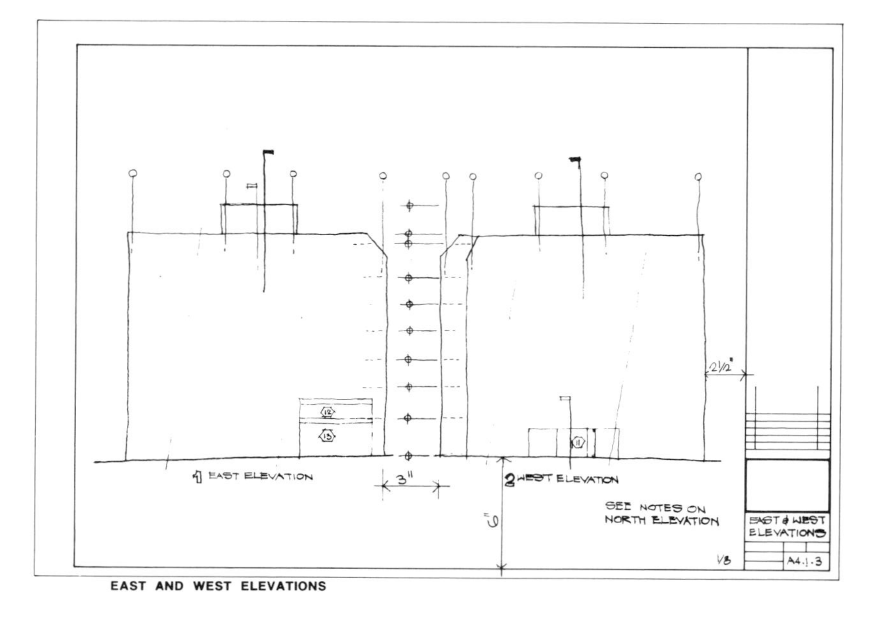

FIGURE 1-13. The Project Mock-Up (cont.)

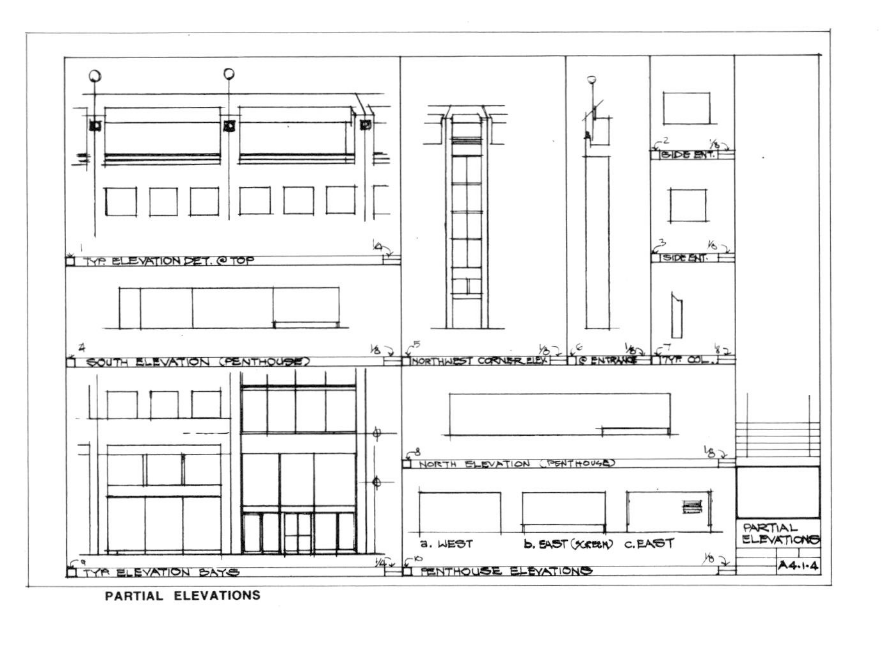

PARTIAL ELEVATIONS

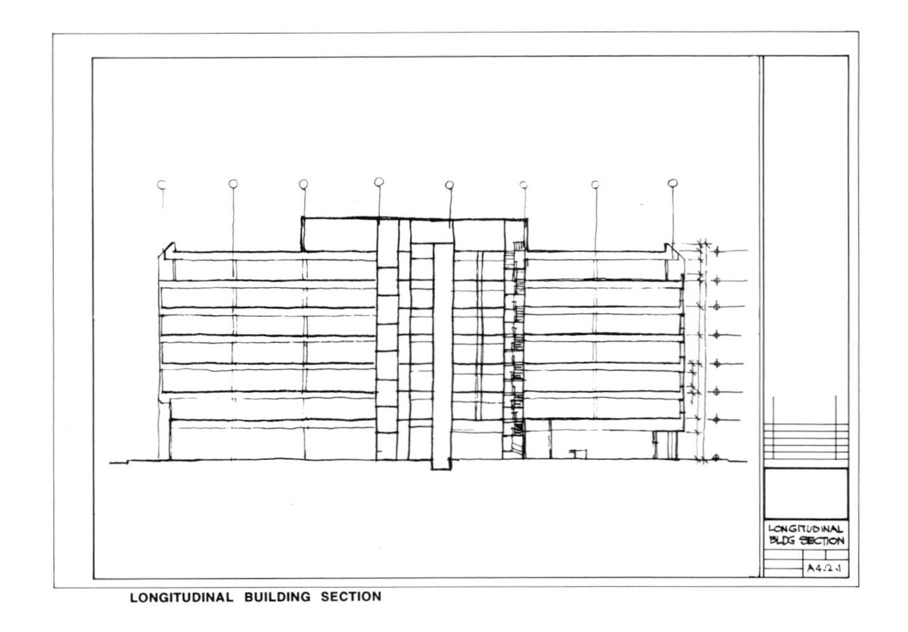

LONGITUDINAL BUILDING SECTION

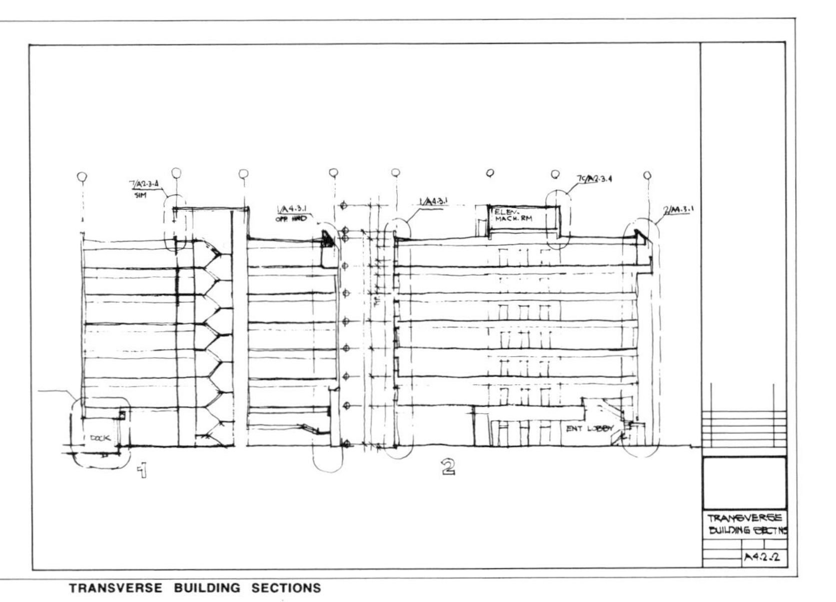

TRANSVERSE BUILDING SECTIONS

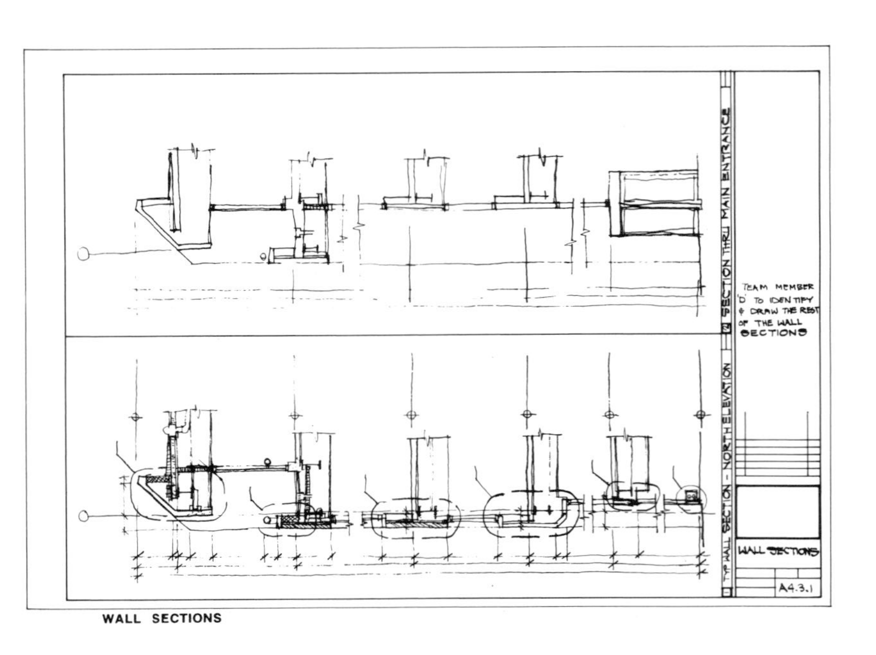

WALL SECTIONS

FIGURE 1-14. The Project Mock-Up (cont.)

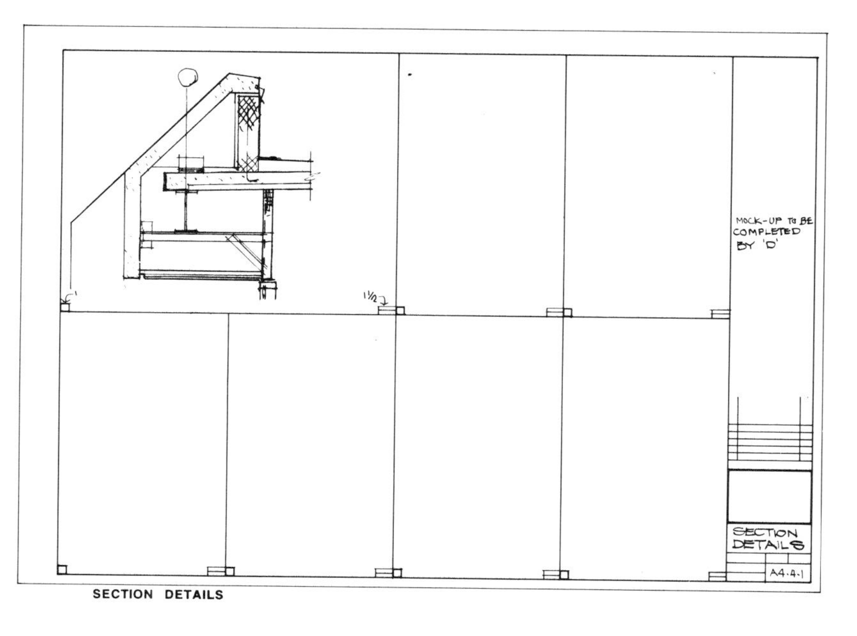

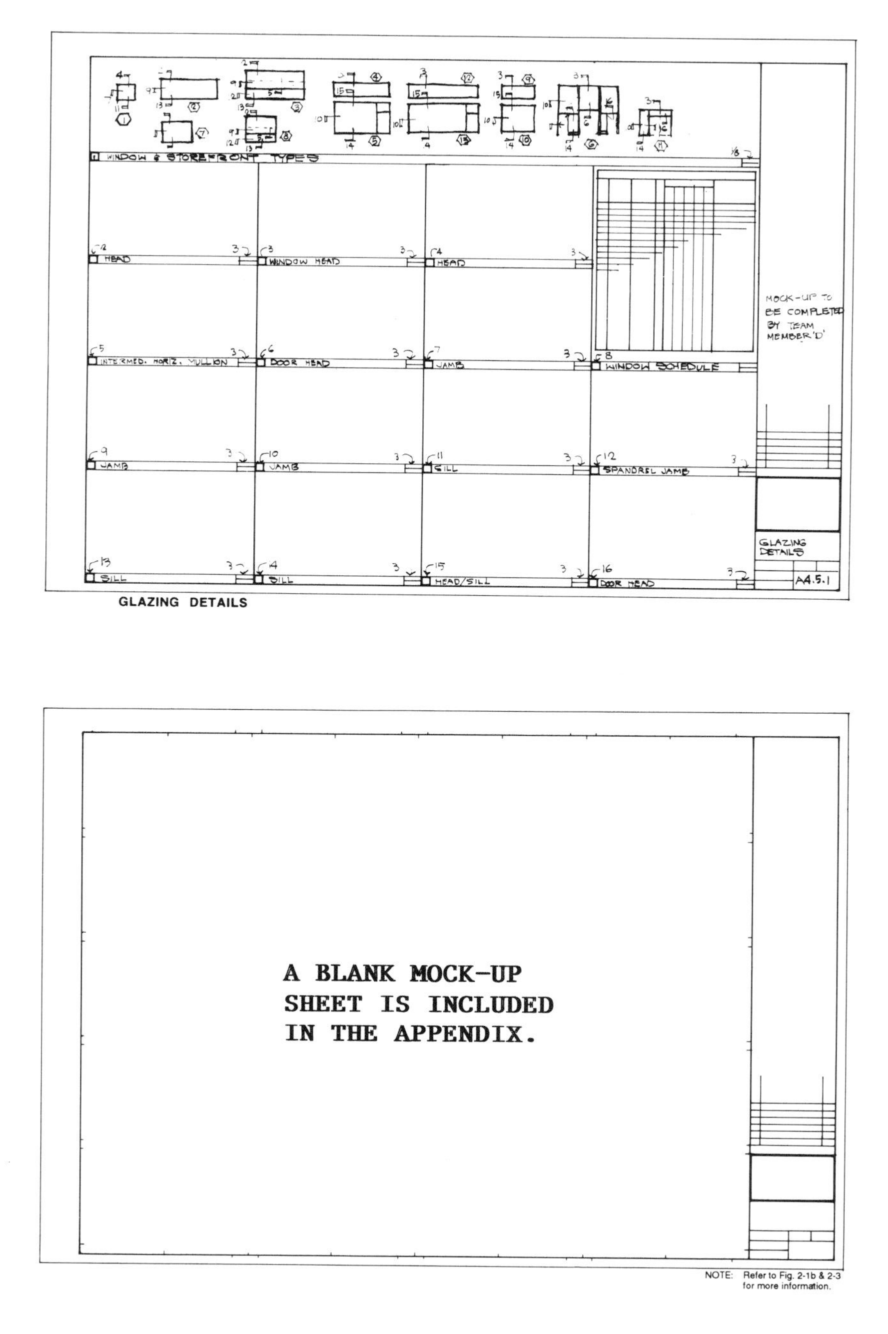

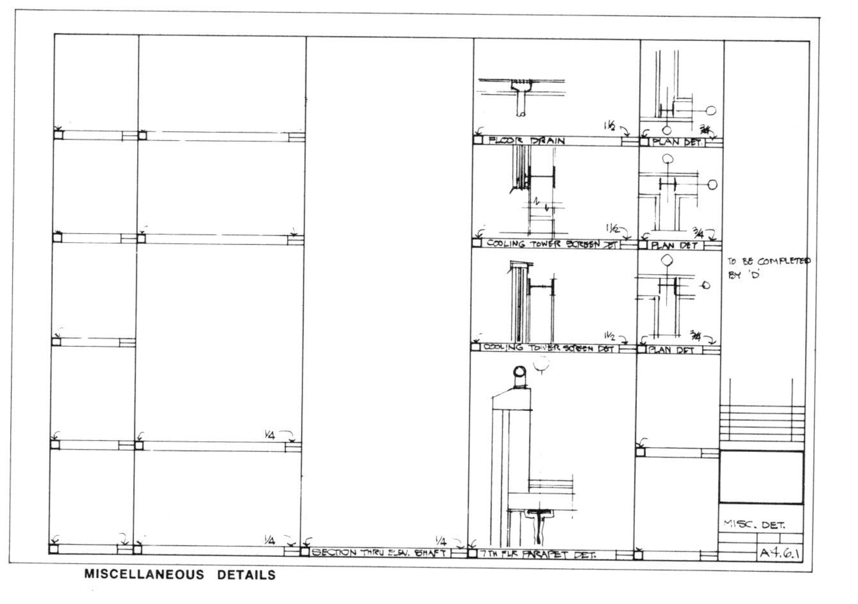

FIGURE 1-15. The Project Mock-Up (cont.)

hand and as a resource for future reference.

Finally, a project directory is started by the PM. It contains the names, addresses, and phone and fax numbers of each major participant in the project. The Team Leader should make copies to distribute to the team at the kick-off meeting and instruct them not to contact the client except through the PM. It is also advisable to channel all questions directed to the consultants through the TL.

These are the main start-up activities required to plan the project and give direction to the team. If done properly, the team leader's job becomes much easier. The members of the team will conduct their tasks efficiently and expeditiously, in most cases, without needing too much guidance.

If a staff member is a new employee, the TL should schedule a one-on-one "get acquainted" meeting to explore the employee's potential and find out what kind of work that person would like to do. Matching experience and interest fosters job satisfaction and professional growth.

1.3.3 The Main Effort

It is recommended that team members be located in close proximity to one another to increase the perception that they are a team, to allow them to exchange information more easily, and to facilitate their helping each other. Experience has proven that this arrangement saves time and raises the morale of the participants. Some offices assign each team member a taboret on wheels to place personal belongings in. Each time a new team is formed, the office is rearranged to allow the new team members to be located near each other. Team members should be chosen with skills compatible with the four task assignments.

The Team Leader, having determined the list of drawings for the CD phase, examines the design development (DD) drawings to determine what changes need to be done to each sheet to convert it into a construction drawing. In most cases, these sheets, unlike schematic design drawings, were drawn

Suggested Project File Headings

1. Owner's Information
   - Memos
   - Meeting Minutes
   - Design and Programming Information
   - Correspondence (original in Central File)
   - Project Directory
   - Transmittals
2. Team Management
   - Sheet Status Reports
   - Progress Graphs
   - Team Task Assignment Schedule
   - Mock-Up
   - Memos to and From Team Members
   - Team Meeting Minutes
3. Product Information
   - Product Reports
   - Cut Sheets
   - Brochures
   - Correspondence and Site Visits
   - Outline and Final Specifications
4. Calculations and Other Data
   - R-Values of Walls and Roof
   - Area Calculations
   - UL Design Equivalent Beams and Columns
   - Gutter and Downspout Sizing
   - Lintel Calculations
   - Glazing Calculations
   - Other
5. Code Issues
   - Code analysis
   - UL Design Selections
   - Code Updates
   - Zoning Requirements
   - Exit Calculations
6. Consultants
   - Telecon
   - FAX
   - Meeting Minutes
   - Transmittals
7. Cost Estimates
   - Preliminary
   - D.D.
   - Final
8. Contractor Data
   - Addenda
   - Change Orders
   - Field Visit Reports
   - Photographs
9. Tenant Correspondence

Note: All bulky items such as soils report, details from power company, research data, specs, owner's construction standards, cost estimates, etc. should be bound separately and placed on a shelf in the team leader's cubicle. After construction is completed, all file folder contents are bound in loose leaf folders. Each section outlined above is tabbed and the binder stored. Hanging file folder headings remain to receive material for a new project.

FIGURE 1-16. Project File Headings

with the intention of conversion to CD's. CADD-generated drawings, on the other hand, can be converted from phase to phase with relative ease.

After the start-up effort is completed and the team members are committed to the project by the PM in consultation with the TL, the latter schedules a kick-off meeting for the project and prepares an agenda. At the meeting, the PM introduces the TL and explains the goals and objectives for the project, the client's importance for the firm, distributes the project directory, and leaves the meeting. It is important that the team members understand that, although the PM is in charge of overall management, the TL is in charge of managing the team.

The TL then distributes copies of the mock-up to each member as well as emphasizes the need to apply the office drafting standards. An example of an initial meeting agenda is shown in figure 1-18.

To prepare for the team meeting on Monday, the TL checks the status of each sheet on Friday afternoon and plots the progress on the Progress Graph. An explanation of this process is included in the following section (1.4). Some TL's play a "guess what your percentage is going to be this week" game to spark interest in the process. They explain how the Project Sheet Status Report and Progress Graph work and ask each member to bring his version to the meeting to compare with the "official version." This is good training, but it may consume time the project could ill afford to lose. Team member D should be trained to participate in the progress-monitoring process in order to be prepared to assume this responsibility when he is promoted to the position of TL.

The Team Leader should schedule a weekly team meeting to discuss the project progress, resolve any problems, give new instructions, provide encouragement, and ask each member about what he plans to accomplish during the week. This meeting is important. Nothing demoralizes a team more than the perception that they are working in a vacuum, that nobody is aware of what they are doing or appreciates their effort, that no one is giving them direction, answering their

Product Information Report

Date:
Product Name:
Company Name:
Represented By:

1. Number of years on the market: *[The longer, the more of a chance that all the bugs have been ironed out.]*
2. Cost per unit (sq. ft., unit, etc.) Installed:
   Material Only:
   Labor Only:
   [Ask if general contractor overhead and profit is included in the price. If not, add an appropriate percentage (approx. 15%)]
3. Product characteristics (including limitations):
   [Ask questions applicable to the product. These may concern wear-resistance, combustibility, applicable standards (ASTM, factory mutual (FM), UL rating, etc.), weatherability, frequency of servicing, compatible sealants, does it require special handling equipment, water absorption, permeance, R-value, available finishes and any other pertinent information. If this information is included in the brochure, let the representative highlight it, attach the brochure to this form and have the secretary type it on the form for filing and future reference.]
4. Address of projects that used the product: *[Visiting these sites may be as informative as the brochure. These visits may highlight any shortcomings or flaws.]*
5. Availability and delivery lead time: *[Check against project schedule.]*
6. Comparable products to be specified "as equal": *[In many cases the representative will name products of higher quality that cost more. the specifier may use this as a clue to research other alternatives.]*
7. Contact for technical assistance:
   Name: Phone:
8. Local architects who used the product:
   Name: Phone:
   Name: Phone:
   [The older the project, the better because if problems have developed since its installation, the architect is sure to inform you about them.]
9. Standard and extended warranty information:
   [See Section 3.8 about roofing for warranty guidelines.]
10. Other information:
   [If the TL's time is very limited, this form should be used before the manufacturer's representative starts to talk at length about how wonderful his product is and, in some cases, how inferior his competitors products are. After the form is filled, he/she should politely but firmly inform the rep. that this was all the time available at present.]

FIGURE 1-17. Product Information Report (Blank form included in Appendix)

questions, or listening to their concerns. The weekly meeting prevents most of these potential problems from occurring.

It is suggested that these meetings take place each Monday between 11 A.M. and noon. This prevents the meeting from dragging on and wasting time. The TL should compliment the group if the work is satisfactory, and individually if some outstanding effort is done by a member. This must be done with total sincerity if it is to be taken seriously. The leader should exhort them to increase their effort if they are lagging behind. The TL should take notes during the meeting, conduct the meeting according to an agenda, and follow through during subsequent meetings on all open or unresolved issues.

Next, the TL must set target deadlines for securing the approval of the authorities having jurisdiction over the project, such as the fire marshal, the building department, the department of public works, or other entities identified by the PM. To prepare for that meeting, a set of drawings marked to show code conformance is prepared and discussed with the official. Any changes requested by that individual are implemented or a variance is requested if mitigating circumstances warrant it.

1.3.4 Coordination and Conclusion

The last two or three weeks of the CD phase must be set aside for coordination. This activity has a two-pronged goal—to iron out any conflicting information within the architectural drawing set and to resolve any conflicts between the architectural drawings and the drawings done by the consultants. Also, incomplete sheets are finished and the specifications are checked against the drawings during that period. If the method suggested by ConDoc (Chapter 2) is used, the task of checking drawings against specs becomes automatic and does not require time at this stage.

If drawings are done conventionally, the TL must point out that the drawings provide a graphic description of materials, relationships, and locations. This includes dimensions and thicknesses. (For example, if the thickness of a particular type of insulation varies at different locations of the project, these locations must be identified in the drawings rather than in the specs.) The specifications indicate quality, desired workmanship, and general contract requirements.

The TL must also emphasize that materials are to be identified by their generic name rather than by their trade

Initial Meeting Agenda

1. Introduction:
   - a. Introduction of new employees if added to the team
   - b. Project description
   - c. Goals
   - d. Design objectives
2. Monitoring Progress
   - a. Project Status Form (Section 1.4)
   - b. Progress Graph - deadlines (Section 1.4)
   - c. Weekly meetings
3. Assignments
   - a. Team Task Assignment Schedule (Fig. 1.8)
   - b. Invitation for each member to mark a preference as to the tasks he/she would like to do
4. Resources
   - a. Office standards folder
   - b. Standard details
   - c. Consultants - Project Directory (by PM)
   - d. Product information report (Fig. 1.17)
   - e. Project mock-up set (Section 1.3.02)
   - f. A layout of the office showing where each team member will be seated
5. Drafting Standards
   - a. Use plastic leads on mylar
   - b. Lettering, sizes
   - c. Check sets of drawings
   - d. Explanation of overlay drafting if applicable (see Chapter 2)
6. Vacation
   - a. Schedule
   - b. Substitutions
7. Quality Control
   - a. Check sets: Yellow for approved information
Red for corrections
Green for questions
   - b. Daily walk through to answer questions
   - c. Adherence to applicable specification sections and copies of the code form highlighted for each task.
8. Team Outings
   - a. Monthly
   - b. At the end of the project
9. Evaluation of Performance
10. Feedback and Comments

Note: Team members are to keep this agenda for future reference.

FIGURE 1-18. Example of a Kick-Off Meeting Agenda for the Architectural Team

name. For instance, gypsum wallboard should be referred to as GWB rather than Sheetrock—a trade name of United States Gypsum Company (USG). Lengthy descriptions belong in the specs not on the drawings, where they may conflict with the text in the specifications.

Checking Methodology

A common and efficient way to check the drawings uses the following method:

1. Make a print that includes the drawings from all disciplines. Mark the cover with the notation, FINAL CHECK SET, and write the date.
2. Use a yellow highlighter to mark everything you agree with. This forces you to read every word and check every dimension. This way, if you are interrupted during this tedious task, you do not have to start from the beginning. Everything that is not highlighted has not been checked yet.
3. Mark all corrections in red. Instruct the team to use a yellow highlighter over the red notations once the correction is made. After the corrections are made, retrieve the check set and make sure that all the red markings are covered with yellow.
4. Use a green pencil to write notes, ask questions, and to draw attention to items that need further investigation or coordination. This way, notes like "Why is this 20′-0″ partition unbraced?" will not accidentally be added to the final drawings, causing embarrassment. Check these notes also to make sure that something was done about them.
5. Use a light table to overlay the prints over plans from the other disciplines, to make sure that there is no conflict between them.

Coordination With HVAC Drawings

Check heating, ventilation, and air-conditioning (HVAC) drawings to make sure that:

1. There is enough clearance for ducts to avoid interference with structural beams.
2. Clear floor openings for vertical shafts are large enough to contain all the ducts, pipes, electrical bus bars, and other mechanical items with enough clearance. Also check if any of these items need access for maintenance. If so, make sure that access panel locations shown on the mechanical drawings are architecturally acceptable.
3. Ducts passing through fire-rated walls have fire dampers to protect the wall opening if required by code.
4. All louvers shown on the HVAC set are also shown on the elevations and that the type of louver specified provides the same free area required by the engineer.

Coordination With Plumbing

1. Check clearance for horizontal runs, making sure that no pipes penetrate through beams unless approved by the structural engineer and that no pipes pierce through ducts. Allow for the slope of horizontal roof leaders as well as for roof slopes when checking the depth of the space above the top-floor ceiling.
2. Make certain that vertical pipe risers are enclosed in shafts or furred-out spaces, that vertical roof leaders are located in shafts or wet columns (Fig. 3-26), that their location does not conflict with beam locations, that fire risers and fire hose cabinets are at the same locations shown on architectural plans, and that risers fit within the detailed furring (Fig. 4-68) without conflicting with beam locations.

Coordination With Electrical

1. Verify that there is no conflict between reflected ceiling plans and lighting plans.
2. Make sure that all items using power are indicated on the power plans (see Power Requirements Worksheet in Chapter 2).

Coordination With Structural

1. Ascertain that all column locations are identical on structural and architectural plans.
2. Check beam depths for clearances.
3. Check top of slab elevations and top of steel elevations, especially at the roof, and make sure that roof slopes are identical to those shown on the roof plan if the structure is sloped toward the drains.
4. Check the lintel schedule and structural stud sizes (if designed by the structural engineer) against the wall sections.

These are but a few examples of items to be aware of. These examples are to be used as a supplement to a more methodical and thorough checking system such as the one included in Volume II of the *Architect's Handbook of Professional Practice* issued by the AIA. The Team Leader must keep the corrected check set as a record until the construction is finished.

Coordination with consultant drawings may require a final meeting with the project managers for the other disciplines to resolve any inconsistencies. One method used on large projects involving joint-venture firms is for each discipline to fill out a form similar to the one shown in figure 1-19. This method reduces the items to be discussed at the meeting to the ones that are not resolved in the response column.

Resolving contradictions and correcting errors must be taken very seriously, must have enough time allocated for it, and must be completed before the drawings are issued for bids.

1.4 MONITORING PROGRESS

1.4.1 General

Different firms use different ways to monitor progress. There is no "one correct and proper way" to gauge the percentage of completion of the work. Small offices use the simple method of "gut feel." The Project Manager reviews a set of the most recent prints and, based on his experience, says, "This looks like 45 percent complete." Oddly enough, this estimate in most cases is not too far from the actual percentage because the project is relatively small and the individual is usually experienced in that type of work.

On medium and large projects, this approach is not recommended. There are too many people involved and any mistake would result in costly missed deadlines and under-billing or overcharging the client.

The better-organized firms use tools to measure progress that are similar to the Project Sheet Status Report (Fig. 1-20 and 1-21) and the Progress Graph (Fig. 1-22).

1.4.2 The Project Sheet Status Report

This form is used to track the progress made on each sheet during the course of the Construction Drawing Phase. Together with the Progress Graph, which plots progress graphically, they form a good progress-monitoring system. The TL must bear in mind that determining the progress of a project is an imprecise process. With time and experience, he becomes more adept at estimating the percentage of completion for each sheet. Time must be allocated for corrections, coordination, and referencing details. It is advisable that the TL set aside a contingency to cover the possibility of adding sheets to the set of drawings.

ABC, Inc.
Architects & Planners

Project Name:
Project No. :
Phase :

Sht._ of __

Sht. No.	Comment No.	Review Comments By: Date:	Designer Reply By: Date:	Action By: Date:

FIGURE 1-19. Drawing Coordination Form

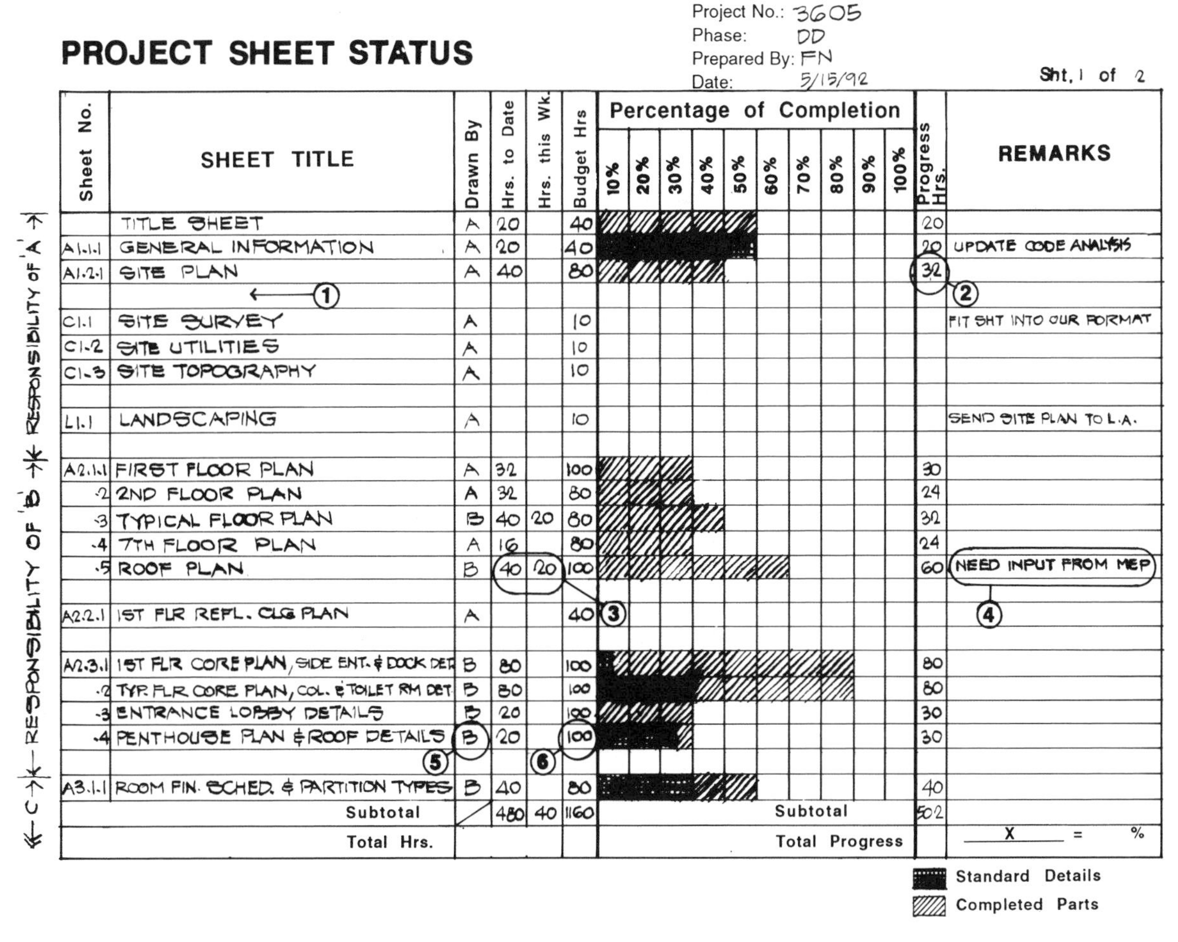

PROJECT SHEET STATUS

Project No.: 3605
Phase: DD
Prepared By: FN
Date: 5/15/92
Sht. 1 of 2

Sheet No.	SHEET TITLE	Drawn By	Hrs. to Date	Hrs. this Wk.	Budget Hrs	10%	20%	30%	40%	50%	60%	70%	80%	90%	100%	Progress Hrs.	REMARKS
	TITLE SHEET	A	20		40											20	
A1.1.1	GENERAL INFORMATION	A	20		40											20	UPDATE CODE ANALYSIS
A1.2.1	SITE PLAN	A	40		80											32	
	← (1)															(2)	
C1.1	SITE SURVEY	A			10												FIT SHT INTO OUR FORMAT
C1.2	SITE UTILITIES	A			10												
C1.3	SITE TOPOGRAPHY	A			10												
L1.1	LANDSCAPING	A			10												SEND SITE PLAN TO L.A.
A2.1.1	FIRST FLOOR PLAN	A	32		100											30	
.2	2ND FLOOR PLAN	A	32		80											24	
.3	TYPICAL FLOOR PLAN	B	40	20	80											32	
.4	7TH FLOOR PLAN	A	16		80											24	
.5	ROOF PLAN	B	40	20	100											60	NEED INPUT FROM MEP
																	(4)
A2.2.1	1ST FLR REFL. CLG PLAN	A			40 (3)												
A2.3.1	1ST FLR CORE PLAN, SIDE ENT. & DOCK DET	B	80		100											80	
.2	TYP. FLR. CORE PLAN, COL. & TOILET RM DET	B	80		100											80	
.3	ENTRANCE LOBBY DETAILS	B	20		100											30	
.4	PENTHOUSE PLAN & ROOF DETAILS	B	20		100											30	
	(5)			(6)													
A3.1.1	ROOM FIN. SCHED. & PARTITION TYPES	B	40		80											40	
	Subtotal		480	40	1160	Subtotal										502	
	Total Hrs.					Total Progress											X = %

FIGURE 1-20. Project Sheet Status Report (A modified version of a form used by F&S Partners, Dallas, TX)

NOTES:

1. Leave gaps after each group of sheets to allow for additional sheets.
2. These numbers represent the actual progress achieved on each sheet. To arrive at that number, multiply the number of budgeted hours by the number of shaded squares and divide by 10. For Sheet A1.2.1 (Fig. 1-20), $80 \times 4 \div 10 = 32$ hours.
3. These numbers are taken from each team member on Friday afternoon. If the total is much higher than the number indicated in the progress column, say 80 instead of 60, a private conversation with Team Member B is called for to determine the cause and devise remedies.
4. These are planning notes indicating what needs to be done. Pursue them from week to week and erase them when accomplished.
5. See Team Task Assignment Schedule (Fig. 1-8) for a week-by-week allocation of time for each sheet. This schedule is related to the target deadlines on the Progress Graph (Fig. 1-22). Each team member is assigned a group of related sheets that he or she is responsible for.
6. On average, 80 hours per sheet are allowed for CD drawings. For more complicated sheets, more time may be allocated. These numbers may also be affected by the negotiated fee. In this example, more hours are allocated to the first-floor plan because the other plans repeat much of the work that goes into that first sheet. The repetitious parts—the core, columns and grid, outline of exterior walls, and exterior dimensions—were drawn and reproduced for the second, typical, and seventh-floor plans on wash-off Mylar (by Du Pont) and then the parts representing differences on each plan were drawn to complete them. Allow more time for detail sheets. Depending on the degree of complexity, the skill of the person entrusted with them, and the number of details per sheet, the time allocated ranges from 100 to 120 hours in most cases.

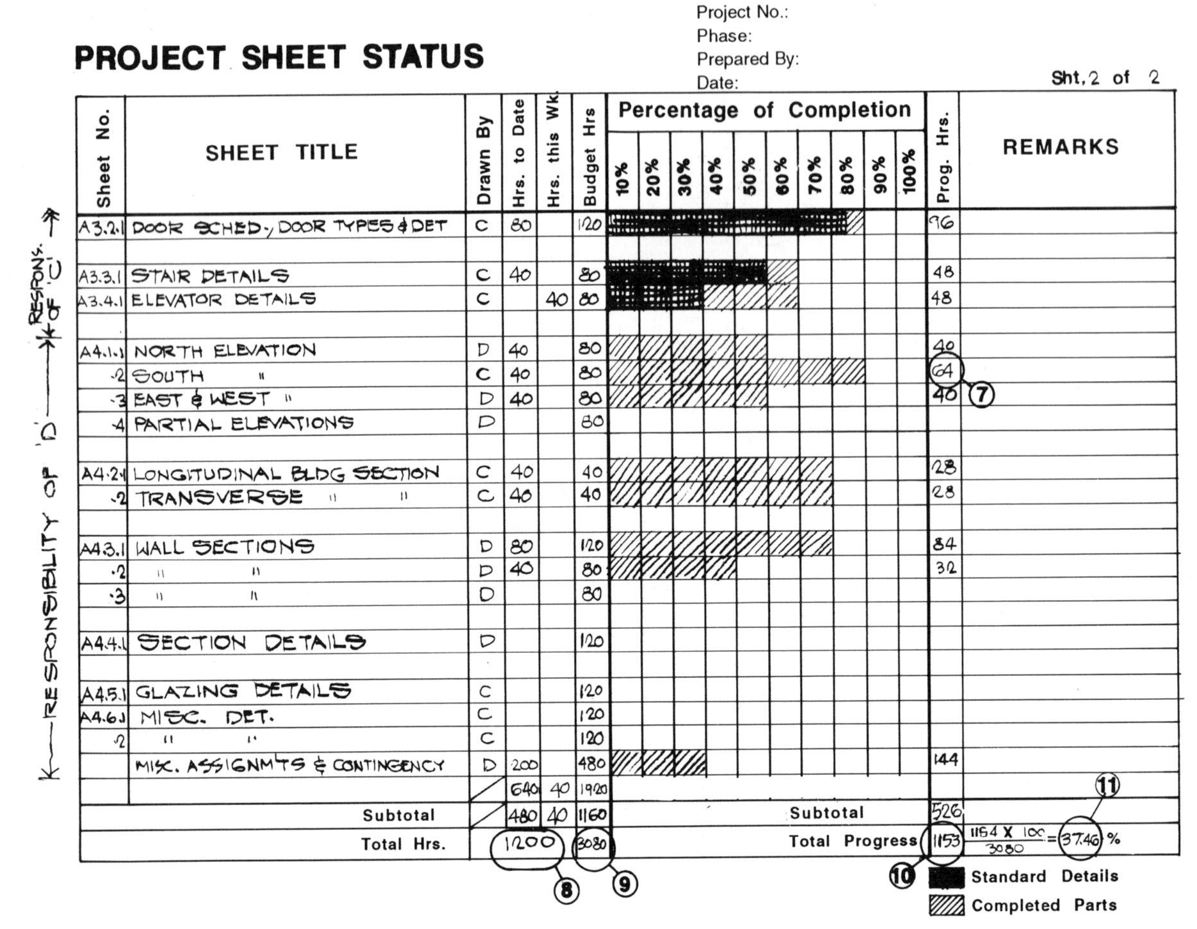

PROJECT SHEET STATUS

Project No.:
Phase:
Prepared By:
Date:

Sht. 2 of 2

Sheet No.	SHEET TITLE	Drawn By	Hrs. to Date	Hrs. this Wk.	Budget Hrs	Percentage of Completion (10%–100%)	Prog. Hrs.	REMARKS
A3.2.1	DOOR SCHED., DOOR TYPES & DET	C	80		120		96	
A3.3.1	STAIR DETAILS	C	40		80		48	
A3.4.1	ELEVATOR DETAILS	C		40	80		48	
A4.1.1	NORTH ELEVATION	D	40		80		40	
-2	SOUTH "	C	40		80		64	(7)
-3	EAST & WEST "	D	40		80		40	
-4	PARTIAL ELEVATIONS	D			80			
A4.2.1	LONGITUDINAL BLDG SECTION	C	40		40		28	
-2	TRANSVERSE " "	C	40		40		28	
A4.3.1	WALL SECTIONS	D	80		120		84	
-2	" "	D	40		80		32	
-3	" "	D			80			
A4.4.1	SECTION DETAILS	D			120			
A4.5.1	GLAZING DETAILS	C			120			
A4.6.1	MISC. DET.	C			120			
-2	" "	C			120			
	MISC. ASSIGNM'TS & CONTINGENCY	D	200		480		144	
			640	40	1920			(11)
	Subtotal		480	40	1160	Subtotal	526	
	Total Hrs.		1200 (8)		3080 (9)	Total Progress	1153 (10)	1154 X 100 / 3080 = 37.46 %

FIGURE 1-21. Project Sheet Status Report (cont.)

NOTES:

7. Team member C has accomplished 64 hours of progress in 40 hours. He should be complimented at the weekly team meeting.
8. This number represents the total of the four numbers above it. Compare it to the total shown under the progress column (1,153 in this case). This indicates that this team is lagging behind and needs to be urged to exert extra effort. Another possibility is that the Team Leader has underbudgeted the project. If underbudgeting is the problem, he should pitch in to help expedite the work. Each pay period, check these numbers against the actual hours charged to the project. This information is readily available from accounting.
9. This total is related to the percentage of the fee allocated to finishing the drawings. It should not include time spent on the project by the Team Leader, the Project Manager, or the Partner-in-Charge.
10. This number represents the number of hours normally required to achieve the progress shown. Because standard details were used, approximately 258 hours, representing 8.3 percent of the project, were saved. Details from Chapter 4 were indicated in this example. Adding exterior standard details can add 3 to 5 percent to that figure, a very substantial savings.
11. This is the number to plot on the progress graph. This must be done each Friday to be ready for distribution to the team during the Monday morning meeting.

1.4.3 The Progress Graph

The Progress Graph is a tool used to track the weekly progress of a project during the CD phase. On some projects such as fast-track, the DD phase is also included. The graph has three functions. The first is to monitor progress graphically, the second is to act as a calendar of target deadlines, and the third is to show the team graphically that their effort is bearing fruit. This last function is similar to the bar graph used by organizations showing their progress toward a fund-raising goal. It motivates the team and shows at a glance whether the past week's progress is similar, ahead, or behind the previous week's. It should be an important ingredient of the weekly project meeting between the TL and the Project Manager, and between the TL and the team.

Using the Progress Graph in conjunction with the Project Sheet Status Report introduces a method to control and monitor the progress of the project. After applying it on a couple of projects, the Team Leader will be able to sleep nights knowing exactly how the work is progressing, and he will be able to communicate that information to the Project Manager, who, in turn, will be able to sleep nights.

The graph shown in figure 1-22 is for a hypothetical office building. The following is a step-by-step explanation of how to construct the graph.

1. At the top of the graph, write all the target dates beginning with the most obvious, such as the start and finish dates and other deadlines based on the time needed to deliver drawings to the consultants. Give a copy to the consultants and ask them to fill out the dates they would deliver information such as wall and roof R-values, structural column and beam sizes, etc. After their response is received, check if any of their dates conflict with project deadlines. Consultants have their own agenda dictated by their work load. Unless committed at the start of the project, they may deliver this information later than the time needed to utilize it.

   For example, some mechanical engineers deliver their final drawings just a day or two before the final deadline. This does not give the team any time for coordination. To prevent this from happening, the Team Leader and the Project Manager must get a firm commitment from the consultants to abide by a target deadline at least ten days before the final deadline.
2. Next, draw the baseline by connecting the project start at 0 percent to the project finish at 100 percent.
3. Construct the modified baseline by determining the zones that will show the least progress. These are located at the start (during the planning stage), between the DD and CD phases, and at the finish during the coordination and corrections stages. Connect these points as shown.
4. Plot the actual progress line each week after preparing the Project Sheet Status Report (Figs. 1-20 and 1-21). If the line strays too far below the modified baseline, help is needed. If above—a rare happening—it means there are a few good people on the team (a lucky situation) or the number of hours had been overestimated. In either case, be sure to compliment the team on the progress that is being made.

The Project Sheet Status Report and the Progress Graph may look like a complicated and time-consuming activity. I remember resenting having to do them the first time I employed this method, but as time went by and I became more skilled at it, I found that I could not operate without them. True, it takes some time and effort at the start of the project, but it is well worth it, and it does not take more than an hour at the end of the day on Friday to figure the percentage and record it on the graph.

These two forms are a modification of a system used at F&S Partners, Inc., in Dallas, Texas, one of the better-organized firms for which I worked.

Monitoring progress, in most cases, helps avoid the frantic charrette activity that may be required in some offices that do not use these management tools.

To recap, managing a project may be done by "gut feel" or through a thoughtful, well-controlled process. This chapter concentrated on the latter method. It proceeded on the assumption that the project is a sizable one managed by a Project Manager and a Team Leader. It analyzed the tasks assigned to the team and explained a method of monitoring progress.

Project management is not a precise science. It must be based on the goals of producing a set of well-coordinated construction drawings, documenting all communications, maintaining good relations with the client and the team, keeping track of the schedule, and staying within the budget.

The Team Leader describes the task, provides as many sources of information as possible *to each team member,* sets the deadlines, and leaves that person to devise ways to accomplish it. General George S. Patton described this process admirably when he said, "Never tell people how to do things. Tell them what to do and they will surprise you with their ingenuity."

1.4.4 Sources of More Information:

Burstein, David, and Stasiowski, Frank. 1982. *Project Management for the Design Professional.* New York: Whitney Library of Design.

DePree, Max. 1989. *Leadership Is an Art.* Garden City, NY: Doubleday & Company, Inc.

Haviland, David, ed. 1988. *The Architect's Handbook of Professional Practice*, 11th ed., 4 vols. Washington, D.C.: The American Institute of Architects.

Haviland, David. 1981. *Managing Architectural Projects: The Effective Project Manager.* Washington, D.C.: The American Institute of Architects.

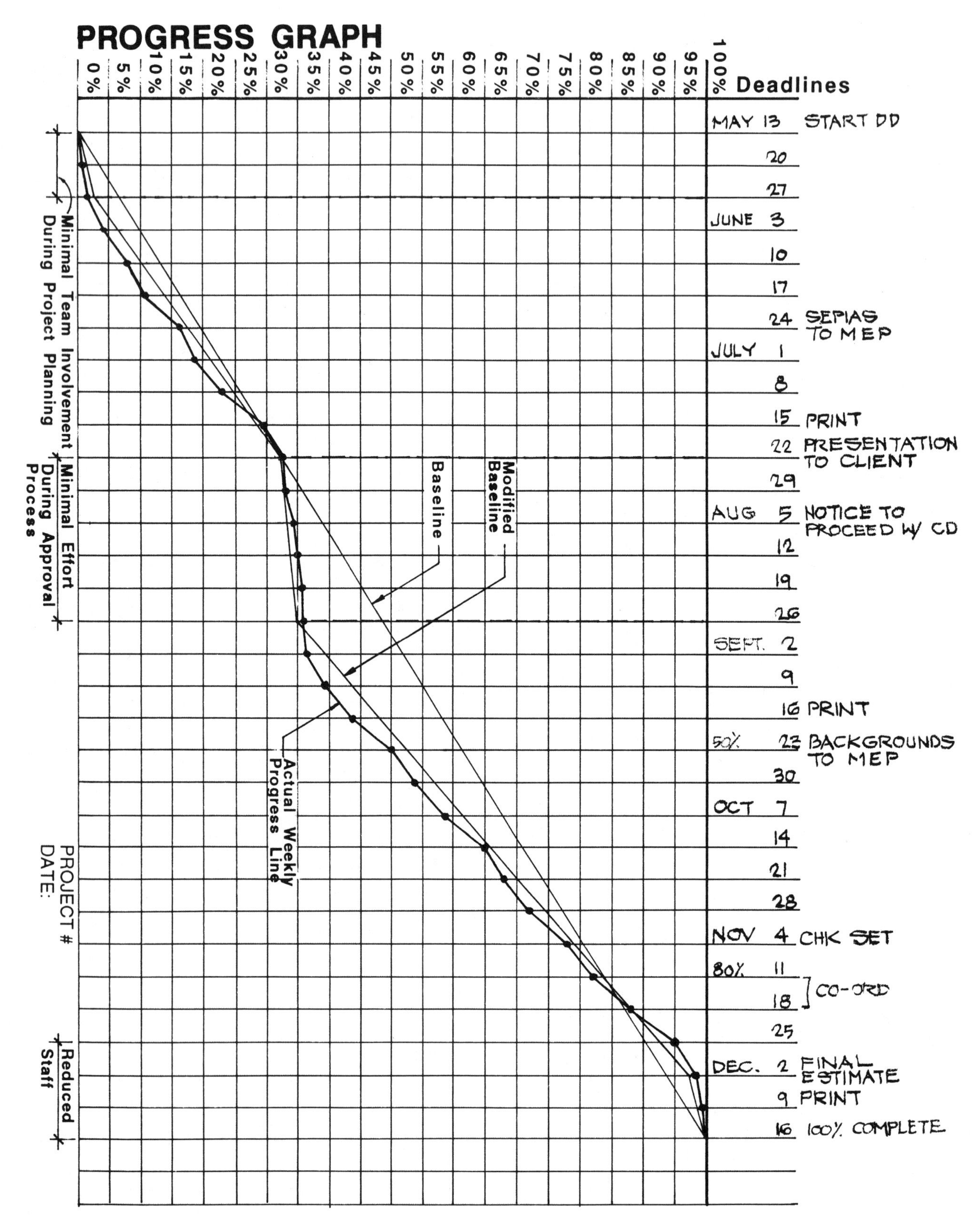

FIGURE 1-22. Progress Graph (Based on a form used by F&S Partners, Dallas, TX)

Haviland, David. 1981. *Managing Architectural Projects: The Process*. Washington, D.C.: The American Institute of Architects.

Haviland, David. 1984. *Managing Architectural Projects: The Project Management Manual*. Washington, D.C.: The American Institute of Architects.

Haviland, David. 1981. *Managing Architectural Projects: Three Case Studies*. Washington, D.C.: The American Institute of Architects.

Nigro, William T. 1987. *Redicheck Interdisciplinary Coordination*. The Redicheck Firm: Stone Mountain, GA.

References:

Burstein, David, and Stasiowski, Frank. 1982. *Project Management for the Design Professional*. New York: Whitney Library of Design.

Haviland, David, ed. 1988. *The Architect's Handbook of Professional Practice*, 11th ed., 4 vols. Washington, D.C.: American Institute of Architects.

Haviland, David. 1981. *Managing Architectural Projects: The Process*. Washington, D.C.: The American Institute of Architects.

OFFICE STANDARDS

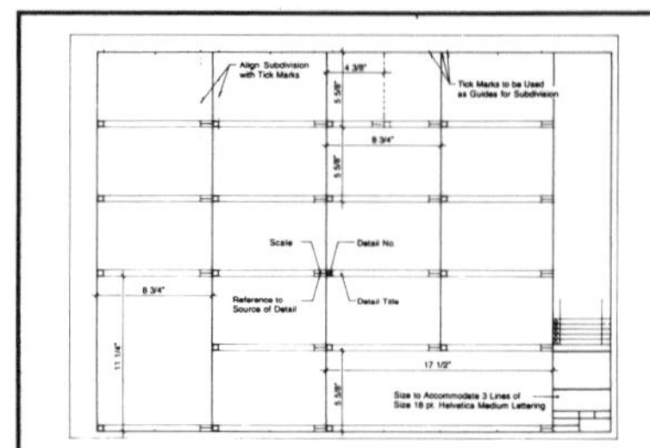

Office staff members have different levels of experience and varying educational backgrounds. Some may still be attending school and working during the summer vacation to gain experience. Some are interns preparing to take the registration test and a few may be transplants from other firms that have a different philosophy toward construction drawings. To bring all these people with different backgrounds into harmony to produce a well-organized, easy to comprehend set of construction drawings and to complete it in the shortest time possible requires a set of rules that each team member must follow. These rules are referred to as office standards.

Many large- and medium-size offices have elaborate, well-organized manuals that are taken seriously by the team and applied diligently to produce a good set of construction documents. Some offices, however, issue manuals that fall far too short of providing the appropriate framework for that purpose. The inevitable result is that almost everybody ignores these so-called standards and follows standards he or she brought from other offices. While this may result in a good set of drawings, one must consider the following side effects: Junior team members will become disoriented every time they are assigned to a new Team Leader who uses different standards; time will be lost while each member of the team gets acquainted with different standards every time a new project is started; and quality control will be hard to enforce.

A well-organized construction drawings production manual should contain the following components: drafting guidelines, sheet format and subdivision, sheet numbering, overlay drafting, lettering (manual drafting only), dimensioning, schedules and work sheets, and general conventions.

Each section provides a fast reference for the team members to use in organizing the project drawings. Firms that use CADD (computer-aided design and drafting) may add a section containing guidelines to the particular system they use.

The office standards manual should be easy to add to. A three-ring binder with the office logo on the cover and tabbed indices marking each section seems to be the most practical kind. A copy should be placed at each workstation. The following sections describe each of the eight components listed above.

2.1 DRAFTING GUIDELINES

2.1.1 General

This section of the manual provides answers to the most frequently asked questions by newcomers to the office. It provides hints on how to organize each component of the drawings. The intent is to illustrate one of the many ways to do the drawings properly. The main goal should be to produce a clear set of drawings and avoid duplication of information.

2.1.2 Plans

1. Orient plans with the north arrow pointing toward the top of the sheet whenever possible. Enlarged plans drawn at ¼-inch scale and plan details drawn at ¾ or ½ inch should have the same orientation as the floor plan unless they relate to a section on the same sheet. An example is a stair plan placed under a stair section which is projected from it. Place a north arrow on all plan sheets. If the

sheets are to be reduced to half size, as happens in most projects, a graphic scale should be drawn on the sheet.

2. For manual drafting, use plastic lead instead of regular lead on polyester film (commonly know as Mylar, a trademark of DuPont) to prevent smearing. It is hard for those accustomed to using lead to switch over to plastic lead because it seems to break more often and offers less control over line thickness to those unaccustomed to using it. Using .5-mm pencil holders for drafting and .9-mm for lettering seems to work for most people. Ink column grid lines on the back of plan sheets and draw the plan at 1/8-inch scale.
3. If the plan is too large to fit on the project sheet, divide the plan into segments, identify each by an alphabetical letter, and place a match line at the border between it and the other segments (Fig. A-17). Locate these match lines where they will not interfere with other information shown on the plan. A key plan must be placed on each plan, showing the area of the segment indicated by shading.

   Each segment should be continued a couple of inches beyond the match line. This is customary in architectural drawings. Engineers, in many cases, prefer to end the plan at the match line. This is done to prevent estimators from making errors by counting items such as columns, equipment, etc., twice—once beyond the match line (where it should not be counted) and once in the segment. Large projects with multiple match lines may require a sheet containing a set of reduced plans showing the whole project at the start of the group of plans to give an overall view of the project at all its levels.
4. Equipment, furnishings, and items that will be installed in the future, but which form no part of the contract, should be indicated by a dashed line and noted as "NIC" (not in contract).
5. Information shown on large-scale plans should not be repeated on floor plans. This applies to dimensions (Sec. 2.6), equipment, and casework. Toilet room fixtures should be shown on both because the floor plan is used as background by the plumbing engineer.

   The reason for this rule is that if a change is made on the large-scale plan, the change would have to be made on the floor plan also if both show the same information, an unnecessary duplication of effort and a waste of time. Partition types should be shown on floor plans only. It is easier for the subcontractor to refer to only one drawing for information pertaining to all partitions.

   These rules do not apply to CADD-generated drawings where both the large plan and the floor plan layers belong to the same active file. This means that changing one, automatically changes the other.

   While most specifications stipulate that large-scale drawings have precedence over smaller-scale drawings, in case of conflict between the two, it is better to prevent the conflict by avoiding the duplication of information in the first place (manual drafting only).
6. Items to be shown on most floor plans are column grid lines and circled designations; dimensions; exterior walls; partition types; room names and numbers; door numbers; expansion joints; drinking fountains, plumbing fixtures, fire hose cabinets (if required); floor pads, drains; casework; and equipment.

   Circled references must be indicated for large-scale plans and plan details, building sections, wall sections cuts (these may be shown on the elevations), interior elevations, and interior window types. Some offices show exterior window and louver types on the plans. Others prefer showing them on the elevations.

   Team members entrusted with the plans may use this as a checklist. Other items may be required by different projects.
7. Reflected ceiling plans should identify areas with exposed structure, show ceiling grid, registers for supply and return air, light fixtures, speakers, and coves or special designs. The ceiling plans should include notations identifying each of these items. Draw door heads but not door swings.
8. Roof plans may be drawn either at ⅛ inch or 1/16 inch and should include drains, scuppers or interior overflow drains, expansion joints, major roof penetrations, mechanical equipment supports and screens (if required), skylights (if included), window-washing davit supports (high-rise projects only), traffic pads, scuttles or smoke vents, slope arrows and rate of slope, and areas of tapered insulation.
9. Refer to Section 2.4 for information on overlay drafting for plans.

2.1.3 Elevations

1. Draw elevations at ⅛-inch scale. If 1/16-inch scale is used for tall buildings or very large, uncomplicated elevations, typical parts of the elevation must be shown at a larger scale.
2. Some offices prefer to show the window and louver types on the elevation. If this is done, do not repeat that information on the plans. There are two advantages to showing this information on the elevations. The first is that the symbols can be accommodated more readily on an otherwise empty drawing. This avoids the congestion that sometimes happens when exterior dimensions vie for space with these symbols. The second is that windows and louvers are sometimes placed one on top of the other in one floor height. This is easier to show on the elevation than on the plan.
3. Identify materials and show the limit of each. Also show relevant column grid lines.
4. Show expansion and control joints, roof scuppers and downspouts (if required), handrails, copings, and other features.
5. Relate the ground line to the actual ground elevations

taken from the site plan, spot elevations, or design contours.

6. Write the floor name next to a partial dashed line indicating the finish floor location. Detailed dimensions and floor elevations belong on the building section, wall section, and section detail sheets. Add a note to that effect on the elevation sheets.
7. Show building and wall section cuts. Some offices may prefer to show the building sections on the plans only and the wall sections on the elevations only, to prevent duplication of information.
8. Show column centerlines and grid numbers above the elevations to provide reference points.

2.1.4 Building Sections

1. Building sections are the first of a four-component section group composed of building sections, wall sections, section details, and glazing details (Fig. 2-15). Locate section cuts where they will provide the most information. Indicate on the plan by a partial line at the exterior walls and at locations where the line changes direction or offsets. Building sections should pass through stairs, elevator shafts, atria, and other features. Column centerlines and designations must be shown on the building section for orientation.
2. See Section 2.6.2 for guidelines to dimensioning.
3. Because the building section is sometimes used by the mechanical engineer to plan duct runs during the initial phases of the project, some architects show the floor slab and the main girders on a print of the building section and send it to the mechanical engineer. Many offices prefer to show the finish floor line and the ceiling only. There is much to be said in favor of that approach. It saves time and it avoids problems that may occur if the depth of certain beams is changed in the latter part of the project.
4. Refer exterior walls to the corresponding wall section.

2.1.5 Wall Sections

1. Wall sections are usually drawn at ¾- or ½-inch scale. Sections drawn at a ½-inch scale may require more section details to clarify the intent. Wall sections set the R-value of the exterior envelope, they detail the air barrier and vapor retarder, and they may be designed according to the "rain-screen" principle (see Sec. 3.5). They require the technical knowledge and the feel for cost implications of the most experienced person on the team working hand-in-hand with the specifications writer and the estimator.
2. Multistory buildings may require the wall section to be foreshortened to fit on the sheet. A natural location for this shortening is at the window. For identical floors, show only one floor and designate as typical.
3. Determining the location and number of wall sections is important. The goal must be to draw the least number of sections, avoid or minimize repetition, and cover every condition so that the contractor is not left to his own interpretations. Typical full-height wall sections should be drawn first. Partial sections showing different areas of the ground floor or the building top should be drawn and a reference to the full-height section is written to relate the partial section to the rest of the building. Repeating the full section because one floor is different does not serve any purpose. If the sheet format or the team leader's preference requires only full-height sections to be shown, draw the different part fully and draw only the outline of the identical part and refer it to the typical section.
4. Remember that besides detailing the wall construction, the wall section must provide durability and protection against the elements. Flashing must be detailed in a way that leaves no doubt about its location, where it starts, and where it ends. Materials must be chosen carefully and minimum clearances observed.
5. Refer to Section 2.6 for dimensioning.
6. In masonry construction, show number of brick courses and relate to the floor-to-floor height. Be sure to identify soldier courses and allow for lintels and shelf angles.

2.1.6 Section Details

Section details are selective enlargements of parts of the wall sections. They are usually drawn at 1½-inch scale and must include complicated areas of the wall assembly, especially where flashing occurs. The easiest way to draw them is to enlarge that part of the wall section on the office copying machine, redline the additional information, then trace it on the project sheet. This step, of course, can be done on the monitor if the office uses CADD.

Figure 2.15 shows an example of a section detail. Note the clearance between the beam fireproofing and the stud wall. Although the GWB adjacent to the beam seems to be inaccessible for fastening to the studs, in this particular case, sections of the stud wall were pre-assembled, including the top part of the GWB, then set in place. This demonstrates the importance of acquiring knowledge about methods of construction when addressing this task. Also note the following:

- Clearance between the stud wall and the floor slab to allow for deflection. (This applies to non load-bearing walls only).
- Clarification of the extent of flashing.
- Air barrier and vapor retarder (see Sec. 3.4).
- Brick dimensions, weep holes, and anchorage (see Sec. 3.10).
- R-value of insulation (see Sec. 3.3).
- Cast stone sill (see Sec. 3.11).
- Dimension between column centerline and exterior face of the building was determined by both section requirements and column plan. One must not arrive at this crucial dimension before detailing both the column plan and this section.

- Stud size, spacing, and gauge are determined by wind loads. The specs should state that the stud manufacturer (in projects with exterior studs) must size the studs to withstand code-mandated loads, define the maximum deflection allowed and require that shop drawings be signed by a professional engineer registered in the state.
- Beam sprayed-on fireproofing requirement is determined by the code (see Sec. 1.1). The thickness of the spray is determined by the U.L. design (see Sec. 1.2).
- Although, as a general rule, horizontal dimensions are usually shown on the plan, it is advisable that detailed horizontal dimensions be shown also on section details due to the complexity of the assembly.

This example illustrates the need for an experienced person to make the decisions that determine the quality, buildability, and cost related to this part of the project. These parameters must be observed with the least modification to the design. Figure 2-15 is only an illustration. Each project detailing must be based on its particular circumstances.

2.1.7 Glazing Details

This is the fourth part of the section sheets that includes the building and wall sections, the section details, and the glazing details also known as window details. The last named must include the head, jamb, and sill details of every window, storefront or curtain wall in the project. They are drawn on a sheet that includes window type elevations. This sheet must include the glass types to be described in the specifications. Some offices organize this information in a window schedule (Fig. 2-18). Certain glass types are mandated by codes and state or federal regulations depending on their proximity to entrances and must be checked with the glass manufacturers' representatives.

2.1.8 Miscellaneous Guidelines

1. Refer to Chapter 4 for information about partition and door types, toilet room details, stair and elevator details and Section 2.7 for schedules.
2. It is important to remember that drawings show locations and extent of materials and that specifications describe quality of workmanship and general requirements. Materials should be identified by their generic names rather than by their trade names. For example, use EIFS (which stands for exterior insulation and finish system) instead of Dryvit, a trademark of one of the systems.
3. It is not necessary to refer to every beam or column on the drawing with a notation stating "refer to struct." A general note stating that fact is sufficient. The various bidders know where to find their information. Only items not shown on structural drawings need to be described.
4. The team leader should distribute to each member of the team the section of the specifications that is applicable to the work he or she is doing with instructions to read it carefully to prevent any conflict.
5. Projects that include a core area containing stairs, toilet rooms, elevators, etc., should have an enlarged core plan rather than a separate plan for each.
6. Manufactured products such as window mullions need not be drawn in detail because the low bidder may opt to choose a different product of equal quality. Just include in the drawings enough information to indicate its relationship to other parts of the assembly such as the way the product is attached to the structure and how it is flashed. Shop drawings will be presented by the contractor and checked for conformance.
7. Do not use the terms "by others"—use "NIC" instead; "as required"—this leaves the decision in the hands of the contractor and is ambiguous at best; and "as specified"—all items must be specified.

2.1.9 Overdrafting

1. Use the smallest size project sheet that can accommodate the plans. When one considers that it takes between 9 and 15 hours to draw each square foot of a project sheet, one can see the consequence of using oversize sheets.
2. Use the smallest scale that will convey the information required. Full-size details are almost never needed. The smaller the scale, the less time and effort are required to finish the drawing. Refer to recommended scales mentioned in the subsections preceding this one.
3. Minimize hand lettering and substitute typing. Use keynotes if possible. This approach can save a sizable amount of time without detracting from the legibility of the drawings.
4. See Section 4.1 for stair drafting recommendations.
5. Show partial material indications on elevations and sections. Hatching a whole brick wall section or stippling a whole elevation area to show plaster is a waste of time.
6. Lengthy descriptions belong in the specifications. Notations should identify the materials in the briefest possible way. Use the same nomenclature used in the specs.
7. Do not draw a site location map if it can be photocopied from a road map and sticky-backed on the general information sheet.
8. Acquire a standard details library and use it as much as possible on every project. Chapter Four is an example of the nucleus of such a library. This can be one of the most timesaving factors for any project.

2.2 SHEET FORMAT

2.2.1 General

Project sheet organization (or format) differs from firm to firm. Many offices use a title block, place a frame around the sheet, and leave the sheet organization up to each team member. While this approach may give some individuals the feeling that they are in charge, it wastes time and effort because it does not provide a structured framework within which one can organize the drawings. Time has to be spent by each team member to invent an organization for each sheet. Details are drawn side by side and, in some cases, overlap, causing confusion and giving a chaotic look to the sheet.

Other firms use a more planned approach. The Northern California Chapter of the AIA (since 1981 known as AIA, San Francisco) issued recommendations to its membership to use a project sheet based on an 8½ × 11-inch module. Figures 2-1a and 2-1b show a sheet layout based on these recommendations. Subdivided detail sheets have the following advantages:

- Details reproduced on sticky-back usually create a shadow on diazo prints. These shadows may be of different shades. Dividing lines between details tend to de-emphasize these differences.
- Each detail sheet can hold a maximum of forty 4⅜ × 5⅝-inch details for a 30 × 42-inch sheet or twenty-eight details for a 24 × 36-inch sheet. To estimate the percentage of completion for the sheet, count the number of 4⅜ × 5⅝-inch rectangles occupied by the details (regardless of the size of each detail) and divide by 40 or 28 as the case may be, then multiply by 100. For example, if a 30 × 42-inch detail sheet contains six fully drawn details occupying eight 4⅜ × 5⅝-inch rectangles, the percentage of completion for that sheet = 8 × 100 ÷ 40 = 20 percent.
- The reference to source of detail shown on the detail module (Fig. 2-2) will reduce the time spent by the person or persons reviewing the project, be it for checking or bidding or code review. Being unfamiliar with the project, these individuals will waste their valuable time hunting through several likely sheets to try to find where the detail originated from if this reference is missing.
- Detail sheets can usually hold more details if subdivided sheets are used.
- It is easier and faster to draw a project mock-up using this system.

Figure 2-3 demonstrates the ability of the system to accommodate many detail sizes.

2.2.2 The ConDoc System

ConDoc, which stands for Construction Documents and is described as "The New System for Formatting and Integrat-

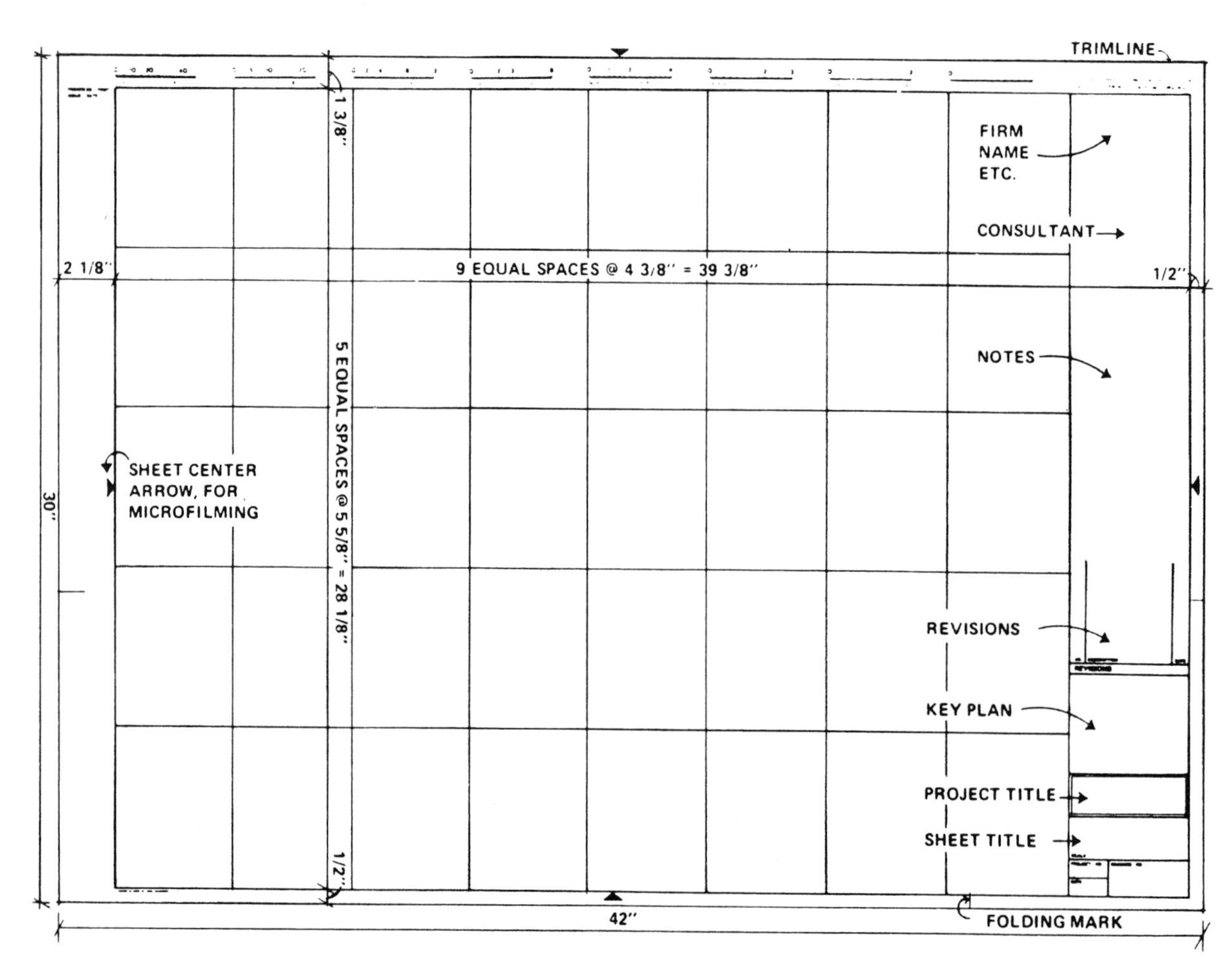

FIGURE 2-1a. The Project Sheet Organization. The format recommended by the Northern California Chapter of the AIA (since 1981 known as AIA, San Francisco)

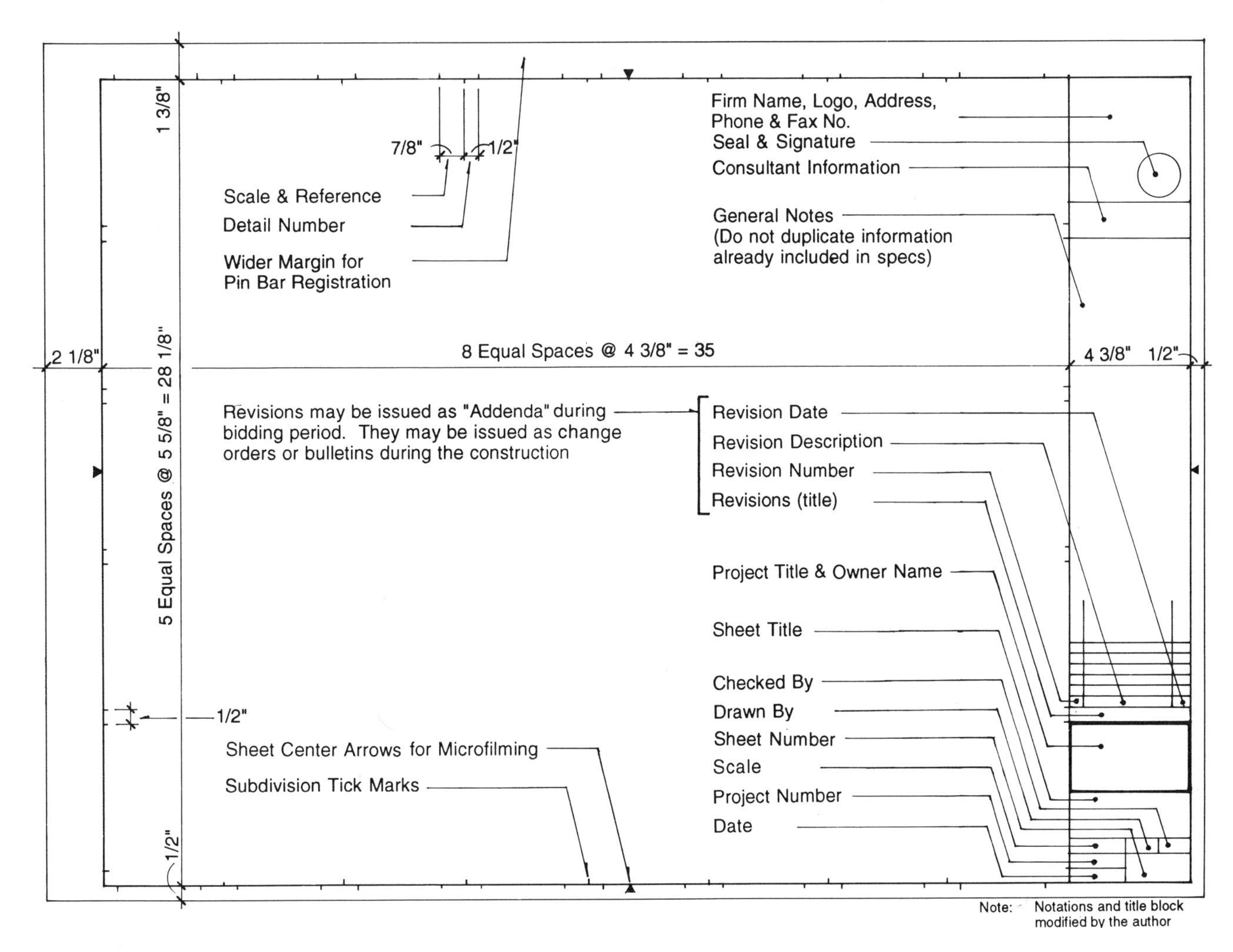

FIGURE 2-1b. Explanations of The Project Sheet (see Fig. 2-1a)

ing Construction Documentation," evolved from an article written by Onkal K. (Duke) Guzey for the fall 1985 issue of *Architectural Technology* magazine. Mr. Guzey, along with James N. Freehof, AIA, conducts seminars nationwide on the subject, under the auspices of the American Institute of Architects.

This approach to sheet organization has the following advantages: It ties specifications to the detail drawings, ensuring that no material used in a detail will be omitted from the specifications (this is done by using keynotes utilizing spec numbers. Notations are typed in a space set aside for that purpose in the title block); sheet subdivision is tied to a 2 × 2-inch module that can accommodate any detail size; and the detail module is divided into zones that separate the dimensions from the keynotes to prevent conflicts.

Using keynotes, whether tied to specifications or not, is advantageous. Repetitious notes are typed only once and keyed to several details where they apply. For both manual drafting and CADD this can save quite a bit of time spent on hand lettering or typing.

2.3 SHEET NUMBERING

2.3.1 General

In the old days, project sheet numbering was very simple. The first sheet was numbered A-1 followed by A-2, etc. This method had a major drawback; every time a sheet had to be added to a group of drawings such as the plans, all the sheet numbers coming after it had to be changed. If this is done late in the project, all the references to the changed numbers had to be hunted down and changed to conform to the new numbers, a lengthy and hazardous procedure that resulted in missed changes that confused everybody. The following is a description of different methods used in numbering project sheets.

2.3.2 Methods of Numbering Sheets

Method 1

One digit is given to the sheet grouping—plans, elevations, etc.—followed by a decimal point and another digit for each

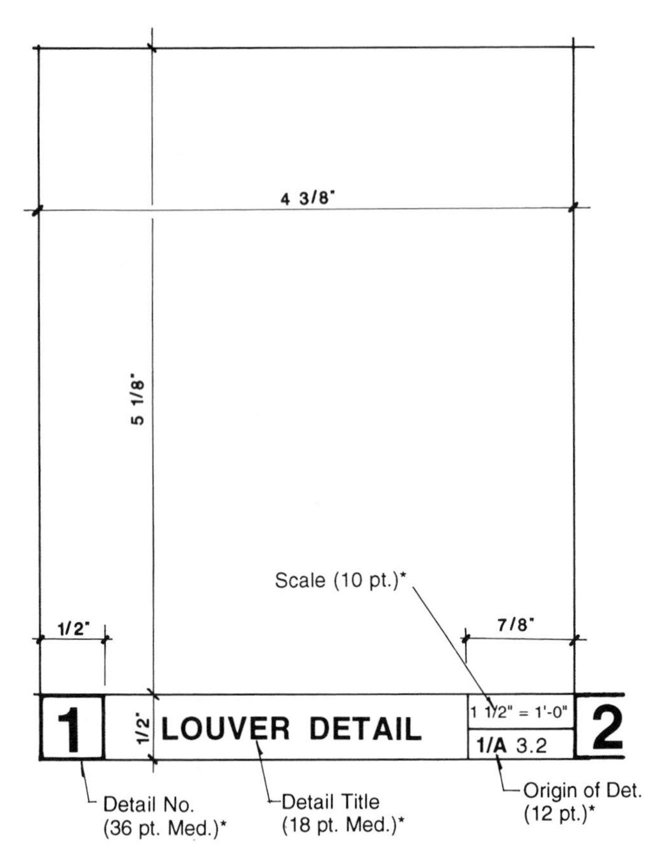

FIGURE 2-2. Detail Module

sheet (Fig. 2-4). For example, if the elevation grouping is designated A3, the second sheet in the group would be numbered A3.2. This way, if the team leader wanted to add a sheet showing, say, partial elevation to the group of elevation sheets ending with sheet A3.3, he or she could easily designate the new sheet A3.4 without affecting the following group, which starts with sheet A4.1. In these examples, the prefix "A" stands for Architecture. A listing of prefixes follows. Other prefixes may be added as needed.

A Architecture
S Structural
M Mechanical
E Electrical
P Plumbing
C Civil
L Landscape architecture
F Fire protection
Q Equipment

This is one of the most common methods used for numbering sheets.

Method 2
Details are drawn on 8½ × 11-inch sheets and bound into a book similar to the specifications to form part of the project manual. These details are numbered according to the same groupings described in Method 1, except that the prefix used is D for detail instead of A. The full-sized sheets are numbered as described in Method 1. This method has the following advantages:

- Reproduction on the office copier is cheap and fast.
- Modifying the details to reissue them as addenda or change orders is faster and much easier.
- Office standard details can be reproduced on standard Xerox paper, saving the time spent on applying sticky-back reproductions on the project sheet.
- Handling the detail booklet in the field on windy days is easier than a full- or half-size set of drawings.

There are also a few disadvantages:

- City plan checkers may find it difficult to check a set of drawings organized so radically different. The trick is to alert them early, explain the system, and get their approval.
- Some details may require a larger size than 8½ × 11 inches. This can be overcome by using 17 × 11-inch sheets exclusively. Each sheet would contain either two details or one large detail. Bigger details would be included in the large drawings set.
- This format is usually unacceptable for use on state and federal projects. These projects require a special predetermined format.
- It is hard to find contradictions, repetitions, and mistakes at a glance during the checking process when each detail is isolated on a sheet by itself. To remedy this shortcoming, a copy of the details should be attached to full-size project sheets with quick-release tape and sent out to be copied (using cheap "Shacoh" prints), then attached to the check set of drawings.

Method 3
The set is organized according to the sequence of construction instead of plans, elevations, building sections, etc. This system has advantages for the contractor, especially for fast-track projects where the contract for the foundations is awarded independently and is followed by other contracts following the sequence of construction. Not too many offices follow this system.

Method 4
ConDoc uses an alphanumeric system of numbering architectural drawings organized around three groupings—general project information, site work, and architectural drawings.

These are four of the methods used for numbering project sheets. Offices that opt to use the four-task team approach, described in Chapter 1, may consider using a numbering method organized to complement this approach. Figure 2-5 explains the numbering system.

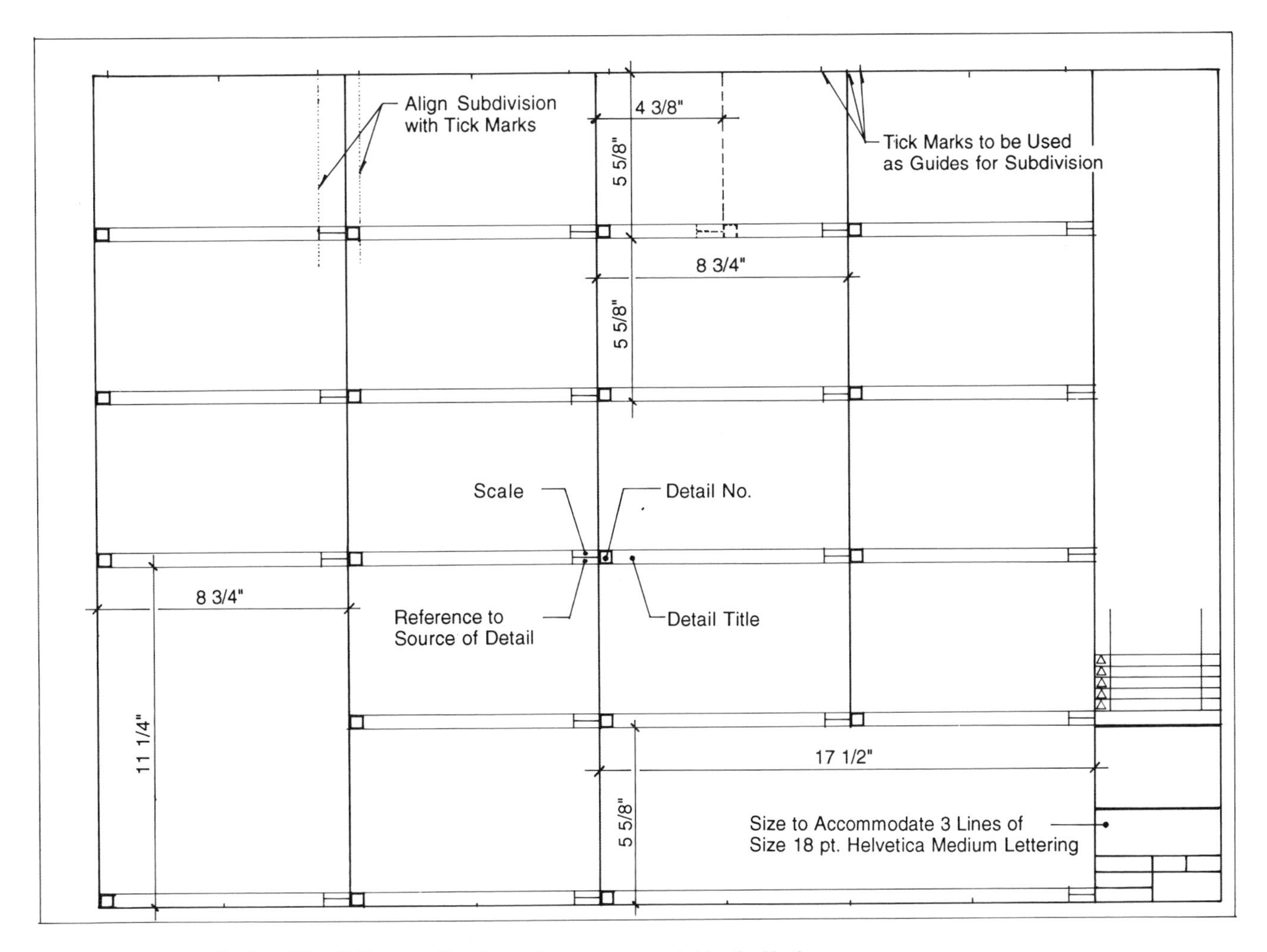

FIGURE 2-3. Subdivision of Detail Sheets (Based on a format recommended by the Northern California Chapter of the AIA, since 1981 known as AIA, San Francisco)

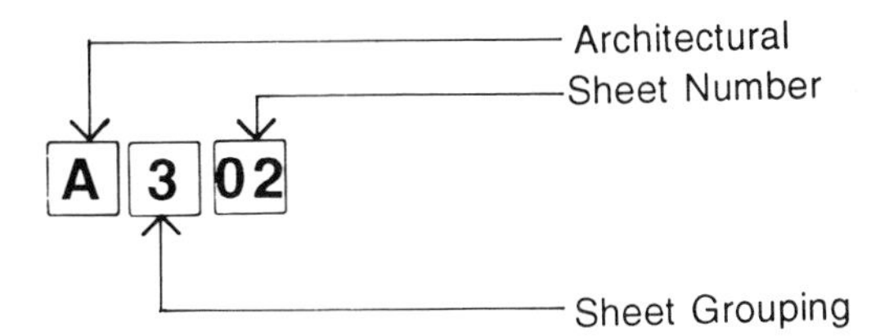

FIGURE 2-4. Common Numbering System

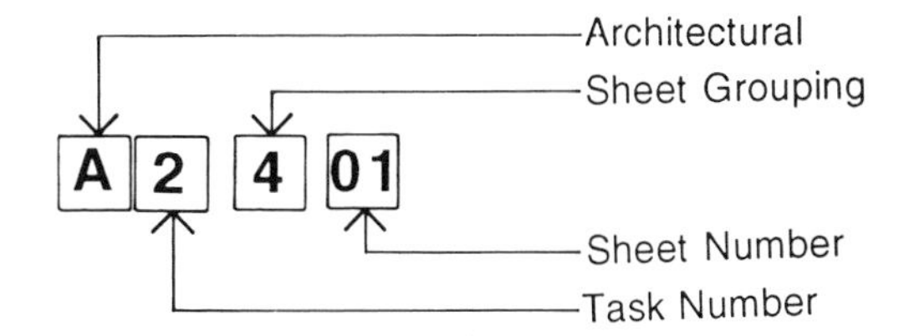

FIGURE 2-5. Numbering System Based on Tasks

In all these methods, plans done by the other disciplines are numbered in a similar fashion.

2.4 OVERLAY DRAFTING

2.4.1 General

Overlay drafting was introduced in architectural projects in the 1960s to reduce the amount of repetitious work done on multistory plans. Prior to its introduction, and even in some offices today, draftsmen had to draw column grids, columns, exterior walls, interior cores, and partitions on each plan sheet. This entailed a lot of repetitious, time-consuming drafting. After completing this task, all work on the plans had to be interrupted while sepia prints were made for the mechanical, electrical, and plumbing engineers (MEP) to be used as backgrounds to draw their ducts, lighting and power, and plumbing plans on. Pressure was brought to bear on the architect to finalize the plans very early in the Construction Drawings phase so that the engineers could start their work. Changes to the background were inevitable and the engineers had to erase the backgrounds (four copies of each plan), draw the changes, and make the necessary changes to their input as a result of the architectural changes. When this occurred two or three times, it brought howls of protests from the engineers and a possible request for a fee adjustment to compensate for the added drafting time.

With pin-registered overlay drafting, drafting time for both the architectural and engineering groups is reduced appreciably. The idea behind overlay drafting is to draw the plans in layers. Some layers, such as the title block and column grid layers, are used on all the floor plans. Some, such as the exterior walls and core layers, are used on the majority of the plans. This depends on how many changes occur to these two items over the height of the building. Finally, some layers, such as the one containing interior notations, dimensions, and sheet number are unique to each plan. Each plan is assembled from layers representing these three categories and composited to produce one print on Mylar representing the completed floor plan.

2.4.2 The System

The term "pin-registered" is descriptive of the hardware used to align the sheets to provide a perfect match between the layers of drawings required to composite each plan. The system employs a stainless steel bar with ¼-inch pins (Fig. 2-6) set at intervals corresponding perfectly to matching holes prepunched in the Mylar sheets. When sheets are overlaid, a perfect match of the layers or "registration" results.

There are four methods used to produce a composited floor plan or, to a lesser degree, a section. These methods are:

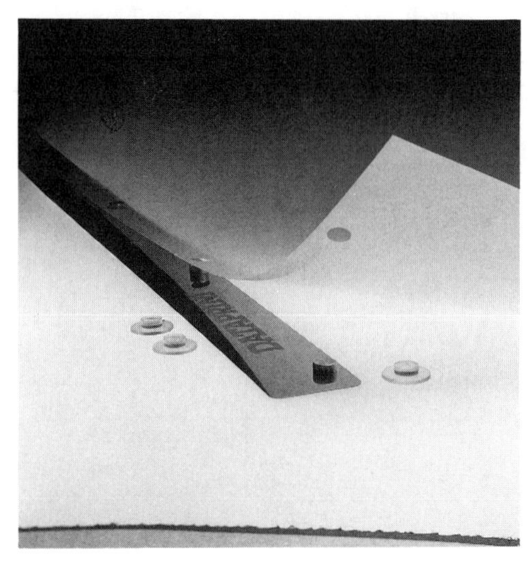

FIGURE 2-6. Pin Bar, Prepunched Film, and Clear Pins (Courtesy of Dataprint)

1. Assemble the layers using rubber pins and run them through the diazo machine to produce an in-house check print. Because of slippage in the machine, this print provides an imperfect image not suitable for review by other than in-house personnel. Up to four layers may be composited using this method.
2. All the layers needed for a plan are assembled on a pin-bar and reproduced as a composite sepia Mylar using a vacuum frame (Fig. 2-7). This can be done economically and fast in-house. This method is used to produce "throwaways" or "slicks" sent to the engineers to be used as background for their engineering drawing layers. A throwaway is a layer to be thrown away when a change to the plan is made. When this happens, the new throwaway is sent by the architect, and the engineer makes changes only to his layer if change is required to conform to the new architectural layout. The term "slick," used to refer to this same layer, describes the surface, which is slick or non-matt on both sides.
3. All the layers are sent to the print shop with instructions in the margin of each notation layer (base sheet) for each plan, listing which layers are to be composited to form the final plan. These instructions also stipulate which layers are to be reproduced in halftone (30 to 50 percent screen) to be combined with the engineering drawings. This makes the engineering layouts of ducts, pipes, conduits, lighting, etc., stand out. It also prevents the architectural lines from interfering with that information while providing the necessary location reference.

   This method of reproducing prints is relatively costly and should be used only after all the coordination and corrections have been done and the drawings are ready to be issued for bids.
4. Reproduction on a CADD plotting machine is the latest development in overlay drafting.

FIGURE 2-7. Vacuum Frame

2.4.3 The Layers

As mentioned above, each plan is made up of several layers. At an early stage in the project, the Team Leader writes a list of layers required to composite the throwaways—the floor plans, the ceiling plans, the equipment plans (if required), the interiors background, the landscape architecture backgrounds, etc. Some of these composites usually require different layer assemblies. In addition, the Team Leader must determine which layers of the exterior walls and cores or interior partitions need to be drawn on more than one layer to reflect changes in the building exterior outline or core configuration. For instance, if the outline of the exterior wall changed at the corners only, three layers are required. The first is drawn showing the exterior walls without the corners. The second shows the corners only and a third layer shows the modified corners (Fig. 2-8). Combining the first and second layers produces the exterior wall for the first plan, and combining the first and third layers produces the second plan.

The following is a listing of layers normally used to produce one composite plan and the reason behind the need for each. These are suggested combinations described for illustration only so that the reader may have a better grasp of the rationale used to determine the content of each layer. The possibilities are almost limitless and the decisions are determined by the ingenuity and experience of the Team Leader. As can be seen from Figure 2-9, the suggested layers are:

Layer 1

Project sheet format, including the title block. This layer is used on all the project sheets. When compositing plans for check sets, it may be omitted.

Layer 2

Structural framing grid lines with circled column designations. The grid forms a reference network for all dimensions and is used on all the plan sheets. This layer may also include the exterior dimensions—overall and between column center lines. The third string containing exterior wall dimensions is drawn on the exterior wall layer.

Layer 3

The slab edge around the perimeter of the exterior and around floor penetrations, such as those located at mechanical shafts, is drawn on this layer. This layer is combined with Layer 2 to produce a Mylar for the structural engineer to draw his framing plan on. A print showing the dimensions of the floor openings is also sent.

This layer is then converted to a base sheet showing all fixed elements, including exterior walls, interior permanent full-height partitions, stairs, elevators, toilet room fixtures, etc. These slicks are produced by combining this layer with Layer 2 and are sent to the MEP engineers to be used as background for mechanical, power, and plumbing drawings. This layer will be reproduced in halftone for the MEP consultant for the final set of drawings. To produce the architectural floor plan, this layer will be combined with Layers 1, 2, and 4 (notations).

Layer 4

This layer contains all the relevant information normally shown on the floor plan—door swings, partition types, keying symbols, dimension notes, door numbers, room names and numbers, floor drains, general notes or keying index, and title block information. It is combined with Layers 1, 2, and 3 to form the plan for each floor. A key plan sticky-back may be added to this layer if each floor plan is shown on more than one sheet. It should be attached to the front of the sheet for better photo reproduction. Other layers showing equipment, removable partitions, furniture, etc., may be added as needed.

Layer 5

A ceiling grid is drawn on this layer. Door heads are added at each door opening. This layer is combined with Layers 2 and 3 to produce a background slick for the electrical engineer to overlay the Mylar on and draw the lighting plan. This composite will be reproduced as a halftone for the final set. It will also be combined with a layer containing notations and

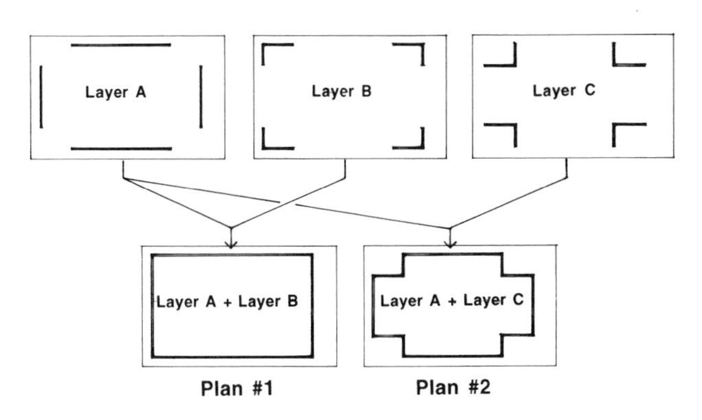

FIGURE 2-8. Combining Layers

PRINCIPAL LAYERS

Layer No.	Layer Name	Content	Layer Required to Produce the Composite	Composite
1	Title Block & Border 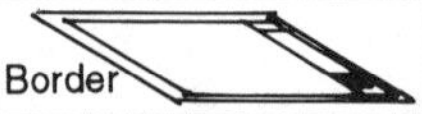	Title Block, Project Name & No., Stamp	N/A	All Project Sheets
2	Column Grid 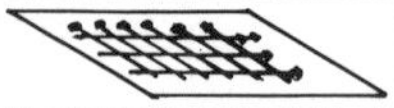	Col. Centerline Col. Numbers	N/A	All Plans
3	Base Sheet 	Ext. Walls Permanent Partitions	1+2+3+10+13+14+15+21	Floor Plans
			1+2+3+5+6+7+8+9+13	Reflected Ceiling Plans
			1+3+13+14+15+21	Leasing Plans
			1+2+11+19+21	Roof Plans
			2+3+5+13	HVAC Back-grounds
			2+3+5+13	Lighting Back-grounds
			2+3+12+14+21	Power Back-grounds
			2+3+10+13+14*+21	Plumbing Back-grounds
			2+3+13+14+15+21	Interiors Back-grounds
			2+3 (Ground Flr.) +20+21	Landscaping & Site Backgrounds
			1+2+3+10+12+13+14+15+21	Equipment Plans
4	Notations 	Dimensions, Door Numbers, Symbols, Sht. Title & No., etc.	N/A	All Plans

*This layer may be composited without screening

SUPPLEMENTARY LAYERS

Layer No.	Layer Name	Layer No.	Layer Name
5	Ceiling Grid	14	Toilet Rm. Fixtures & Partitions
6	Ceiling Plan Notations	15	Low Partitions (from Interior)
7	Light Fixtures (from Elect.)	16	Changes in Exterior Walls
8	A/C Registers (from Mech.)	17	Changes in Core Plans
9	Sprinkler Heads (from Plumb'g)	18	Landscape Overlay (from L. Arch.)
10	Floor Drains (from Plumb'g)	19	Roof Parapet, Slopes, Notations, etc.
11	Roof Drains (from Plumb'g)	20	Site Boundary
12	Equipment Layout	21	Door Swings
13	Structural Columns (from Struct.)		

Note: Layers 16 & 17 may not be required

Screened to produce a halftone print (HVAC, Lighting, Power and Plumbing Backgrounds)

FIGURE 2-9. Architectural Layer Designations

keys to ceiling details, Layers 1, 2, and 3, to produce the architectural reflected ceiling plan.

To summarize, five layers are required to produce the floor and reflected ceiling plans. These layers are: 1. The Title Block, 2. The Grid, 3. The Fixed Walls, 4. The Notations, and 5. The Reflected Ceiling Grid. The first one is used on all the project sheets. The second is used on all the floor plans. The third may be applicable to all floor levels or may require additional layers to modify it (Fig. 2-8). The fourth layer is combined with the preceding layers to create the floor plans, and the fifth layer, which is exclusive to each ceiling plan, is combined with Layers 1, 2, and 3. Base sheets are, in essence, the conventional floor plans without the notes (Layer 4) and title block (Layer 1). They are the basis for the backgrounds sent to the MEP consultant in the form of slicks updated periodically during the preparation of the drawings. For the final set to be issued for bids, they are photographed in halftones to be combined with the MEP overlays to produce composite wash-off Mylars. They are combined full tone with selected architectural layers to produce the floor and reflected ceiling plans. Some offices prefer to show the duct layout on a halftone reflected ceiling plan rather than on a floor plan. Some prefer to poshé rated partitions to alert the mechanical engineer to the need for fire dampers and other requirements to prevent fire from spreading through penetrations in these walls.

2.4.4 CADD

CADD overlay drafting uses basically the same approach as pin-bar drafting, utilizing electronics to manipulate the layers. If the consultants use compatible hardware, the information or background layers are transmitted directly to their monitors by modem or disk. The layers are displayed in different colors for clarity. Figures 2-10 and 2-11 are reproduced from *CAD Layer Guidelines* by permission from the AIA.

2.4.5 The Advantages

The following is a description of some of the advantages associated with overlay drafting:

1. Because each line on each layer is drawn only once, drafting time is reduced significantly. The drudgery associated with repetitious activities is also eliminated. Initially, as the system is first introduced in the office, there will be a learning period that will cost the office time and money. As the staff members gain experience, work will proceed faster and projects will be finished more economically.
2. The final set of drawings is better coordinated, has fewer mistakes, and is of better quality, which gains the respect of all those who review it.
3. CADD-generated layers are stored on disks that require very little space and are easy to organize, retrieve, and plot. Likewise, negatives generated by the wash-off Mylar process are easy to store, retrieve, and develop. Negatives are produced on 8½ × 11-inch sheets, which should be stored in a fire-rated cabinet or vault.

   This contrasts very favorably with the conventional method of storing full-size original drawings, which are heavy, require a large and costly space to warehouse, and are not easy to locate if stored at the bottom of a stack of drawings in a remote warehouse. This is usually the case when an old drawing is added to or modified. If either CADD or wash-off is used, the originals can be warehoused for three or four years, then either destroyed or presented as a gift to the client. A copyright clause should be included in the title block to prevent illegal use of these documents.
4. Using CADD enables each participant to work on a different layer simultaneously using individual work stations. In a similar fashion, pin-bar drafting utilizes base sheet background slicks, to enable one person to draw the ceiling grid while another works on the notations and a third draws the equipment or the dimensions, etc. for the same floor plan.
5. "Leasing plans" for commercial projects can be produced by combining the base sheet with a new layer containing information more suited for marketing.
6. Other disciplines benefit because they do not have to make changes on the architectural backgrounds.

2.4.6 Sources of More Information:

ConDoc Seminars conducted by Onkal Guzey, AIA, and James N. Freehof, AIA, a professional development program under the auspices of The American Institute of Architects.

Schley, Michael K., ed. 1990. *CAD Layer Guidelines*. Washington, D.C.: The American Institute of Architects.

Stitt, Fred A. 1980. *Systems Drafting*. New York: McGraw-Hill, Inc.

Woods, Frank, and Powell, John. 1987. *Overlay Drafting*. Stoneham, MA: Butterworth Architecture.

2.5 LETTERING

Hand lettering is fast becoming a lost art as computers become more universally used. Even before the advent of CADD, some offices realized that the cost of producing a set of drawings could be reduced appreciably by writing the notations in longhand and having them typed. Some offices utilized a special typewriter that enabled the typist to type directly on the project sheet.

Since many offices still use hand lettering and may continue to do so for a period of time, I have included a sample (Fig. 2-12). The following rules must be observed: Always use guidelines (the least distance between these guidelines must be ⅛ inch. This makes the letters legible if the drawing is reduced to half size; a printed grid can be slipped under the sheet for this purpose instead of drawing the guidelines); experiment with leads until you find the one you are comfortable with. Use plastic lead in a 9-mm holder when writing on Mylar; and use a Kroy, Merlin, or similar lettering machine for sheet and detail titles (this gives the set of drawings a crisp, well-organized look).

Hand lettering is still a good skill to acquire even if one is using CADD exclusively.

2.6 DIMENSIONING

2.6.1 General

Dimensioning a drawing is not a simple matter of drawing a line and writing numbers corresponding to lengths. The

MASTER LAYER LIST WITH MODIFIERS

General Information

The layer list is divided into eight major groups. Within each group, building information layers are listed first, followed by drawing information layers.

The following modifiers may be used with any building information layer:

Long Format Layer Name	Short Format Layer Name	Layer Description
*-****-IDEN	***ID	Identification Tag
*-****-PATT	***PA	Cross-hatching and Poche
*-****-ELEV	***EL	Vertical Surfaces (3D Drawings)
*-****-EXST	***EX	Existing to Remain
*-****-DEMO	***DE	Existing to Be Demolished or Removed
*-****-NEWW	***NW	New or Proposed Work (Remodeling Projects)

For example, A-WALL-DEMO would be used to designate walls to be demolished.

The following modifiers may be used with any drawing information layer:

Long Format Layer Name	Short Format Layer Name	Layer Description
*-****-NOTE	***NO	Notes, Call-outs and Key Notes
*-****-TEXT	***TE	General Notes and Specifications
*-****-SYMB	***SY	Symbols, Bubbles, and Targets
*-****-DIMS	***DI	Dimensions
*-****-PATT	***PA	Cross-hatching and Poche
*-****-TTLB	***TT	Title Block Sheet Name and Number
*-****-NPLT	***NP	Nonplot Information and Construction Lines
*-****-PLOT	***PL	Plotting Targets and Windows

Read-Me Layer

The following layer is common for all major groups:

X-RDME	XRD	Read-Me Layer, Not-to-Be-Plotted, Information on File Organization

FIGURE 2-10. Master Layer List with Modifiers

Architecture, Interiors, and Facilities

Long Format Layer Name	Short Format Layer Name	Layer Description
		Building Information Layers
A-WALL	**AWA**	**Walls**
A-WALL-FULL	AWAFU	Full Height Walls, Stair and Shaft Walls, and Walls to Structure
A-WALL-PRHT	AWAPR	Partial Height Walls (Not on Reflected Ceiling Plans)
A-WALL-MOVE	AWAMO	Movable Partitions
A-WALL-HEAD	AWAHE	Door and Window Headers (Shown on Reflected Ceiling Plans)
A-WALL-JAMB	AWAJA	Door and Window Jambs (Not on Reflected Ceiling Plans)
A-WALL-PATT	AWAPA	Wall Insulation, Hatching and Fill
A-WALL-ELEV	AWAEL	Wall Surfaces (3D Views)
A-DOOR	**ADO**	**Doors**
A-DOOR-FULL	ADOFU	Full Height (to Ceiling) Door: Swing and Leaf
A-DOOR-PRHT	ADOPR	Partial Height Door: Swing and Leaf
A-DOOR-IDEN	ADOID	Door Number, Hardware Group, etc.
A-DOOR-ELEV	ADOEL	Doors (3D Views)
A-GLAZ	**AGL**	**Windows, Window Walls, Curtain Walls, Glazed Partitions**
A-GLAZ-FULL	AGLFU	Full Height Glazed Walls and Partitions
A-GLAZ-PRHT	AGLPR	Windows and Partial Height Glazed Partitions
A-GLAZ-SILL	AGLSI	Window Sills
A-GLAZ-IDEN	AGLID	Window Number
A-GLAZ-ELEV	AGLEL	Glazing and Mullions (Elevation Views)
A-FLOR	**AFL**	**Floor Information**
A-FLOR-OTLN	AFLOT	Floor or Building Outline
A-FLOR-LEVL	AFLLE	Level Changes, Ramps, Pits, and Depressions
A-FLOR-STRS	AFLST	Stair Treads, Escalators, and Ladders
A-FLOR-RISR	AFLRI	Stair Risers
A-FLOR-HRAL	AFLHR	Stair and Balcony Handrails and Guard Rails
A-FLOR-EVTR	AFLEV	Elevator Cars and Equipment
A-FLOR-TPTN	AFLRP	Toilet Partitions
A-FLOR-SPCL	AFLSP	Architectural Specialties (Toilet Room Accessories, Display Cases)
A-FLOR-WDWK	AFLWD	Architectural Woodwork (Field-Built Cabinets and Counters)

Architecture, Interiors, and Facilities (continued)

Long Format Layer Name	Short Format Layer Name	Layer Description
A-FLOR-CASE	AFLCA	Casework (Manufactured Cabinets)
A-FLOR-APPL	AFLAP	Appliances
A-FLOR-OVHD	AFLOV	Overhead Skylights and Overhangs (Usually Dashed Lines)
A-FLOR-RAIS	AFLRA	Raised Floors
A-FLOR-IDEN	AFLID	Room Numbers, Names, Targets, etc.
A-FLOR-PATT	AFLPA	Paving, Tile, and Carpet Patterns
A-EQPM	**AEQ**	**Equipment**
A-EQPM-FIXD	AEQFI	Fixed Equipment
A-EQPM-MOVE	AEQMO	Movable Equipment
A-EQPM-NICN	AEQNI	Equipment Not in Contract
A-EQPM-ACCS	AEQAC	Equipment Access
A-EQPM-IDEN	AEQID	Equipment Identification Numbers
A-EQPM-ELEV	AEQEL	Equipment Surfaces (3D Views)
A-FURN	**AFU**	**Furniture**
A-FURN-FREE	AFUFR	Freestanding Furniture (Desks, Credenzas, etc.)
A-FURN-CHAR	AFUCH	Chairs and Other Seating
A-FURN-FILE	AFUFI	File Cabinets
A-FURN-PNLS	AFUPN	Furniture System Panels
A-FURN-WKSF	AFUWK	Furniture System Work Surface Components
A-FURN-STOR	AFUST	Furniture System Storage Components
A-FURN-POWR	AFUPO	Furniture System Power Designations
A-FURN-IDEN	AFUID	Furniture Numbers
A-FURN-PLNT	AFUPL	Plants
A-FURN-PATT	AFUPA	Finish Patterns
A-FURN-ELEV	AFUEL	Furniture (3D Views)
A-CLNG	**ACL**	**Ceiling Information**
A-CLNG-GRID	ACLGR	Ceiling Grid
A-CLNG-OPEN	ACLOP	Ceiling and Roof Penetrations
A-CLNG-TEES	ACLTE	Main Tees
A-CLNG-SUSP	ACLSU	Suspended Elements
A-CLNG-PATT	ACLPA	Ceiling Patterns
A-ROOF	**ARO**	**Roof**
A-ROOF-OTLN	AROOT	Roof Outline
A-ROOF-LEVL	AROLE	Level Changes
A-ROOF-STRS	AROST	Stair Treads and Ladders
A-ROOF-RISR	ARORI	Stair Risers
A-ROOF-HRAL	AROHR	Stair Handrails, Nosings, and Guard Rails
A-ROOF-PATT	AROPA	Roof Surface Patterns (Hatching)
A-ROOF-ELEV	AROEL	Roof Surfaces (3D Views)

FIGURE 2-10. *Continued*

Architecture, Interiors, and Facilities (continued)

Long Format Layer Name	Short Format Layer Name	Layer Description
A-AREA	**AARE**	**Area Calculation Boundary Lines**
A-AREA-PATT	AARPA	Area Cross-hatching
A-AREA-IDEN	AARID	Room Numbers, Tenant Identifications, and Area Calculations
A-AREA-OCCP	AAROC	Occupant or Employee Names
A-ELEV	**AEL**	**Interior and Exterior Elevations**
A-ELEV-OTLN	AELOT	Building Outlines
A-ELEV-FNSH	AELFN	Finishes, Woodwork, and Trim
A-ELEV-CASE	AELCA	Wall-Mounted Casework
A-ELEV-FIXT	AELFI	Miscellaneous Fixtures
A-ELEV-SIGN	AELSI	Signage
A-ELEV-PATT	AELPA	Textures and Hatch Patterns
A-ELEV-IDEN	AELID	Component Identification Numbers
A-SECT	**ASE**	**Sections**
A-SECT-MCUT	ASEMC	Material Cut by Section
A-SECT-MBND	ASEMB	Material beyond Section Cut
A-SECT-PATT	ASEPA	Textures and Hatch Patterns
A-SECT-IDEN	ASEID	Component Identification Numbers
A-DETL	**ADE**	**Details**
A-DETL-MCUT	ADEMC	Material Cut by Section
A-DETL-MBND	ADEMB	Material beyond Section Cut
A-DETL-PATT	ADEPA	Textures and Hatch Patterns
A-DETL-IDEN	ADEID	Component Identification Numbers

Architecture, Interiors, and Facilities (continued)

Long Format Layer Name	Short Format Layer Name	Layer Description
		Drawing Information Layers
A-SHBD	ASH	Sheet Border and Title Block Line Work
A-SHBD-TTLB	ASHTT	Project Titleblock
A-SHBD-LOGO	ASHLO	Office or Project Logo
A-PFLR	APF	Floor Plan
A-PLGS	APL	Large Scale Floor Plan
A-PCLG	APC	Reflected Ceiling Plan
A-PROF	APR	Roof Plan
A-PXFU	APX	Fixtures and Furniture Plan
A-PEQM	APE	Equipment Plan
A-PMFN	APM	Materials and Finish Plan
A-PDEM	APD	Demolition Plan
A-PARE	APA	Area Calculations
A-POCC	APO	Occupancy Plan
A-P***	AP*	Other Plan Drawings
A-ELEV	AEL	Interior and Exterior Elevations
A-SECT	ASE	Building and Wall Sections
A-DETL	ADE	Details
A-SCHD	ASC	Schedules and Title Block Sheets
A-****-NOTE	A**NO	Notes, Call-outs, and Key Notes
A-****-TEXT	A**TE	General Notes and Specifications
A-****-SYMB	A**SY	Symbols, Bubbles, and Targets
A-****-DIMS	A**DI	Dimensions
A-****-PATT	A**PA	Cross-hatching and Poche
A-****-TTLB	A**TT	Title Block Sheet Name and Number
A-****-NPLT	A**NP	Nonplot Information and Construction Lines
A-****-PLOT	A**PL	Plotting Targets and Windows

FIGURE 2-10. *Continued*

locations of these lines must be planned beforehand to prevent conflict with other elements on the plan or duplication of dimensions shown on another string of dimensions. The person entrusted with this task should make a print of each plan, sketch all the other elements usually shown on the plan, such as room name and numbers, reference symbols, equipment or furnishings, notations, etc., then draw the dimension lines through this obstacle course, making adjustments as needed. The tracing would then be placed over the print and the information traced to provide a legible, uncluttered drawing.

Although this mock-up approach may seem like a waste of time to some, it is not. Omitting this step is like omitting the project mock-up (Chapter 1). It takes a relatively short time to plan the sheet but saves time spent on drawing, erasing, and redrawing, and the frustration associated with this activity is eliminated if this step is taken.

CADD-generated drawings must be reviewed on the monitor. Overlay all the layers and check for overlapping, inconsistencies, duplications, etc., and make adjustments before plotting.

2.6.2 Guidelines

1. If a part of the plan such as a toilet room is enlarged, only the overall dimension of the enlarged area is shown on both plans. All the detailed dimensions are shown on

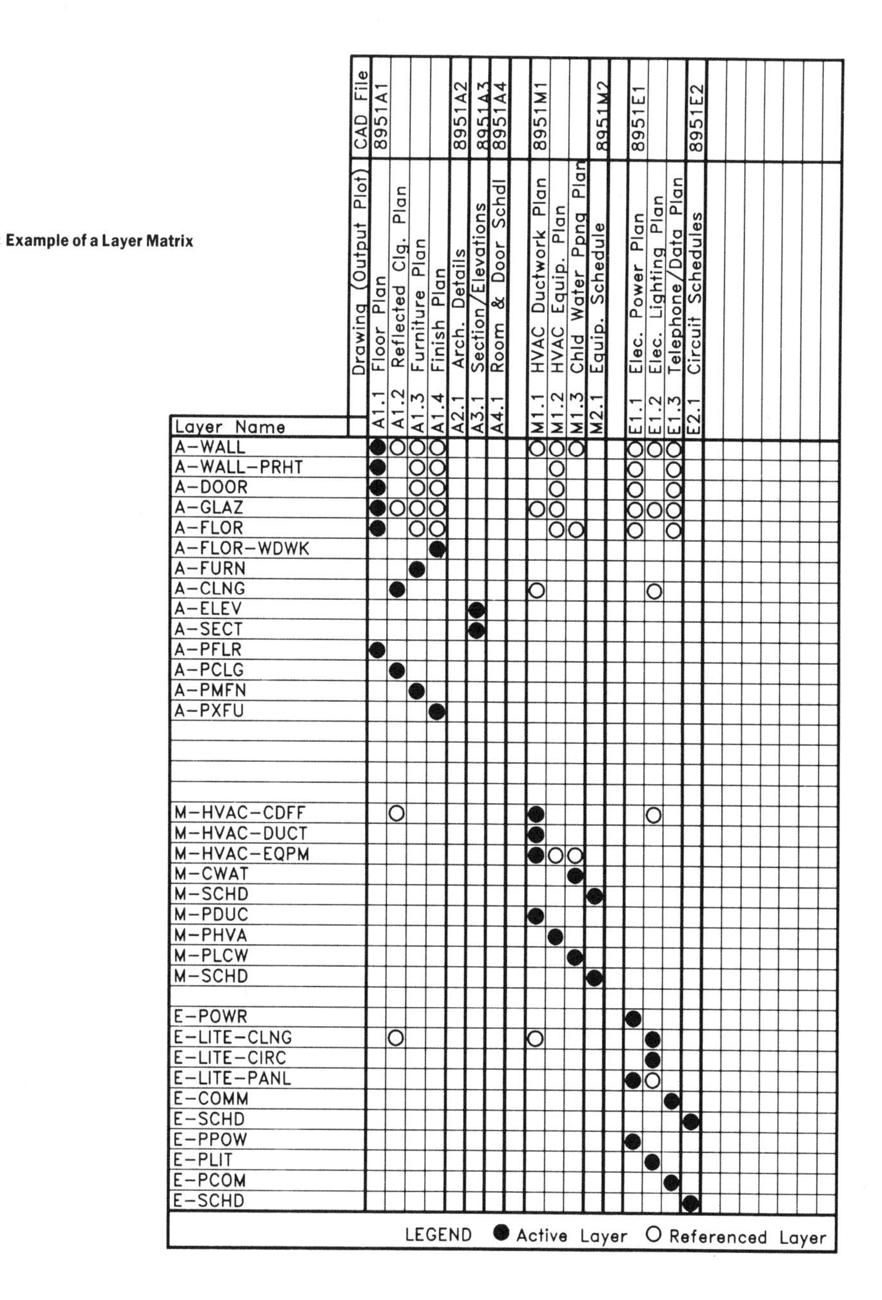

Example of a Layer Matrix

Layer Name	A1.1 Floor Plan	A1.2 Reflected Clg. Plan	A1.3 Furniture Plan	A1.4 Finish Plan	A2.1 Arch. Details	A3.1 Section/Elevations	A4.1 Room & Door Schdl	M1.1 HVAC Ductwork Plan	M1.2 HVAC Equip. Plan	M1.3 Chld Water Ppng Plan	M2.1 Equip. Schedule	E1.1 Elec. Power Plan	E1.2 Elec. Lighting Plan	E1.3 Telephone/Data Plan	E2.1 Circuit Schedules
CAD File	8951A1				8951A2	8951A3	8951A4	8951M1			8951M2	8951E1			8951E2
A-WALL	●	○	○	○				○	○	○		○	○	○	
A-WALL-PRHT	●		○	○					○			○		○	
A-DOOR	●		○	○					○			○		○	
A-GLAZ	●	○	○	○				○	○			○	○	○	
A-FLOR	●		○	○					○	○		○		○	
A-FLOR-WDWK				●											
A-FURN			●												
A-CLNG		●						○					○		
A-ELEV						●									
A-SECT						●									
A-PFLR	●														
A-PCLG		●													
A-PMFN			●												
A-PXFU				●											
M-HVAC-CDFF		○						●					○		
M-HVAC-DUCT								●							
M-HVAC-EQPM								●	○	○					
M-CWAT										●					
M-SCHD											●				
M-PDUC								●							
M-PHVA									●						
M-PLCW										●					
M-SCHD											●				
E-POWR												●			
E-LITE-CLNG		○						○					●		
E-LITE-CIRC													●		
E-LITE-PANL												●	○		
E-COMM														●	
E-SCHD															●
E-PPOW												●			
E-PLIT													●		
E-PCOM														●	
E-SCHD															●

LEGEND ● Active Layer ○ Referenced Layer

FIGURE 2-11. Example of a Layer Matrix

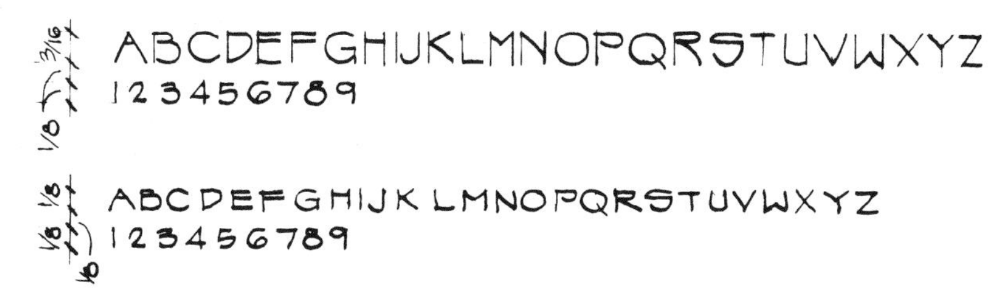

FIGURE 2-12. Hand Lettering

the enlarged plan only (Fig. 4-24 and 4-57). This ensures that if a change is made on the floor plan and is not reflected on the enlarged plan, the person doing the final checking will be able to catch the discrepancy. This does not apply to CADD, where the correction is done simultaneously on both plans.

2. Tie all dimensions to a fixed reference point. In steel or concrete frame structures, the reference point is the column grid line. For bearing wall construction, it is the face of the bearing wall.
3. Exterior plan dimensions are usually written on three lines (Fig. 2-13). The one nearest to the wall is tied to wall openings and offsets. The second string of dimensions indicates column grid lines. If bearing walls are used instead of columns, this line is either omitted or used to dimension major features such as building wings. The outermost line shows the overall dimension.
4. Repetitious dimensions should not be shown individually. Either draw one line for the group and write the numbers of the segments and the dimensions of each (e.g., 5 EQ. PARTS @ 2′-6″), or subdivide the line and write EQ. on each subdivision and show the overall dimension (Fig. 2-14).
5. The principle of writing dimensions only once also applies to vertical dimensions. Building section, wall section, and section detail dimensions should be written as shown in Fig. 2-15.

   Do not repeat the same dimensions on the elevation sheets. If you must show dimensions on the elevations, just show the floor-to-floor heights and identify each floor.
6. If the length of the line representing the dimension is not the same length as the written dimension, write NTS (not to scale) beside or under the dimension. This should be used as a last resort; drawings should be accurate. Always double-check the written dimension against the measured length of the dimension line.
7. Strings of dimensions stretching between the same column grid lines must each add up to the same amount. If CADD is not used, a special inch/foot calculator is a time-saver that helps in this task.
8. Always allow for adequate clearance between construction elements. For example, if a wall panel is to be attached to the edge of a slab, allow a gap for shimming between the panel and the slab. In this imperfect world, the slab edge may not be perfectly straight throughout the facade or it may not be perfectly parallel to the slabs above and below it. Check the specifications for the allowable tolerances.

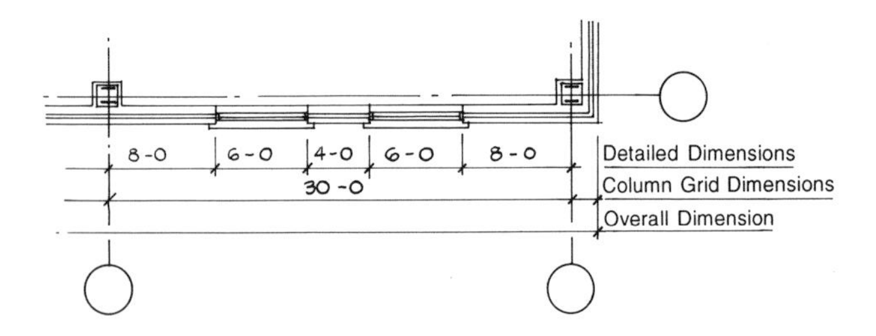

FIGURE 2-13. Exterior Wall Dimensions

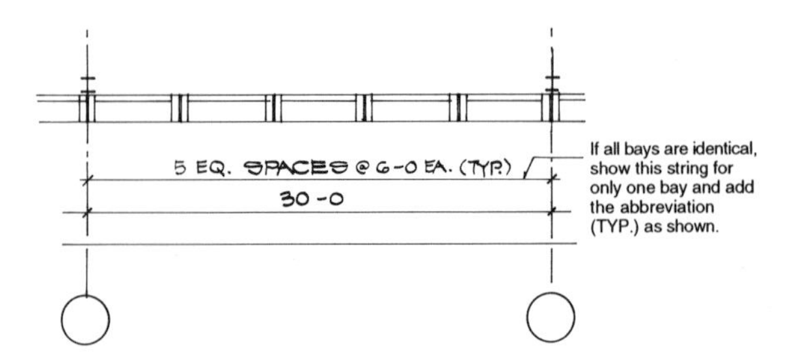

FIGURE 2-14. Repetitive Dimensions

9. The distance between stud partitions should be shown either to the face of the stud or to the wall finish. Both methods are correct. The first method is easier for the erector. The second is easier for the architect and ensures that critical code-mandated dimensions are done correctly. A general note describing which method is used should be included.

   Do not dimension to partition centerlines. If the erector marks the slab at that point, the stud runner will hide it when it is placed on top of it and the location will not be accurate.
10. Fractions of an inch should not be less than ⅛ inch.
11. Dimensions for projects with angles other than 90 degrees require calculations based on trigonometric formulas. The *Manual of Steel Construction* issued by the American Institute of Steel Construction, Inc., contains these formulas and other mathematical data.

These are but a few of the things to observe when one is entrusted with dimensioning. The name of the game is to check and double-check the computed dimensions against the measured dimension lines until you are absolutely sure that all the dimensions are correct.

2.6.3 Sources of More Information

American Institute of Steel Construction, Inc. Latest edition, *Manual of Steel Construction*. One E. Wacker Drive, Chicago, IL 60601. 312-670-2400.

Liebing, Paul. 1977. *Architectural Working Drawings*. New York: John Wiley & Sons, Inc.

Stitt, Fred A. 1980. *Systems Drafting*. New York: McGraw-Hill, Inc.

2.7 SCHEDULES AND WORK SHEETS

2.7.1 Schedules

Schedules are a means to provide the greatest amount of information in the least amount of space in an organized, easy-to-read fashion. Room finish schedules (Fig. 2-16) and

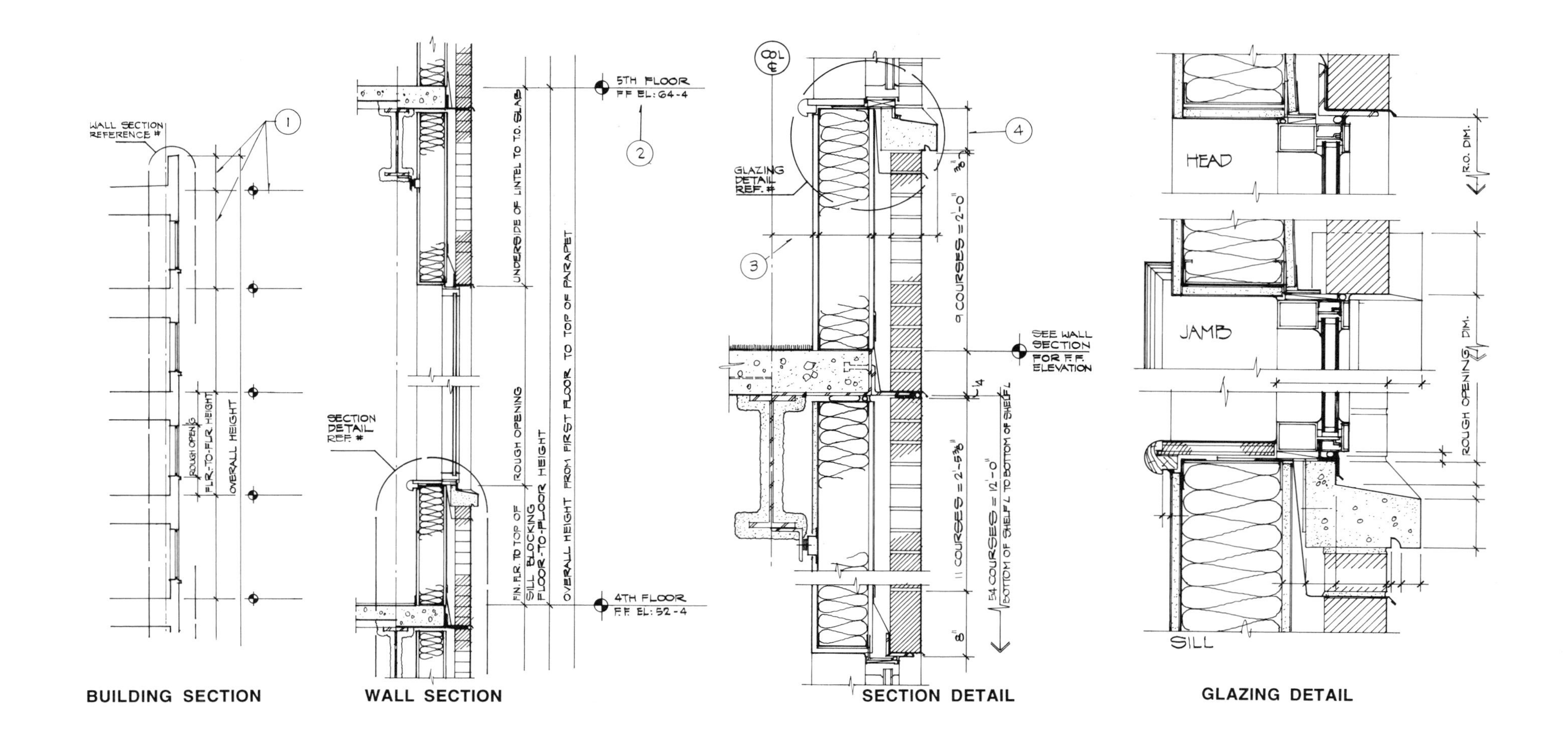

Guidelines:

1. If roof edge slopes along the perimeter of the building to direct water toward the drains, write the word "varies" for the T.O. slab elevation and the dimensions indicated and show spot elevations on the roof plan.
2. Some offices prefer to designate the first floor elevation as 100' or ±0' and relate all elevations to that number. The actual elevation is shown between brackets or a note stating that ±0' (or 100' as the case may be) = the elevation from the survey. If, for any reason, the ground floor elevation is modified, only the note will have to be changed.
3. Although, as a general rule, horizontal dimensions should be shown only on the plans, these dimensions may be shown here if plan details do not show them in that level of detail.
4. Show overall height of the sill. Detailed dimensions are shown on the Window Detail.

FIGURE 2-15. Hierarchy of Vertical Dimensions

ROOM FINISH SCHEDULE*											
ROOM #	ROOM NAME	FLOOR	BASE	WALLS				CEILING	CEILING HT.	REMARKS
				NORTH	EAST	SOUTH	WEST			
304	WOMEN	D 6	A 6	A 5	A 5	A 6	A 5	I 1	8-0	SEE LAVATORY ELEVATION
305	VEST.	D 12	A 19	A 5	A 5	A 5	A 5	I 1	8-0	

LEGEND

* Title may be placed at the bottom (See Detail Module Fig. 2-2)

SURFACES
A. Gypsum Wallboard
B. Concrete Masonry Units
C. Metal
D. Concrete
E. Plaster on Exp. Mtl.
F. Glazing
G. Wood Subfloor
H. Exposed Overhead Struct.
I. Acoustic Tile Type 1
J. Acoustic Tile Type 2

FINISHES
1. Prefinished
2. Sealer
3. Epoxy Paint
4. Paint
5. Vinyl Wall Covering
6. Ceramic Tile
7. Wood Paneling
8. Acoustic Panels
9. Plaster Coat
10. Marble
11. Granite
12. Carpet
13. Terrazzo
14. Wood
15. Vinyl Composition Tile
16. Rubber
17. Resilient Base
18. Entrance Mat
19. Vinyl base
20.

FIGURE 2-16. Room Finish Schedule

door schedules (Fig. 2-17) are included in almost every architectural project.

Door schedules identify each door by its number shown on the floor plan. It is recommended that, unless requested otherwise by the owner, door numbers be the same as the room number into which they open. If more than one door opens into a room, add a letter to the number. For instance, if the room number is 204, the first door is numbered 204, the second is numbered 204A, the third 204B, etc.

Because different doors have different hardware, such as lock set, threshold, hinge type, closer, gaskets, security measures, etc., the schedule identifies each hardware set by a number keyed to the description in the specifications. These hardware sets are usually prepared by a hardware consultant or a manufacturer's representative. The architect must check each set for conformance to code requirements and security measures required by the client.

The schedule enables the door subcontractor to install the right hardware on each door and install each door in the right location. Refer to the guidelines under the schedule (Fig. 2-17) for more pertinent information.

Other schedules such as the window and column fireproofing schedules (Fig. 2-18 and 2-19) are sometimes included in the project. While ensuring that the fireproofing is properly done, the fireproofing schedule is time-consuming to develop. It should be considered for small projects only. For larger projects and multistory buildings, indicate the code-mandated fire ratings on the general information sheet or in the specs. An alternative method is to define the rating required by the code for the steel columns and specify the acceptable kind of fireproofing and leave it up to the contractor to execute the required thicknesses based on the manufacturer's recommendations. A partition schedule may also be included (Fig. 2-20).

Blank schedules are included in the Appendix for easy reproduction on sticky-back and application to the back of the project sheet. The lines can then be extended to accommodate the data for any project size.

2.7.2 Work Sheets

Work sheets fulfill a function similar to the schedules. However, they are only used in special cases. For this reason, they are included in the Appendix. Unlike schedules, which are used as a means of communication between the architect and the client's representatives as well as between the architect and the MEP engineers. Here is how they work. Let us assume for example that we are dealing with a complicated university research lab project. After the preliminary or schematic design is done, the Project Manager meets with each department head. A ¼-inch plan of each lab conveniently superimposed over a ¼-inch grid is given to the department head. Each professor may draw the preferred lab layout allocated to him or her, including equipment as well as casework. In conjunction with the plan, work sheets such as those shown in figure A-13 and figure A-14 are also used to communicate the lab requirements to the architect. Manufacturers' representatives will assist each professor in drawing the layout and filling out the work sheets.

After the Project Manager receives this information, he or she gives a copy to the team leader and sends a copy to the MEP engineers to review and ultimately to use in sizing the required utilities and exhaust duct work. Blank work sheets are also included in the Appendix.

DOOR SCHEDULE*

DOORS						FRAMES				HRDWR	RATING (4)	REMARKS
NO.	H	W	T	TYPE (1)	MAT'L	HEAD (2)	JAMB	THR'D	TYPE (1)	SET NO. (3)		
2.07	7-0	3-0	1¾	A	WD	3/A4.2.1	4/A4.2.1	5/A4.2.1	a	14	1 HR	
											/	

Guidelines:

1. If all types are the same, write the type designation on the first line and draw an arrow extending to the bottom of the schedule. If the elevation is the same for a hollow metal and a wood door, they have the same type designation. The door "Material" column indicates the difference.
2. Write the detail number under the first three headings. If the jamb detail is different for each side of the door, write the other detail number in the "Remarks" column and identify which side.
3. This column is usually filled out by the hardware consultant and checked by the architect. The hardware set is listed in the specifications.
4. Write the time rating mandated by the code (20 min., 1 hr., 2 hr., etc.). If no rating is required, cancel the square as shown on the second line.
5. Show glass type on door-type elevations. Show louvers in the Remarks column.

Notes:

1. Door types and details are shown on Sht. A4.2.1 of the Mock-up (Chapter 1).
2. Double doors are indicated under the "Type" heading thus: A x 2 ("A" is an example, show actual door type).
3. Refer to the project specifications for hardware description.
4. All rated doors are installed in frames having the same rating.
5. All frames are hollow metal unless stated otherwise in the Remarks column.

* Table title may be placed in the detail title block instead of above the table.

Abbreviations:

HM	Hollow Metal Door
WD	Wood Veneer Door
PL	Plastic Laminate Veneer Door
HDW	Hardware
H	Height
W	Width
T	Thickness

FIGURE 2-17. Door Schedule

WINDOW SCHEDULE*

WINDOW TYPE	ROUGH OPENING (W X H)	OPERATION	GLASS TYPE	DETAILS					REMARKS
				HEAD	JAMB	SILL	RAIL	MULLION	
1	3'-4" x 6-0	FIXED	A/B	5/A3.5.1	6/A3.5.1	7/A3.5.1	2/A3.5.2	1/A3.5.2	SEE WINDOW TYPE FOR GLASS AREAS
2									
3									
4									

Guidelines:

1. If window details are drawn on only one sheet, delete the sheet number and add a note stating: Refer to Sheet A_____ for window details.
2. Glass types must conform to code requirements for safety requirements, energy savings, and structural strength to withstand wind pressure. Glass types are described in the specifications.
3. Window types and glazing details should be drawn adjacent to the schedule and keyed to the elevations and section details.

* Table title may be placed in the detail title block instead of above the table. (See Fig. 2-2)

FIGURE 2-18. Window Schedule

COLUMN FIREPROOFING SCHED.

COLUMN				UL. DESIGN			
NUMBER	SIZE	W/D RATIO	FLOOR NO.	NUMBER	FIRE RATING	W/D RATIO	DETAIL NO.
A-10	W12x53	.84	1 TO 7	X517	2 HR	.83	

This number must be equal to or greater than the U.L. design number.

COL. SIZE	PERIMETER*	W/D RATIO
W 10 x 49**	(9.98 x 2) + [(10 x 4) - (2 x .56)] = 58.84	49 ÷ 58.84 = .83
W 14 x 228**	(16 x 2) + [(15.568 x 4) - (2 x 1.045)]= 61.37	228 ÷ 61.37 = 3.72
W 12 x 53†	(12.06 x 2) + [(9.995 x 4) - (2 x .345)] = 63.41	53 ÷ 63.41 = .836

* Perimeter = Column Depth x 2 + Flange Width x 4 - 2 x Web Thickness

** These column sizes are the basis for most UL. designs.

† This column is the one shown on the example above.

Notes:

1. UL.designs listed in the table show intent. Other UL.designs may be substituted if they provide the same fire-rating.
2. W/D ratio = Column Weight ÷ Column Perimeter

FIGURE 2-19. Column Fireproofing Schedule

PARTITION TYPE SCHEDULE*

USED / NOT USED	TYPE	WALL THICKNESS	STUDS: SIZE	STUDS: SPACING	FURRING: HAT	FURRING: ZEE (DEPTH)	FURRING: RESILIENT	BOARD TYPE THICKNESS: GWB	BOARD TYPE THICKNESS: GWB, TYPE"X"	BOARD TYPE THICKNESS: GWB, MR	BOARD TYPE THICKNESS: CEMENT BD.	CMU THICKNESS	INSULATION: MINERAL FIBER	INSULATION: FIBERGLASS	INSULATION: RIGID	INSULATION: OTHER**	T.O. WALL DET.	FIRE RATING	FIRE TEST	STC RATING	REMARKS
	1	4 7/8	3 5/8	16"				5/8					3				3.1	0	N/A	49	MR AT WET WALLS
	2	24	1 5/8	16						5/8			1 1/2				1.1	0	N/A	N/A	INSULATION STAPLED TO ONE SIDE ONLY
	3 (1" ... 1/2")	3 1/2	2 1/2	24					1/2 1								7.1	1	UL#U469	N/A	
	4	9 1/2			24			5/8				8					2.2	2	UL#U906	N/A	CMU BEARING THE UL CLASSIFICATION MARKING

Notes:

1. Partition types shown at right indicate main categories. They may be further subdivided to indicate different designs within each category, for example 1.1, 1.2, 1.3 . . . etc. (Type 1 includes examples shown in Fig. 4-26 thru 4-34.) These sub-categories may also include different size studs, different stud spacing, addition of insulation, different top of wall conditions, etc.
2. Fill schedule with the types most used by the office. Then reproduce to be used on all projects. Add types as needed.
3. For complicated partitions, identify layers on the sketch shown under the "Type" column and write "shown" under the "Board" column.
4. Refer to Sub-section 4.2.8 for more information.

* Schedule title may be placed in the detail title block.

** Identify in the remarks column.

PARTITION TYPES

1. REGULAR GWB WALLS
2. GWB CHASE WALLS
3. SHAFT WALLS
4. CMU WALLS
5. PARTY WALLS
6. OTHER

ABBREVIATIONS:

GWB	GYPSUM WALL BOARD
MR	MOISTURE RESISTANT
BD	BOARD
CMU	CONCRETE MASONRY UNIT WALL
T.O.	TOP OF
N/A	NOT APPLICABLE

FIGURE 2-20. Partition Type Schedule (Example) (Blank schedule included in the Appendix)

Other work sheets, such as those shown in figure A-15, may be useful in remodeling or school projects. Another work sheet shown in figure A-16 is used to convey to the electrical engineer the power for equipment such as dock levelers, motorized roll-up doors, projection screens, etc. These work sheets may be modified or new work sheets may be custom-designed for special projects.

2.8 GENERAL CONVENTIONS

2.8.1 Graphic Symbols

These symbols (Fig. A-17) provide a means to reference items shown on the plans to other drawings or schedules for more detailed or complementary information. A few, such as the floor and spot elevation indicators, apply also to elevation and section sheets. I have chosen symbols that require the least amount of time to draw. Please note that the method of dimensioning is indicated in this set of symbols. This saves the Team Leader from having to add a general note for that purpose.

The "alternate number" symbol is used on projects that contain alternate methods of construction. For instance, if two methods of construction, such as precast and GFRC (glass fiber reinforced concrete), are deemed to be equal by the architect and one of them requires different profiles, panel sizes, or joint locations, the architect prepares drawings for one and calls it the base bid. He then draws the GFRC alternate, places it in a frame on the same sheet, and places that symbol with a number to signify that it is an alternate. The contractor prices all the alternates (there may be several alternates in the project) for the owner and architect to evaluate and decide whether the price difference warrants a change from the base bid to the alternate.

The revision symbol is used when changes to the contract documents are required during the bidding phase or during construction (see Fig. 2-1b for explanation).

2.8.2 Materials Symbols

These symbols are used to identify materials used in an assembly, especially in sections and details. They enable a person to identify the limits of a material such as brick at a glance. Figure A-18 shows an example of a materials symbol index. This sheet and the preceding one may be reproduced on sticky-back and affixed to the general information sheet at the front of the construction drawings set. Space is provided at the bottom to add symbols as needed.

2.8.3 Abbreviations

The Team Leader should emphasize that abbreviations must be kept to a minimum. If, due to lack of space or multiple repetition, an abbreviation is absolutely unavoidable, it must be chosen from the abbreviations included in the general information sheet. The same abbreviations should be used for both the drawings and the specifications.

References:

Guzey, Onkal, and Freehof, James N. 1985. "The Ongoing Revolution in Contract Documents," *Architectural Technology*, Fall,: 27–31.

Schley, Michael K., ed. 1990. *CAD Layer Guidelines*. Washington, D.C.: The American Institute of Architects.

3 TECHNICAL INFORMATION

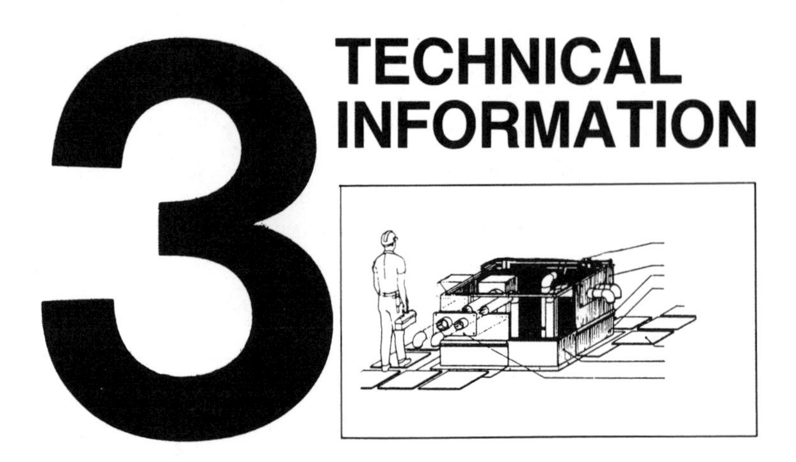

In the past, there were far fewer systems and products used on architectural projects. Most projects were not as big or complicated as projects being built today. The architect was able to acquire a certain expertise in a relatively short time without recourse to seminars or visits to conventions.

Today conditions are much different. The array of products, systems, and building types is mind boggling. One has only to look at *Sweet's Catalog* to get an idea about the magnitude of the choices being offered. The architect must educate himself to make the most appropriate choices to prevent problems from occurring.

A sizable amount of time used on a project is spent on research for background information, such as code analysis, fire protection, wall and roof systems, insulation to satisfy energy code requirements, sealants, and other items necessary to base the details and Construction Drawings on. To research products and systems, many architects follow the fast and relatively easy method of calling two or three product representatives (reps, for short) to get as much information in the shortest time possible. This process is acceptable if done by an experienced person knowledgeable on the subject being investigated, that is, one who is able to separate fact from claims by the salesman designed to sell the product he or she represents. While most product reps are conscientious people, one has to be aware that they are in the business of selling their product. This, in some cases, may mean downplaying similar products that may be as good or better than those being represented. Product reps should not be confused with impartial advisers. So the architect must prepare for these meetings by reading background material to be able to ask intelligent and pertinent questions designed to separate fact from fiction.

To acquire technical knowledge, architects attend seminars offered by universities or product associations such as the Brick Institute of America (BIA), the National Roofing Contractor's Association (NRCA), the Precast/Prestressed Concrete Institute (PCI), and similar organizations. Some attend the yearly AIA convention to hear the speakers and see exhibits of the latest products and systems. These functions are rather costly and are attended mostly by more senior staff. This leaves junior people, who do the actual work, wasting time hunting for answers.

Offices use different methods to provide information. They range from the grossly inadequate to the highly developed approach. The following is a partial listing and evaluation of these methods.

1. The minimal approach is a "library" consisting of *Sweet's Catalog, Architectural Graphic Standards, The Architect's Handbook of Professional Practice* (four volumes), and a copy of the building code. This approach relies almost entirely on information gathered from product reps. As mentioned previously, this is not the best way to gather information. Unfortunately, many offices still use this method.
2. The next step up is an office with a library that contains, in addition to the items listed above, a few useful books such as:

    - *Architectural Precast Concrete* issued by the Precast/Prestressed Concrete Institute
    - *Standard Pratice in Architectural Sheet Metal Work* by the Sheet Metal and Air Conditioning Contractor's National Association (SMACNA)

- *The National Roofing Contractor's Association Roofing & Waterproofing Manual* issued by the National Roofing Contractor's Association
- *Manual of Steel Construction* published by the American Institute of Steel Construction
- *Fire-Resistance Directory* (usually referred to as the U.L. book)
- *Technical Notes on Brick Construction* by the Brick Institute of America

A few other publications and journals may also be included. While this is a distinct improvement over the first library, any "library only" approach falls short of the mark. There is not enough time to read in depth about a subject before making a decision. More often than not, the team member walks to the library, spends time searching for the book, finds out that somebody else has checked it out, gets interested in a design journal, and spends time perusing it.

3. The third approach, in addition to an adequate library, is to designate a senior staff member as the technical adviser or resident guru. Theoretically, this should work perfectly. It simulates the old apprentice approach that supplied old societies with fine craftsmen. In practice, however, this task is added on top of this person's other duties, such as project management or specification writing. In many cases, junior staff hesitate to approach a person at that level lest their question be construed as a display of ignorance that may be a factor in evaluating their performance. The project Team Leader may become offended that the team member bypassed him and got his answers from the "guru," or the Team Leader may have a different answer that leaves the team member in a quandary about which answer to use.

These are but a few of the methods used to disseminate information. As can be readily seen, they fall short of providing a fast reference that saves time and effort for the junior members of the team. Devising a way to find dependable information in a hurry without wasting time should be the goal. The following integrated resource system has the potential for achieving that.

- A source of information in digest form written in a concise, easy-to-understand language. This can be in the form of an office manual containing information used on every project. The contents of Chapter 3 of this book can form the nucleus of such a manual. These can be updated and added to on a yearly basis by an experienced staff member or associate given that responsibility.
- Periodic, preferably monthly, seminars given by a reputable rep representing several companies manufaturing similar products.
- A well-organized office library with a sign-out sheet to track who has which book. An index of its contents and a brief description of each subject should be available at each workstation.
- Team leaders to serve as designated advisers so they can be accessible to the team. The function of guru, in this case, is covered by the specification writer, the library, and seminars.

Many offices incorporate one or more of the features listed above. In my opinion, the Technical Information Manual can be a valuable asset and a real timesaving tool that can help interns acquire knowledge to prepare them for becoming the team leaders of tomorrow.

3.1 CODES

3.1.1 General

At the start of the schematic design, a code analysis should be done by the Team Leader. This analysis must be updated at each phase of the project and discussed with the authority having jurisdiction. Most city codes are modeled after one of the national model codes.[1] These codes are similar in organization except that each puts emphasis on conditions peculiar to the region it applies to. For instance, codes applicable in California emphasize earthquake resistance, while codes in the northern half of the nation contain sizable sections on snow loads and energy conservation. Other codes stress other local requirements.

Codes ensure that buildings function properly, are sturdy enough to resist forces of nature, are fire protected, allow safe evacuation for the occupants in case of fire, are accessible to the physically handicapped, and safeguard the health and comfort of the occupants. The building design must also conform to the plumbing code, which may or may not be part of the building code, and the zoning code, which, in addition to identifying the use of buildings constructed in different zones of the city (e.g., industrial, medical, etc.), may limit the height and area of the building, designate the number of parking spaces, and, in some instances, require that the appearance of the building be compatible with neighboring structures. In addition, OSHA (the Occupational Safety and Health Act) regulations apply to certain industrial buildings to ensure safety, noise control, and other factors.

Other codes and regulations may also apply to the projects. For instance, if the project is above a certain height and is located in the flight path for an airport, it is subject to FAA (Federal Aviation Administration) height limitations and warning light requirements. Elevator codes must be consulted for buildings served by elevators. The Project Manager must make sure that the building is in conformance with all applicable codes. Large cities sometimes have their own codes. If the project is outside city limits, it may be subject to the state building code, the Life Safety Code issued

[1]The BOCA National Building Code, published by the Building Officials and Code Administrators International, Inc., the Uniform Building Code (UBC), published by the International Conference of Building Officials, and the Standard Building Code (SBC), published by the Southern Building Code Congress International, Inc.

by the National Fire Protection Association (NFPA), or one of the model codes.

The annotated edition of NFPA 101 gives insights on the rationale behind the egress and fire protection requirements in most building codes. I recommend that each member of the team read NFPA 101 to be more knowledgeable about code application. Also, if there is any doubt about the meaning of a word, consult the "Definitions" section. For instance, there is quite a difference between exit, way of exit, and exit discharge.

The code analysis form included at the end of this section illustrates the methodology of building code review. This form is not based on any specific code. It addresses only the architectural sections and a few structural requirements that may have a bearing on the architectural design. Let me emphasize that this is only an example that must be modified to conform to the specific code applicable to the project at hand. Each office should have a timesaving code analysis form conforming to the code applicable to the area in which most of their projects are located. The code reference column must also be filled out, stating the section number opposite each item, to aid in double-checking. With this step-by-step guide, almost any member of the team will be able to fill out the form.

3.1.2 Code Analysis Guidelines

The following is a sequential method of code analysis related to the form in the following section.

1. Determine the main occupancy (educational, residential, etc.) of the building and identify any ancillary occupancies and determine the fire separation requirements between them. In some cases, incidental uses that occupy less than 10 percent of the floor area are allowed to conform to the provisions of the main occupancy, provided that the ancillary occupancy is not classified as hazardous.
2. Compute the area per floor and total height of the building and compare to the maximum allowed. Most codes allow substantial increases for added street frontage and the presence of a fire-suppression system. Choose the type of construction based on the area and height allowed. If the floor area exceeds the area allowed, the building may be subdivided by one or more fire walls. In that case, each segment is considered to be an independent building. Be sure to check the zoning ordinance also.
3. Refer to the fire-resistance ratings table and find the ratings required to protect the structure and the means of egress for the type of construction chosen in Step 2. Check the table of contents at the beginning of the code for other fire-resistive requirements.
4. Determine the occupancy load based on the number of square feet per person tabulated for that occupancy (the code defines whether the area is net or gross). Find corridor, door, and stair width by translating the occupancy load to units of egress width, usually 22 inches (if that is the basis used in the code). The number of persons per unit can be found in the egress capacity table.
5. Find the minimum number of toilets required for each sex using the occupancy load number found in Step 4. In some cities, this information may be found in a separate plumbing code. If necessary, modify this number by adding fixtures to satisfy the convenience of the occupants. Remember, these are minimums. (see Sec. 4.3)
6. Choose a suitable fire-protection system. The systems listed in most codes are sprinkler, wet or dry standpipe (the last is used where uninsulated water pipes are subject to freezing such as in a parking structure), foam, carbon dioxide, and halogenated and dry chemical systems. The choice is influenced by the occupancy. For instance, aircraft hangars require a foam system. It is also influenced by the type of fire (electrical, gasoline, etc.) and whether the building contents are too valuable to be damaged by water (museum, specialized libraries, mainframe computers, etc.)

   Fires are usually defined as Class A, B, or C. After taking your best shot and consulting with manufacturer representatives, check with the fire marshal and find out what is acceptable to him before specifying the system. After all, he is the one responsible for fighting these fires. Besides, it is a requirement.
7. Determine the requirements for the physically handicapped, to conform to both the local and federal code.
8. Check the code chapters dealing with mechanical and structural design. Although these chapters are in the domain of the consultants, the architect is responsible for making sure that they are applied. This does not mean checking the calculations usually done by other professionals. Check for things like masonry wall thickness, wind pressure data for selecting windows, fire damper locations, etc.
9. Refer to the code table of contents and check for sections you may have missed. Items such as boiler room requirements, elevators, penthouses, atria, interior finishes, fire doors, plastics, energy conservation, etc., must conform to the code.

A building code analysis form is included in the following section. After modifying it to conform to the code applicable to the project, fill it out and take a set of drawings marked with all dimensions, fire hose cabinet locations, means of egress requirements, and the code analysis to the "plan checker" at city hall for review. If you are confronted with a situation where it is impossible to conform to the code, which sometimes happens in remodeling existing buildings, apply for a variance.

3.1.3 Building Code Analysis Form*

Project Name: ______________________ Project No.: ______________________

Project Location: ______________________ Prepared By: ______________________

Applicable Code: ______________________ Checked By: ______________________

Fire Zone: ______________________ Date: ______________________

Code Requirements				Code Ref. #
1. Principal Occupancy:				
Ancillary Occupancies:				
a.				
b.				
c.				
d.				
Occupancy Separations				
Between: a. and:	=	hours		
Between: b. and:	=	hours		
Between: c. and:	=	hours		
Between: d. and:	=	hours		
2. Gross Floor Area (actual)	=			
Net Floor Area (actual)	=			
Maximum Gross Building				
Area Allowed**	=	sq. ft.	sq. ft. (if sprinklered)	
Maximum Height Allowed**	=	floors	ft.	
Construction Type	Allowed:		Used:	
(Based on maximum area and height allowed)				
3. Fire-Resistance Ratings				
Exterior Walls	=	hours		
Fire Walls and Party Walls	=	hours		
Fire Separation Assemblies	=	hours		
Exitway Enclosure (stair)	=	hours		
Exitway Access Walls (corridors)	=	hours		
Exitway Discharge Walls***	=	hours		
Shaft Walls (mech. and elevator)	=	hours		

Code Requirements			Code Ref. #
Structural Frame	=	hours	
Floor Assembly (slab and joists)	=	hours	
or Floor/Ceiling Assembly	=	hours	
Roof/Ceiling Assembly	=	hours	
or Roof Slab	=	hours	
Roofing	=	hours	

4. Occupancy Load

Number of Units of Exit Width = Actual Numer of Persons per Floor ÷

Number of Exits × Number of Persons per Unit = "X" Units @ 22″ = ft.

a. Occupancy: sq. ft. ÷ sq. ft. per person = persons

b. Occupancy: sq. ft. ÷ sq. ft. per person = persons

c. Occupancy: sq. ft. ÷ sq. ft. per person = persons

d. Occupancy: sq. ft. ÷ sq. ft. per person = persons

Total = persons

5. Exit Requirements

Minimum Number of Exits = Provided =

Maximum Dead-end Length =

Maximum Distance to Exits =

Minimum Width of Exit Corridor = Provided =

Stairs

Minimum Width of Door =

Access to Roof Required? ☐ Yes ☐ No

Minimum Width of Flights and Landings =

Maximum Riser =

Minimum Tread =

Maximum Height Between Landings =

Handrail Height =

Maximum Distance Between Rails =

Smoke-proof Enclosure Required? ☐ Yes ☐ No

Guardrail Minimum Height =

Ramps

Maximum Slope =

Code Requirements					Code Ref. #
Handrail Height	=				
Handrail Required on One or Both Sides	=				
6. Toilet Requirements					
Minimum Number of Toilet Stalls for Men****	=	Provided =			
Minimum Number of Toilet Stalls for Women	=	Provided =			
Minimum Number of Lavatories for Men	=	Provided =			
Minimum Number of Lavatories for Women	=	Provided =			
Minimum Number of Drinking Fountains	=	Provided =			
Number of Urinals****	=	Provided =			
7. Fire Protection System					
Sprinkler (required?), Partial or Full?:					
Dry Standpipe (required?):	☐ Yes	☐ No	Size:		
Number of Standpipes:	Location:				
Number of Outlets:	Hose (required?): ☐ Yes	☐ No			
Siamese Connection Required?:	☐ Yes	☐ No			
Wet Standpipe (required?):	☐ Yes	☐ No			
Number of Standpipes:					
Length of Hose:	Length of Throw:				
Number of Fire Hose Cabinets:	Location:				
Fire Extinguishers (required?):	☐ Yes	☐ No	Type:	Location:	
Other (describe):					
8. Structural Requirements (may be filled in by consultant)					
Loads					
Wind Loads —1st floor to floor	=	psf			
— floor to floor	=	psf			
— floor to floor	=	psf			
—Building corners	=	psf			
—Roof	=	psf			

Code Requirements	Code Ref. #
Snowguards (required?): ☐ Yes ☐ No Where:	
Earthquake Requirements:	
CMU Masonry	
Span: Minimum thickness:	
Span: Maximum thickness:	
Cavity walls	
Minimum cavity: Maximum cavity:	
Ties at:	
Metal studs (exterior)	
Span: Size: Gauge: Spacing:	
Span: Size: Gauge: Spacing:	
Other:	
9. Miscellaneous	
Penthouse	
Area limitation =	
Use limitation =	
Height limitation =	
Elevators	
Ventilation (required?): ☐ Yes ☐ No Area:	
Maximum number per shaft =	
Maximum rail span =	
Machine room wall rating: = Hours	
Other requirements:	

*This form must be modified to conform to the code applicable to the project before using.

**Add or subtract any modifications allowed by code.

***Corridor to street or exterior court.

****Most plumbing codes allow up to one-third substitution of urinals for toilet stalls. For example, if the minimum number of toilets is required to be nine, six stalls and three urinals are allowed.

3.2 HOW TO USE THE U.L. BOOK

3.2.1 General

The Underwriter's Laboratories, a nonprofit organization, periodically issues two directories that are applicable to architectural projects. The first is called *Fire-Resistance Directory*. It is commonly referred to as the U.L. book and includes fire-rated designs for beams, floor/ceiling assemblies, roof/ceiling assemblies, walls, partitions, and columns. These designs help in making choices that satisfy code fire-resistance requirements. The second directory is called *Building Materials Directory* and contains surface burning characteristics (flame spread and smoke development ratings) for building products, including ceilings, floor coverings, finishes, building panels, sheathing, adhesives, sealants, glazing, insulation, and other materials. It also includes information about rated doors, frames, and roof coverings. The last category, which occupies a sizable part of the directory, defines class A, B, and C roofing as well as wind uplift ratings for roofs.

The folllowing section relates to the *Fire-Resistance Directory* only. It shows in tabular form, and with explanatory notes, how to choose structural elements that satisfy the code. Wall designs, included in this directory, are relatively few and easier to choose from.

3.3.2 Column Designs

Defining the thickness of a column fireproofing early in the DD phase enables the project architect to determine the distance between the column centerline and the face of the exterior wall. It also aids in developing column details.

Table 3-1 can be used as a tool to facilitate and expedite the task of choosing a U.L.-approved design from the myriad of designs offered. The architect may use a schedule similar to the one shown in figure 2–19 to list the acceptable U.L. designs. A general note, stating that other U.L. designs which satisfy the fire rating and the weight to perimeter ratio (W/D ratio) would be acceptable, may be added if no factors restrict the choice. This note enables the contractor to choose from a wide range of designs. U.L. designs are usually acceptable to the authority responsible for seeing to it that the building code is conformed to. However, the TL must ascertain that this is the case.

The choice of fireproofing design may be affected by whether the column is exposed to view or touch (use lath and plaster or GWB), whether the whole building frame is scheduled to be sprayed, as is usual in multistory buildings (use cementitious, fiber or, in some cases, mastic spray), or other constraints or cost implications.

While it is desirable that the project should be designed to accommodate the thickest fireproofing to provide the contractor with the widest range of U.L. designs to choose from, one must balance this against the fact that thick designs add to slab edge cantilevers. Longer cantilevers require heavier bent plates and, in some cases, outrigger angle supports, adding to the cost.

Because the U.L. book is issued periodically with additions and deletions to the designs, the following table must be updated periodically. The following guidelines should be considered in conjunction with Table 3–1:

1. Column designs are currently identified by an X prefix to the design number. These designs are similar to the XR prefixes; however, the XR ratings are more stringent. They are suitable for areas which may develop fire temperatures at a faster rate than the X rating assemblies. "X" stands for column, "R" stands for Rapid Temperature Rise.
2. The column sizes indicated are the minimum sizes allowed. **Any column having a greater weight to perimeter (W ÷ D) ratio than the ratio of the column under consideration (the design column) is acceptable.**
3. Columns and beams carrying a rated wall should not be rated less than the wall. Check the building code.

3.2.3 Beam, Floor/Ceiling, and Roof/Ceiling Assemblies

Choosing a horizontal separation (floor or roof assembly) is a little more complicated than choosing columns. It involves the following steps:

1. Determine the required fire rating by referring to the building code.
2. Read the code carefully. In most codes, joists have the same rating as the slabs while girders and trusses are part of the structural frame and have the same rating as the columns.
3. Decide whether a slab/beam design will be chosen or a floor/ceiling assembly is preferred. The latter requires dampers on the ducts, enclosures around the lights, clips on the ceiling, etc. Floor/ceiling designs usually do not require fireproofing of the deck or beam. In the final analysis, cost determines which system is chosen.
4. Refer to the table at the beginning of the U.L. book titled *Fire Resistance Ratings (BXUV)*. Narrow down the choice to the structural system used on the project (steel, poured-in-place concrete, precast, wood, etc.).
5. Refer to the chosen system section and select a design that comes closest to satisfying the requirements. If the structure is steel and the beam is sprayed, choose a design that has a beam size equal to or more than the minimum size shown. This is done by comparing the weight ÷ perimeter ratio of the design beam to the weight ÷ perimeter ratio for the beam used in the project.
6. The following guidelines give further helpful hints.

   A. In floor-ceiling assemblies no substitution is allowed between lightweight and normal weight concretes.

TABLE 3-1. Column Fireproofing Design Guide

Ratings	Thickness of Fire Proofing	Column Size	UL Design Numbers: Cementitious Mix. X700 Series	Fiber, Sprayed X800 Series	GWB X500 Series	Plaster X400 Series	Mastic, Fiber Bd., Bldg. Units X000, X300, X600
3/4 HR.	1/16 to 1/2	W 10 x 49	724, 732, 733, 736				618
		W 14 x 228	766				
		4 Ø x .237	771				
		8 Ø x .322	771				
		ST 4 x 4 x .375	771				
		ST 36 x 24 x .5	771				
3/4 HR.	9/16 to 1	W 6 x 16		821, 831			
		W 10 x 49	701*, 702*, 703*, 732, 753, 754, 755, 766				
		W 14 x 228	766				
		4 Ø x .188	771	835			
		4 Ø x .237		835			
		8 Ø x .188	771	835			
		8 Ø x .322		835			
		ST 4 x 4 .188	771	835			
		ST 4 x 4 x .375	771	835			
1 HR.	1/16 to 1/2	W 6 x 16	756, 757				
		W 8 x 35	741				
		W 10 x 49	725, 726, 733, 758, 759, 760, 761, 776, 777, 780, 781		526		203, 601*, 602*, 604*, 605*, 606, 607, 607*, 612, 614, 618, 620
		W 12 x 106		829, 838			
		W 14 x 228	718, 738, 756, 757, 766, 772, 785	823, 824, 838			609
		W 14 x 233	711, 764, 774, 787	829			
		W 14 x 730	711, 764, 771, 772	829			
		4 Ø x .237					615
		8 Ø x .322					603*
		ST 4 x 4 x .375	771				
		ST 36 x 24 x .5	775, 771				
1 HR.	9/16 to 1	W 4 x 13			526		
		W 8 x 10	711				
		W 6 x 15.5			526		
		W 6 x 16	711, 722, 738, 764, 772	821, 829, 831, 838			
		W 8 x 28	723, 738, 772	813, 829, 833, 838			301*
		W 8 x 35		826			

TABLE 3-1. Column Fireproofing Design Guide *(Continued)*

Ratings	Thickness of Fire Proofing	Column Size	UL Design Numbers: Cementitious Mix. X700 Series	Fiber, Sprayed X800 Series	GWB X500 Series	Plaster X400 Series	Mastic, Fiber Bd., Bldg. Units X000, X300, X600
		W 16 x 40	740				
		W 10 x 49	701, 701*, 702*, 703*, 708, 711, 717, 724, 728, 729, 730, 731, 732, 736, 738, 739, 753, 754, 755, 764, 766, 769, 770, 772, 773, 784, 786, 788, 789	823, 824, 829, 838			004, 201*, 301, 301*, 604*
1 HR.	9/16 to 1	W 12 x 106					301*
		W 14 x 228	749, 750				
		W 14 x 233					301*
		4 Ø x .188	771	835			
		4 Ø x .237	751, 768, 771, 775	835			
		8 Ø x .188	771	835			
		8 Ø x .332	771	835			
		ST 4 x 4 x .188	752, 767, 771, 775	835	526		301*
		ST 4 x 4 x .375	752, 767, 775	835			
		ST 8 x 8 x .25			526		301*
1 HR.	1 1/16 to 2	W 6 x 9	772				
		W 8 x 10		829, 837			
		W 10 x 49		825			
	2 1/16 to 3	W 4 x 13			528 (1 + 1 5/8)		
		W 6 x 15.5			528 (1 + 1 5/8)		
		W 10 x 49			528 (1/2 + 1 5/8)		
		ST 4 x 4 x .188			528 (1 + 1 5/8)		
		ST 8 x 8 x .25			528 (5/8 + 1 5/8)		
1 1/2 HR.	1/16 to 1/2	W 10 x 49	716, 719, 725, 726				602*, 604, 604*, 605*, 608, 608*, 613, 614, 616, 620
		W 14 x 228	735, 738, 743, 744, 756, 757, 772, 785	823, 824, 838			
		W 14 x 233	711, 764, 774, 787	829			
		W 14 x 730	711, 764, 772	829			
		ST 36 x 24 x .5	775				
1 1/2 HR.	9/16 to 1	W 6 x 16	756, 757, 764				
		W 8 x 35	741				
		W 16 x 40	740				
		W 10 x 49	701, 701*, 702*, 708, 709, 724, 726, 727, 729, 730, 732, 733,	829			004, 301*, 601*, 603, 604*

TABLE 3-1. Column Fireproofing Design Guide *(Continued)*

			UL Design Numbers				
Ratings	Thickness of Fire Proofing	Column Size	Cementitious Mix. X700 Series	Fiber, Sprayed X800 Series	GWB X500 Series	Plaster X400 Series	Mastic, Fiber Bd., Bldg. Units X000, X300, X600
			736, 738, 758, 759, 760, 761,				
			764, 766, 769, 770, 772, 773,				
			776, 777, 780,781, 788, 789				
		W 12 x 106	718, 766	829, 838			301*
		W 14 x 228	764				
		W 14 x 233	764				301*
		W 14 x 730					
		4 Ø x .237	771				615
		8 Ø x .322	752, 767, 775, 771				603*
		ST 4 x 4 x .375	771				
		ST 36 x 24 x .5					
1 1/2 HR.	1 1/16 to 2	W 6 x 9	772				
		W 8 x 10	711	829			
		W 6 x 16	711, 738, 772	821, 829, 831, 838			301*
		W 8 x 28	738, 772	829			301*
		W 10 x 49	703*, 711, 717, 753, 754, 755,	823, 824, 838			301
			766, 784, 786				
		4 Ø x .188	771	835			
		4 Ø x .237	751, 768, 775, 771	835			
		8 Ø x .188	771	835			
		8 Ø x .322		835			
		ST 4 x 4 x .188	752, 767, 775, 771	835			301*, 307
		ST 8 x 8 x .25					301*
		ST 4 x 4 x .375		835			
	2 1/16 to 3	W 8 x 10		837			
		4 1/2 Ø x .109			531 (1 + 1 3/8)		
2 HR.	1/16 to 1/2	W 10 x 49					601, 602, 605*, 608*, 620
		W 14 x 228	725, 726, 762, 763, 778, 779,	807, 809, 819, 834			610
			782, 783				
		W 14 x 233	711	829			
		W 14 x 730	711, 738, 764, 772	829			
2 HR.	9/16 to 1	W 8 x 35	741				
		W 10 x 49	701*, 702*, 716, 719, 724, 725,			401, 402	002 (9/16 + 1/2), 003, 004,
			726, 732, 733, 736, 758, 759,				201*, 203, 204, 601*, 602*,
			780, 781, 789				605*, 614, 619

TABLE 3-1. Column Fireproofing Design Guide *(Continued)*

Ratings	Thickness of Fire Proofing	Column Size	UL Design Numbers: Cementitious Mix. X700 Series	Fiber, Sprayed X800 Series	GWB X500 Series	Plaster X400 Series	Mastic, Fiber Bd., Bldg. Units X000, X300, X600
		W 12 x 106		829			301*
		W 14 x 228	704, 707, 710, 718, 735, 738,	808, 823, 824, 832, 838			
			743, 744, 749, 750, 756, 757,				
			766, 769, 770, 772, 785, 788, 789				
			764, 774, 787				
		W 14 x 233	738				301*
		W 14 x 730					
		4 Ø x .237					615
		8 Ø x .322					603*
		ST 36 x 24 x I5	775, 771				
2 HR.	1 1/16 to 2	W 6 x 9	772				
		W 8 x 10	711				
		W 4 x 13			526		307
		W 6 x 15.5			526		
		W 6 x 16	711, 722, 738, 756, 757, 764, 772	821, 829, 831, 838			301*
			723, 738, 745, 772				
		W 8 x 28		813, 822, 829, 833, 838			301*
		W 8 x 35	740	826			
		W 16 x 40	701, 703*, 708, 709, 711, 717,				
		W 10 x 49	727, 728, 729, 730, 731, 734,	818, 823, 824, 829, 838	516, 526		002 (9/16 + 1/2), 004, 204, 301,
			737, 738, 739, 746, 747, 753,				301*
			754, 755, 760, 761, 764, 766,				
			772, 773, 776, 777, 784, 786				
2 HR.	1 1/16 to 2	W 12 x 92					606*
		W 12 x 106		838			
		W 14 x 228			523		
		3 1/2 Ø x .216				413	
		4 Ø x .237	751, 771				
		8 Ø x .322	748, 771	835			
		ST 3 1/2 x 3 1/2 x .25				413	
		ST 4 x 4 x .375	752, 767, 775, 771	835			
		ST 4 x 4 x .188			526		
		ST 8x 8 x .25					301*, 307
2 HR.	2 1/16 to 3	W 8 x 10		829, 837			
		W 10 x 49		836	517, 518, 519, 528 (1 1/8 +		305

TABLE 3-1. Column Fireproofing Design Guide *(Continued)*

			UL Design Numbers				
Ratings	Thickness of Fire Proofing	Column Size	Cementitious Mix. X700 Series	Fiber, Sprayed X800 Series	GWB X500 Series	Plaster X400 Series	Mastic, Fiber Bd., Bldg. Units X000, X300, X600
					1 5/8)		
		W 14 x 228			520, 521		
		4 Ø x .188	771	835			
		4 Ø x .237	768, 775	835			
		8 Ø x .188	771	835			
		ST 4 x 4 x .188	752, 767, 771, 775	835			301*
	3 1/16 to 4	W 4 x 13			528 (1 1/2 + 1 5/8)		
		W 6 x 15.5			528 (1 1/2 + 1 5/8)		
		W 8 x 28			525 (1 1/4 + 2 1/4 + 1/4),		
					527 (1 1/4 + 2 1/4 + 1/4)		
		ST 4 x 4 x .188			528 (1 5/8 ± 1 5/8)		106
		ST 8 x 8 x .25			528 (1 5/8 + 1 5/8)		
2 1/2 HR.	9/16 to 1	W 10 x 49					605*, 609*
		W 12 x 106					301*
		W 14 x 233					301*
		ST 36 x 25 x .5	771				
	1 1/16 to 2	W 8 x 28					301*
		W 10 x 49	703*				001* (2 + 1/4),
							005 (1 1/2 + 1/4), 301*
		8 Ø x .322	771				
		ST 4 x 4 x .375	771				
	2 1/16 to 3	W 6 x 16					301*
		4 Ø x .188	771				
		4 Ø x .237	771				
		8 Ø x .188	771				
		ST 8 x 8 x .25					301*
		ST 4 x 4 x .188	771				301*
3 HR.	1/16 to 1/2	W 14 x 228					611
		W 14 x 730	711, 738, 772	829			
	9/16 to 1	W 10 x 49	716, 719, 725, 726			401	605, 609*, 614, 620
		W 14 x 228	704, 707, 710, 718, 725, 726,	807, 808, 809, 819, 832,			302, 307
			735, 738, 742, 743, 744, 756,	834, 838			
			757, 762, 763, 772, 778, 779,				
			782, 783, 785				
		W 14 x 233	711, 764, 774, 787	829			301*

TABLE 3-1. Column Fireproofing Design Guide *(Continued)*

Ratings	Thickness of Fire Proofing	Column Size	UL Design Numbers				
			Cementitious Mix. X700 Series	Fiber, Sprayed X800 Series	GWB X500 Series	Plaster X400 Series	Mastic, Fiber Bd., Bldg. Units X000, X300, X600
		W 14 x 730	764				
3 HR.	1 1/16 to 2	W 6 x 16	756, 757, 764				
		W 8 x 28	723, 772	813, 833			
		W 8 x 35	741				
		W 16 x 40	740				
		W 10 x 49	701, 701*, 702*, 703*, 708, 709, 711, 717, 727, 728, 729, 730, 731, 732, 733, 736, 737, 738, 739, 746, 747, 753, 754, 755, 758, 759, 760, 761, 764, 766, 769, 770, 772, 773, 776, 777, 780, 781, 784, 786, 788, 789	815, 818, 829	508, 526	402, 403, 410	001, 002 (7/8 + 1/2), 004, 203, 204, 301*, 306, 307, 601*
							301*
		W 12 x 106		829, 838			
		W 14 x 228	749, 750, 766	823, 824			
		8 Ø x .322	771				
		ST 4 x 4 x .375	752, 771				
		ST 36 x 24 x .5	771				
3 HR.	2 1/16 to 3	W 6 x 9	772				
		W 8 x 10	711				
		W 4 x 13			526		307
		W 6 x 15.5			526		
		W 6 x 16	711, 722, 738, 772	829, 831, 838			301*
		W 8 x 28	738, 745	822, 829, 838	525 (2 + 2 1/4 + 1/), 527 (2 + 2 1/4 + 1/4)		301*
		W 8 x 35		826			
		W 10 x 49		823, 824, 838	511, 515	411 (1 1/2 + 1 1/4)	001 (2 + 1/4), 301
		W 14 x 228			512, 513, 514		302
		4 Ø x .237	751, 771				
		8 Ø x .322		835			
		ST 4 x 4 x .188			526		
		ST 4 x 4 x .375	767	835			
		ST 8 x 8 x .25					301*
3 HR.	3 1/16 to 4	W 8 x 10		837			

TABLE 3-1. Column Fireproofing Design Guide *(Continued)*

Ratings	Thickness of Fire Proofing	Column Size	UL Design Numbers: Cementitious Mix. X700 Series	Fiber, Sprayed X800 Series	GWB X500 Series	Plaster X400 Series	Mastic, Fiber Bd., Bldg. Units X000, X300, X600
		W 4 x 13			528 (2 1/4 + 1 5/8)		
		W 6 x 15.5			528 (2 1/4 + 1 5/8)		
		W 6 x 16		821			
		W 10 x 49		816	509, 510, 528 (1 7/8 + 1 5/8)		
		4 Ø x .188	771	835			
		4 Ø x .237	775	835			
		8 Ø x .188	771	835			
		ST 4 x 4 x .188	752, 771	835	528 (2 1/2 + 1 5/8)		
		ST 8 x 8 x .25			528 (2 1/2 + 1 5/8)		
4 HR.	1/16 to 1/2	W 14 x 730	711				
	9/16 to 1	W 14 x 228	742				307
		W 14 x 233					301*
		W 14 x 730	738, 764, 772	829			
4 HR.	1 1/16 to 2	W 6 x 16	764				
		W 8 x 35	741				
		W 10 x 49	701*, 702*, 709, 716, 719, 724, 725, 726, 727, 728, 729, 731, 732, 733, 736, 739, 758, 759, 760, 761, 764, 769, 770, 773, 776, 777, 780, 781, 788, 789			401, 402, 403, 405, 406, 408	
		W 12 x 106		829			
		W 14 x 228	704, 707, 710, 718, 725, 726, 738, 756, 757, 762, 763, 772, 778, 779, 782, 783, 785	807, 808, 809, 819, 823, 824, 832, 834, 838			301* 304
		W 14 x 233	711, 764, 774	829			
		ST 36 x 24 x .5	771				
4 HR.	2 1/16 to 3	W 8 x 10	711				
		W 6 x 16	711, 738, 756, 757				
		W 8 x 28	723, 738, 745, 772	813, 829, 833			
		W 8 x 35		826			
		W 16 x 40	740				
		W 10 x 49	701, 703*, 708, 711, 717, 730, 738, 746, 747, 753, 754, 755,	801, 823, 824, 829, 838	502, 526	404, 407	001 (2 5/8 + 1/4), 502, 301*

TABLE 3-1. Column Fireproofing Design Guide *(Continued)*

Ratings	Thickness of Fire Proofing	Column Size	UL Design Numbers Cementitious Mix. X700 Series	Fiber, Sprayed X800 Series	GWB X500 Series	Plaster X400 Series	Mastic, Fiber Bd., Bldg. Units X000, X300, X600
			766, 772, 784, 786				
		W 12 x 106		838			302
		W 14 x 228	766		506 (1 + 1 5/8), 507, 522, 533 (1 + 1 5/8), 534 (1 + 1 5/8)		
		8 Ø x .322	771				
		ST 4 x 4 x .375	752, 771				
4 HR.	3 1/16 to 4	W 6 x 9	772				
		W 6 x 15.5			504, 526		
		W 6 x 16	772	829, 838			
		W 8 x 28		822, 838	525 (2 13/16 + 2 1/2)		
		W 10 x 49			501 (2 1/2 + 1 5/16), 504		
		4 Ø x .237	771				
	4 1/16 to 5	W 8 x 28			525 (2 13/16 + 2 1/2), 527 (2 3/4 + 2 1/2)		
		4 Ø x .188	771	835			
		4 Ø x .237		835			
		8 Ø x .188	771	835			
		ST 4 x 4 x .188	771	835			

Abbreviations:

Mix.	Mixture	ST	Steel Tube
Bd.	Board	Ø	Indicates Pipe
Bldg.	Building	W	Wide Flange column

Notes:

1. "XR" prefix precedes design numbers marked with an asterisk(*). All other design numbers are preceded by an"X" which has been omitted to simplify the table.
2. The following designs were not included in the table because either they do not include column size or the design does not mention fire proofing thickness: 101, 104, 106, 203, 204, 309, 310, 524, 530, 201*. Consult with manufacturer for more information.
3. The thickness of cementitious or fiber spray in some designs is increased when the protection of the contour sprayed column's flange tips are reduced to one half. These designs are 701, 703*, 704, 711, 722, 723, 727, 764, 772, and 829. Refer to the Fire Resistance Directory for this information.
4. This table is based on information included in the Fire Resistance Directory issued in 1989. Because these directories are issued periodically, this table should be updated as soon as new directories are issued. It is the sole responsibility of the architect in charge of the project to verify the accuracy of the information included herein. This table is not intended as a substitute for the U.L. directory but rather as a guide for using it to maintain enough clearances in architectural details.

Hold-down clips are required in all lay-in ceilings to prevent panels from being uplifted by the draft caused by fire, exposing the unprotected structure to the flames. Clips are not required if the panels weigh one pound per foot or more.

B. Composite slabs are those where the deck has indentations that allow the deck to act in unison with the concrete topping (Fig. 3-1). Composite slabs deflect more under fire conditions. Non-composite beams may be substituted for composite, but not vice versa. (Composite beams are beams where a portion of the floor acts as part of the compression flange of the beam. This is accomplished by fusing studs—a type of bolt—to the top of the beam.)

C. Steel bar-joist sizes larger than those specified may be used, provided the accessories are compatible. Substitutions are described in detail in the Directory. The structural engineer should be consulted if there is any doubt.

D. **Beams with the same or greater weight-to-perimeter ratio as the specified beam are allowed.** If a beam substitution is made, the spray thickness shall be adjusted as follows:

$$\text{Thickness of Spray} = \frac{[(\text{U.L. Beam Weight} \div \text{U.L. Beam Perimeter} + .6) \times \text{Spray Thickness}]}{(\text{Substitute Beam Weight} \div \text{Substitute Beam Perimeter}) + .6}$$

Provided:
Minimum Weight ÷ Perimeter is not less than .37
Minimum new spray thickness = 3/8 in. or more

E. Unrestrained beam ratings in the N400, N600, N700, and N800 "beam only" series with a sprayed deck may be used with the D900 series if a 12-inch overspray on each side is provided on the deck beyond the flanges of the beam. The same thickness is required for the beam.

Steel beams are considered restrained if they are welded, riveted, or bolted to framing members, cast-in-place slabs are secured to framing, and prefab systems are secured to framing so that the framing resists thermal expansion.

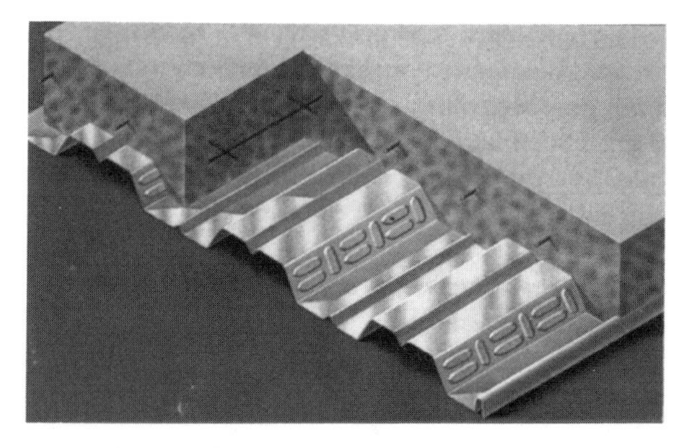

FIGURE 3-1. Composite Deck (Vulcraft 1992).

3.3 INSULATION AND R-VALUES

3.3.1 General

Many design decisions have an effect on energy consumption in buildings. The shape and orientation of the building, the ratio of window area to total wall area, and the type of glass as well as the window treatment (blinds, louvers, etc.) are important factors. The number of occupants, the lighting intensity, and the type of heating, ventilating, and air-conditioning (HVAC) equipment are also important. The type and thickness of insulation protecting the building envelope plays an important role in conserving energy.

One of the initial Design Development tasks is the development of a typical wall section showing the materials, the cavity width, the type of insulation, and part of the roof showing the same kind of information. This drawing contains a wealth of information for the mechanical engineer. Based on this information along with the plans, elevations, and building section, the engineer can size the HVAC system and estimate the preliminary Heat Transfer Coefficients (U-Values) for the walls and roof. These can be transformed into R-values (R stands for Thermal Resistance), since R is the reciprocal of U. The architect then calculates the R-values for the wall and roof assemblies and makes adjustments to confirm that the design satisfies these values. Failure to choose the right amount of insulation will result in occupant discomfort or high energy bills if too little insulation is provided or, unnecessary added cost if too much is installed.

3.3.2 How to Figure R-Values

All building materials have some resistance to heat (or cold) passing through them. Insulating materials have a much better resistance than other materials because they are composed of fibrous or cellular elements that enclose tiny amounts of still air or gas both good insulators.

To find the thermal resistance "R" of a wall or roof assembly, the architect refers to tables listing R-values for building materials and selects the value applicable to each component. Such tables can be found in *Architectural Graphic Standards*, *Time-Saver Standards*, and *The ASHRAE Handbook of Fundamentals*, as well as in manufacturers' literature. Because air films have a tendency to adhere to interior and exterior wall surfaces and provide some resistance to heat passage, they are included in the computation. Figure 3-2 shows an example of a wall assembly to illustrate the method of calculating the R-value. Please note that the air cavity is also included. Table 3-2 gives values for air films and air spaces. (Refer to the latest edition

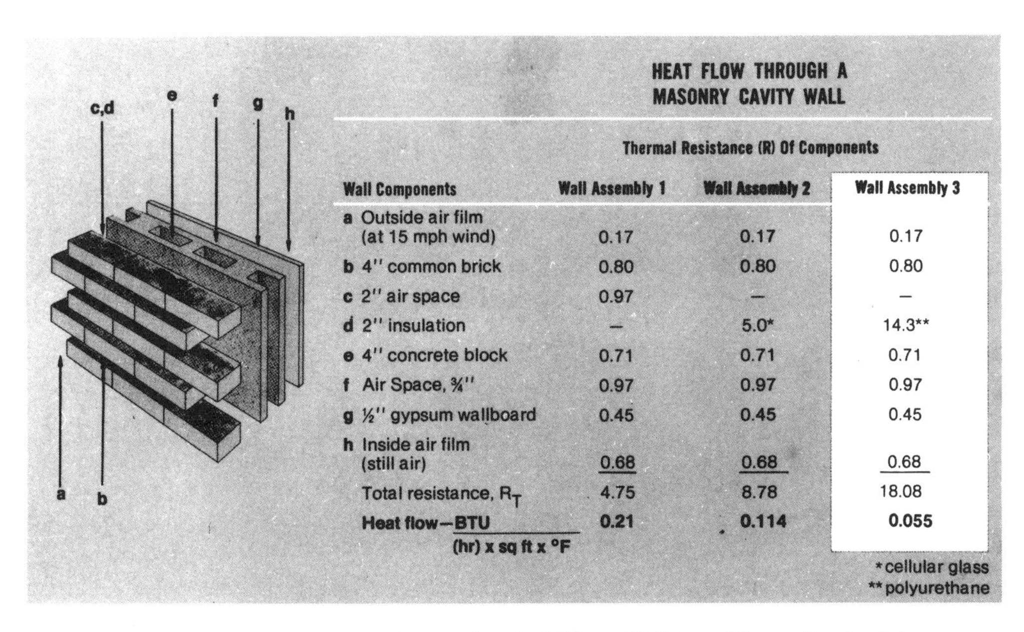

HEAT FLOW THROUGH A MASONRY CAVITY WALL

Thermal Resistance (R) Of Components

Wall Components	Wall Assembly 1	Wall Assembly 2	Wall Assembly 3
a Outside air film (at 15 mph wind)	0.17	0.17	0.17
b 4" common brick	0.80	0.80	0.80
c 2" air space	0.97	—	—
d 2" insulation	—	5.0*	14.3**
e 4" concrete block	0.71	0.71	0.71
f Air Space, ¾"	0.97	0.97	0.97
g ½" gypsum wallboard	0.45	0.45	0.45
h Inside air film (still air)	0.68	0.68	0.68
Total resistance, R_T	4.75	8.78	18.08
Heat flow—BTU / (hr) x sq ft x °F	0.21	0.114	0.055

*cellular glass
**polyurethane

FIGURE 3-2. Comparison of R-Value Computations for Different Wall Assemblies (Reprinted, by permission, from Mobay Chemical Corporation, *Urethane Insulation Energy Saver Manual for Commercial, Institutional, Industrial, and Residential Construction*).

TABLE 3-2. Surface Conductances, Btu/h · ft² · °F, and Resistances, °F · ft² · h/Btu, for Air[a,b,c,d]

Position of Surface	Directions of Heat Flow	Surface Emittance, ε [e]					
		Non-reflective		Reflective			
		ε = 0.90		ε = 0.20		ε = 0.05	
		h_i	R	h_i	R	h_i	h_i
STILL AIR							
Horizontal	Upward	1.63	0.61	0.91	1.10	0.76	1.32
Sloping - - 45°	Upward	1.60	0.62	0.88	1.14	0.73	1.37
Vertical	Horizontal	1.46	0.68	0.74	1.35	0.59	1.70
Sloping - - 45°	Downward	1.32	0.76	0.60	1.67	0.45	2.22
Horizontal	Downward	1.08	0.92	0.37	2.70	0.22	4.55
MOVING AIR	(Any Position)	h_o	R	h_o	R	h_o	R
15-mph Wind (for winter)	Any	6.00	0.17	--	--	--	--
7.5-mph Wind (for summer)	Any	4.00	0.25	--	--	--	--

a No surface has both an airspace resistance value and a surface resistance value. No airspace value exists for any surface facing and airspace of less than 0.5 in.

b For ventilated attics or spaces above ceilings under summer conditions (heat flow down), see Table 5*.

c Conductances are for surfaces of the stated emittance facing virtual blackbody surroundings at the same temperature as the ambient air. Values are based on a surface-air temperature difference of 10°F and for surface temperature of 70°F.

d See Chapter 3* for more detailed information, especially Tables 5 and 6, and see Figure 1* for additional data.

e Condensate can have a significant impact on surface emittance (see Table 3*).

* Refers to other sections of the ASHRAE, 1989 Handbook, Fundamentals from which this table was reproduced by permission.

of *The ASHRAE Handbook of Fundamentals* for more detailed information.)

3.3.3 Definitions

An understanding of a few abbreviations used by mechanical engineers is always helpful. The following are a few basic definitions:

BTU (British Thermal Unit): The amount of heat required to raise the temperature of one pound of water 1°F.

K (Thermal Conductivity): A measure of heat flow per hour through one square foot of material 1-inch thick for every 1°F of temperature difference.

C (Thermal Conductance): Same as K but for any thickness.

R (Thermal Resistance): The resistance of a material to heat transfer (R = 1/c).

U (Heat Transfer Coefficient): The number of BTUs that can flow through a square foot of a wall or roof assembly in an hour for every 1°F of temperature difference. The total R for an assembly = 1/U.

HDD (Heating Degree Days): The sum of degrees that the average outdoor temperature is below 65°F in a twenty-four hour period computed for the whole heating season. Example: If the temperature during a winter day fluctuated from 17° to 21°F (19° average), the HDD for that day is 65 − 19 = 46 degree days.

R × K: Thickness of material to achieve the required rating.

3.3.4 Insulating Materials

When choosing insulation, the R-value is not the only kind of information to look for. The following factors must also be considered:

1. Thermal drift:
   Some kinds of insulation, such as polyisocyanurate and polyurethane, lose some of their insulation transmission resistance over time. This phenomenon is called "thermal drift." One of the theories about the cause of this loss is that these types emit chlorofluorocarbon gas (CFC). The use of this gas as a blowing agent during the manufacturing process is scheduled to be banned by the year 2000 because it depletes the ozone layer that protects the earth from solar radiation. Substitute gases, however, will probably not affect thermal drift.

   On the plus side, this type of insulation has good compressive strength, a good R-value, and is dimensionally stable.
2. Compressive strength:
   Roof or plaza insulation is required to support design live loads including the weight of equipment. Insulation used in these locations must be chosen to satisfy this requirement. Boards with a weak bearing capacity will compress, resulting in loss of some R-value. They will also not support the membrane properly and may eventually cause it to fail.
3. Flammability, smoke generation, and meltability:
   Polystyrene, commonly known as Styrofoam, is flammable. A fire-retardant may be added to the product or an "X" type gypsum board layer may be placed between the insulation and the roof deck. In addition, polystyrene has a melting point in the 250°F–300°F range. If hot asphalt is used to adhere these panels to the roof, melting point temperatures must not be exceeded. Local codes must be consulted in that regard.
4. Moisture resistance:
   In some applications such as "upside-down roofs," where the insulation is placed on top of the roofing (also known as IRMA—Insulated Roof Membrane Assembly), the insulation is almost always wet. Needless to say, the type of insulation chosen must be impervious to water penetration. Likewise, insulation placed on the exterior of walls, whether below grade or in wall cavities, must follow the same criterion.
5. Compatibility with adhesives:
   Manufacturers must be consulted to determine whether adhesives used to attach the insulation to a substrate or attach another layer to the insulation is compatible with it. The same caution also applies to solvents.
6. Impermeability:
   Foil-faced insulation acts as a vapor barrier. If used, it must be placed with the foil facing the warm side (usually the interior of the building in colder regions) to prevent moisture from traveling through the insulation and condensing in the assembly, causing damage (see Sec. 3.4).
7. Compaction:
   Some types of loose-fill insulation have a tendency to become compacted over time. This compaction means that the voids between the particles of insulating material become smaller. Because these voids provide a percentage of the R-value, their compaction results in loss of insulating value.
8. Cost:
   Cellular glass roof insulation withstands high compressive loads, is dimensionally stable, and resists vapor migration. Furthermore, it is not affected by ultraviolet rays or chemicals and has an excellent R-rating. It is used mostly on prestigious projects.

3.3.5 Site Visits

Make sure that the insulation used is as specified or an approved substitute, that the thickness installed conforms to the drawings, and that the method of attachment is adequate. Storage prior to installation must be protected from moisture. It is also important to ascertain that compressible insulation such as fiberglass batts fit in the space between studs without undue compression, which will result in a reduction of R-value. Finally, the integrity of the vapor barrier must be maintained to prevent moisture from damaging the insulation (Sec. 3.4).

To summarize, determining the right amount of insulation for large buildings is a complicated procedure that involves life-cycle costs, heat generated inside the building by lighting, occupants, and equipment, as well as solar gain, window-to-wall ratio, and type of glazing. It is the duty of the Team Leader to provide the mechanical engineer with a set of plans, elevations, building sections, and the crucial typical wall section describing the components of the building envelope as early as possible. This enables the engineer to make his preliminary calculations and supply the Team Leader with the design U-values for the walls and roof. To translate this information into the final wall and roof design, the architect makes a fast calculation based on the method described in Section 3.3.2, after determining the type and thickness of the insulation. This process may result in a change of wall thickness. For instance, if the structural studs are required to be 16 gauge, 4 inch deep to resist deflection caused by wind pressure and the insulation is required to be 6-inch batts to satisfy the design R-value, either a change to a lighter gauge 6-inch stud satisfying the same structural requirements is made or another type of insulation with a higher R-value is used.

Codes must be consulted before finalizing the choice of insulation, especially where flammability, smoke generation, and energy conservation are concerned. Protection from moisture by placing a vapor retarder on the warm side is very important because moisture is a good conductor of heat and can reduce the R-value drastically. The emphasis on energy conservation is important and architects must take the subject seriously.

3.3.6 Sources of More Information:

American Society of Heating. Refrigeration, and Air Conditioning Engineers, Inc., latest edition. *ASRAE Handbook of Fundamentals*, 1791 Tullie Circle, N.E., Atlanta, GA 30329. 404-636-8400.

Ballast, David Kent. 1988. *Architect's Handbook of Formulas, Tables and Mathematical Calculations*. Englewood Cliffs, NJ: Prentice-Hall.

Callender, John Hancock, ed. *Time-Saver Standards*, latest edition. New York: McGraw-Hill, Inc.

Maslow, Philip. 1982. *Chemical Materials for Construction*. New York: McGraw-Hill, Inc.

McQuiston, Faye C., and Parker, Jerald D. 1988. *Heating, Ventilating and Air-Conditioning, Analysis and Design*, 3rd ed. John Wiley & Sons, Inc.

Ramsey, Charles G., and Sleeper, Harold R. *Architectural Graphic Standards*, latest edition. New York: John Wiley & Sons, Inc.

3.4 AIR BARRIERS AND VAPOR RETARDERS

3.4.1 General

Moisture in the form of water vapor is always present inside buildings. It is introduced through the HVAC system and forms an essential part of the air we breathe. Water vapor, however, can cause damage to buildings if it is allowed to condense inside walls, reducing the effect of insulation, creating musty smells, and causing rust or corrosion to the structure and damage to the finishes.

Wind pressure, in the absence of an air barrier, can cause air to penetrate through wall interstices on the windward side and exit the building on the leeward side, creating uncomfortable drafts in winter and wasting energy. Wind can also drive rain into these same access points and cause water damage on the windward side and accelerate water vapor migration to the leeward side. This vapor may condense when it hits cold surfaces in the wall assembly and do similar damage.

As can be readily perceived, vapor and air barriers perform very important functions and must be included in the design of the building envelope. A vapor retarder must also be placed under concrete slab-on-grade construction within the footprint of the building to prevent vapor in the soil from migrating into the building. This section explains, in a simplified manner, the function of these barriers.

3.4.2 Vapor Retarders

Water vapor is a gas. It has the ability to diffuse through seemingly impermeable barriers. It migrates from areas with high moisture content to low moisture areas. Because cold air has less capacity to carry vapor, moisture migrates from the warm interior in winter to the cold exterior. It is during that season that the potential for damage is greatest due to the big difference between interior and exterior temperatures.

The harmful effects of vapor migration can be avoided with the placement of a vapor retarder on the warm side of the walls and, in some cases, the roof. This prevents the bulk of the water vapor from passing through the insulation, condensing, and causing damage. Depending on the location of the building, the "warm" side can be either at the interior or exterior of the building envelope or the vapor retarder may even be omitted (Fig. 3-3).

Materials used as vapor barriers have low water-vapor permeance (perm rating). A perm is the ratio at which water will diffuse through a given area of the material over a specified period of time. To qualify as a vapor retarder, a

Humid climates in the continental U. S. follow southern coastal belts of Gulf of Mexico and Atlantic Ocean. Note that humid-climate area can provide problems for vapor retarder use if precautions are not taken. Fringe climate may need no vapor retarder at all.

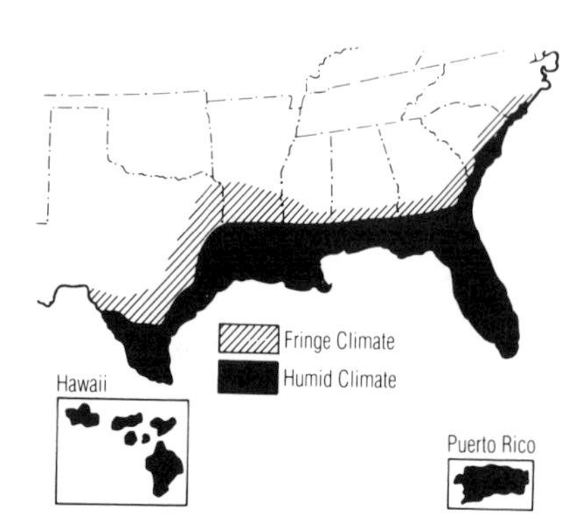

Typical wall for residential and low-rise commercial construction in cold climate illustrates temperature gradient within wall. Note that in this case the inside temperature is 70°F., inside relative humidity is 50%, the outside temperature is 24°F., and the dew point is 51°F. (see Fig. 2). The overlaid graph shows that moisture will condense on the sheathing if no vapor retarder is used on the "warm side" of the condensing plane.

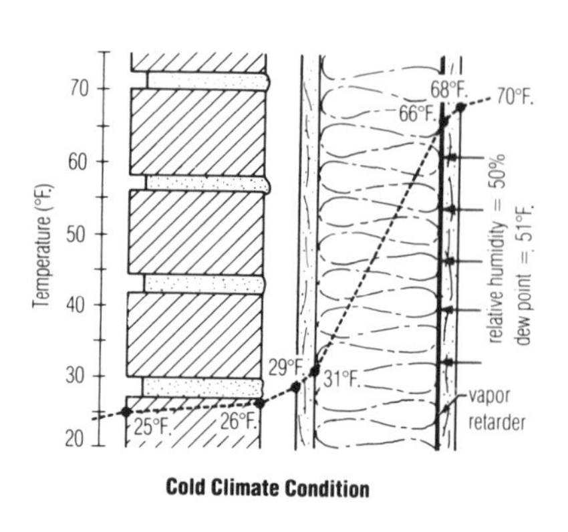

Typical wall for residential and low-rise commercial construction in hot, humid climates illustrates that vapor pressure drive is opposite that of cold climates. Vapor retarder in this example is needed on the exterior side of the insulation. Note that in this case the outside temperature is 86°F., the outside relative humidity is 90%, the inside temperature is 70°F., and the dew point is 83°F. (see Fig. 2). The overlaid graph shows that moisture will condense on the gypsum panel if no vapor retarder is used.

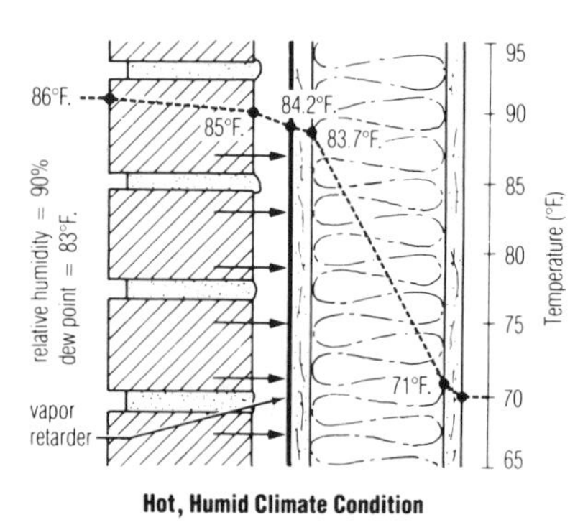

FIGURE 3-3. Location of the Vapor Retarder (Reprinted, by permission, from Shipp and Marchello, *"What You Ought to Know About Air Barriers and Vapor Retarders"*).

material must be rated at less than .5 perm. One of the most prevalent vapor retarders is 4-mil. polyethylene sheeting. Although some materials such as aluminum foil are totally impervious to water vapor (0 perm), it is impractical to expect that, if used, it would form a perfect envelope around the building interior. There will be fasteners going through it, poorly sealed edges, punctures and pinholes. That is why a vapor barrier should be referred to as a vapor retarder. It reduces the ratio at which vapor passes through the building envelope and prevents any substantially harmful condensation.

Wayne Tobiasson, a research civil engineer at the Cold Regions Research and Engineering Laboratory of the U.S. Army Corps of Engineers and a recognized authority on the subject, developed the map shown in figure 3-4. It represents relative humidities (RH) above which a vapor retarder is needed. It is based on an indoor temperature of 68°F, the usual design temperature for HVAC calculations. The graph in figure 3-5 is designed to modify the figures shown in figure 3-4 if the indoor temperature is other than 68°F. The example shown on the graph represents a building in New York (50 percent relative humidity) with an indoor temperature of 75°F. This combination requires a vapor retarder above 40 percent RH instead of the 50 percent RH shown on the map.

Mr. Tobiasson also developed the graph and roof sections shown in figures 3-6 and 3-7 to determine the optimum location of the vapor retarder in roofs. To use this graph, the indoor design relative humidity determined by the HVAC consultant is used in conjunction with the RH determined from figures 3-4 and 3-5. In the example shown in figure 3-5, for a 45 percent RH in a city close to the Canadian border, a maximum 62 percent of the R-value for the roof assembly may be placed under the vapor retarder. Table 3-3 shows some typical indoor relative humidities.

Examples of vapor retarders include all roofing membranes, polyethylene (4 mil. and 6 mil.), and aluminum foil. Refer to *Architectural Graphic Standards* for a listing of materials and their perm ratings.

3.4.3 Air Barriers

Air barriers in exterior walls prevent wind from creating uncomfortable cold air currents and forcing moisture into the building interior. They prevent water vapor from being

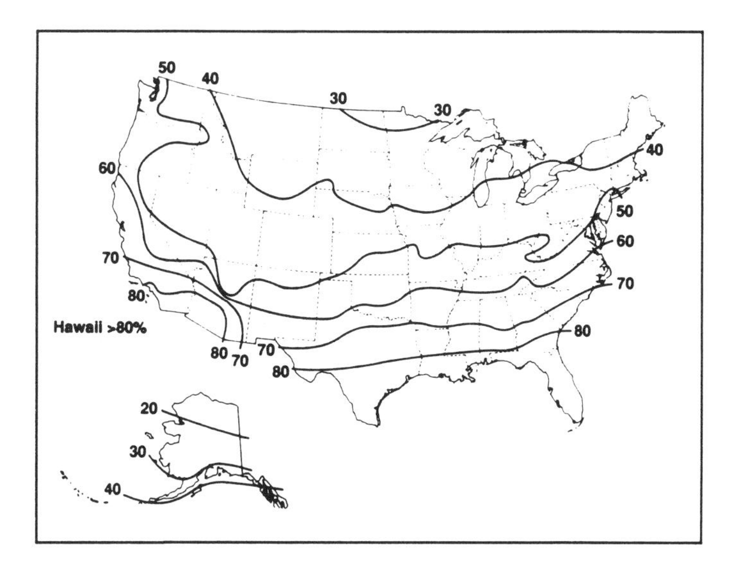

FIGURE 3-4. Indoor Relative Humidities at 68°F, Above Which a Vapor Retarder Is Needed in Membrane Roofing Systems. Use the Graph Shown in Figure 3-5 to Modify These Values for Temperatures Other Than 68°F. (Reprinted, by permission, from Tobiasson, *Vapor Retarders for Membrane Roofing Systems*).

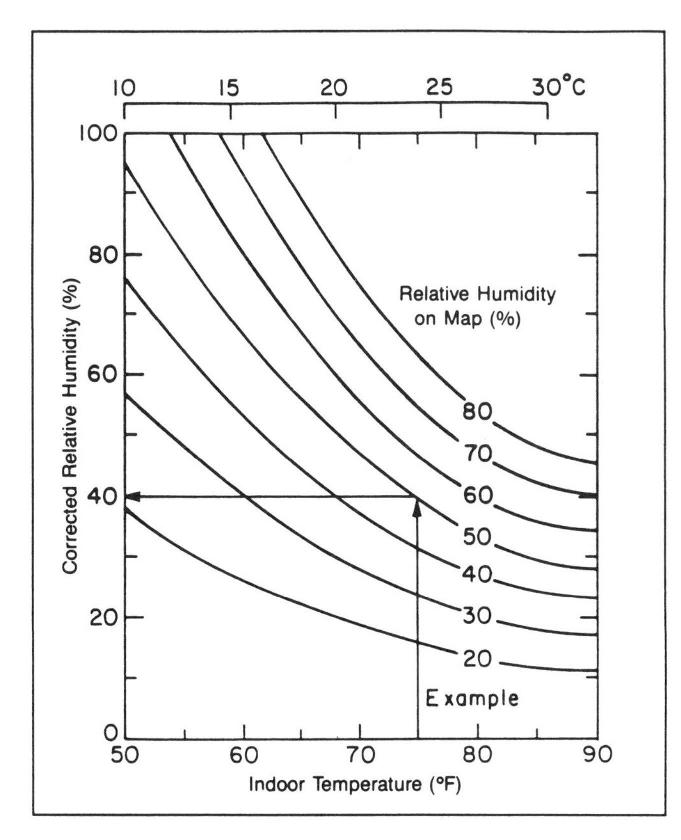

FIGURE 3-5. Graph for Correcting the Map Values in Figure 3-4 for Indoor Air Temperatures Other than 68°F. (Reprinted, by permission, from Tobiasson, *Vapor Retarders for Membrane Roofing Systems*).

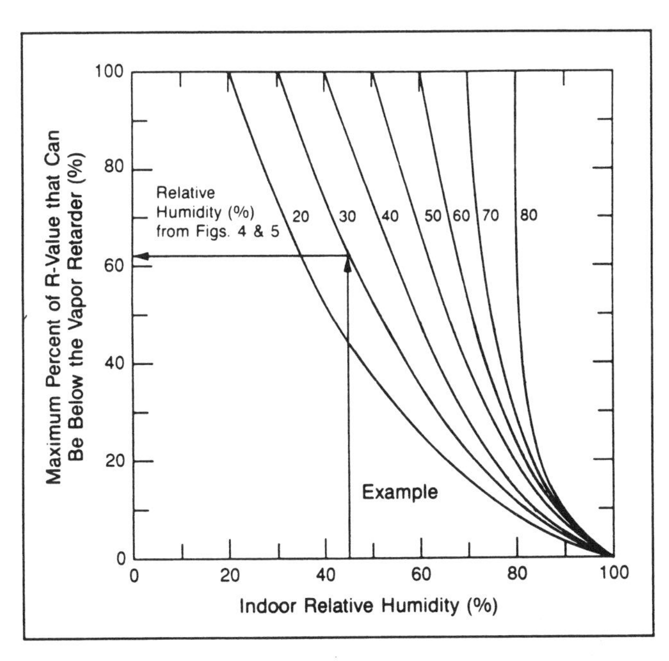

FIGURE 3-6. Graph Used With Figure 3-4 and 3-5 to Determine the Maximum Percent of the Thermal Resistance of the Roof That Can Be on the Warm Side of the Vapor Retarder. (Reprinted, by permission, from Tobiasson, *Vapor Retarders for Membrane Roofing Systems*).

Compact roofing system with the vapor retarder installed between insulation layers

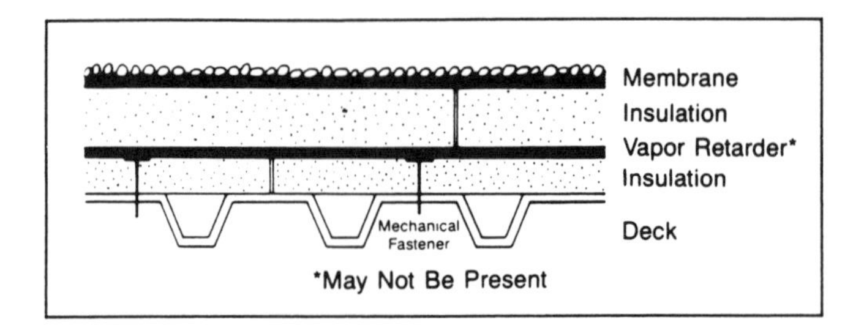

Compact membrane roofing system

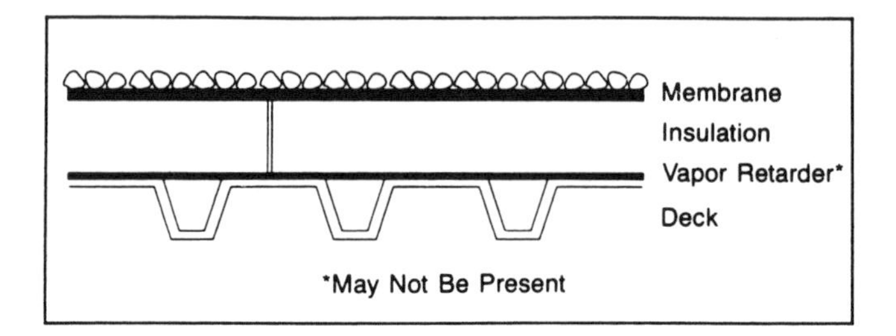

A protected membrane roofing system (PMR)

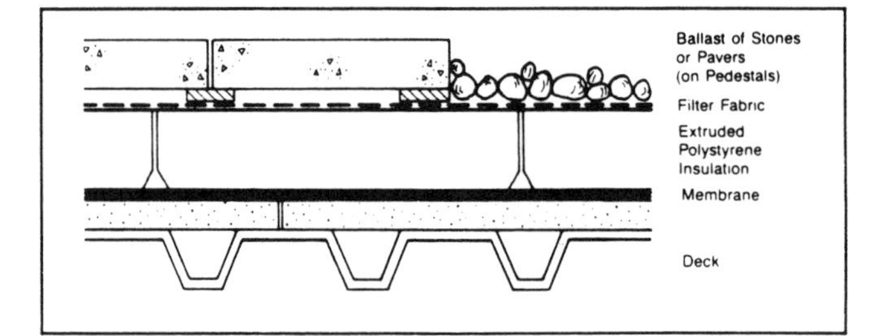

FIGURE 3-7. Optimum Location of Vapor Retarders in Roofs (Reprinted, by permission, from Tobiasson, *Vapor Retarders for Membrane Roofing Systems*).

moved from the interior of the building to condense inside wall cavities, causing damage. They also prevent wind pressure from causing drafts.

Air barriers must be strong enough to withstand the substantial pressure brought to bear against them. They must form an uninterrupted skin enclosing the building. All gaps must be sealed properly at wall-slab junctions and around

TABLE 3-3. Typical Winter Indoor Relative Humidities

Offices	30 - 50%
Hospitals	30 - 55%
Computer Rooms	40 - 50%
Department Stores	40 - 50%
Swimming Pools	50 - 60%
Textile Mills	50 - 85%

(Reprinted, by permission, from Tobiasson, *Vapor Retarders for Membrane Roofing Systems*)

windows to prevent any substantial leakage. If the air barrier is placed on the warm side of the wall, no gaps between it and the insulation should be allowed. If this condition should occur, cold exterior air may find its way through joints between the insulation components and create a cold spot on the face of the GWB, causing condensation and damage.

Air barriers, unlike vapor retarders, do not have to be located at the warm side of the wall. They may be placed anywhere in the assembly. In fact, any part of the assembly may qualify as an air barrier if it provides an uninterrupted sheath around the building. This sheath must be structurally adequate to withstand wind pressure, which can be quite substantial during storms. If the vapor retarder acts also as the air barrier, it must be supported by the GWB panels, which must extend all the way above the ceiling and be sealed to the floor slab or deck above. The membrane must also be able to withstand negative pressure. This may require a thicker or reinforced membrane if the only support is fiberglass batts.

Some experts recommend that if any component of the

building envelope located on the cold side of the vapor retarder is a potential obstacle to vapor migration, it must be, at most, one tenth the perm rating of the vapor barrier or it must allow the vapor to be vented to the exterior.

If the insulation is placed directly on the hanging soffit, vapor migrating from the floor above will condense in the cold soffit area, damage the insulation, and stain the soffit unless this area is vented to the outside. It is almost impossible to place an airtight membrane on top of the insulation if the insulation is located on the soffit.

The air barrier may qualify as a vapor retarder if it is placed on the warm side of the wall and if it has the required maximum perm rating. Most foil-faced insulation boards and GWB are considered to be adequate air/vapor barriers if sealed properly at the seams and around their perimeter and provided with adequate structural support.

Building enclosures are complex structures that perform many functions. They protect the occupants from wind and rain, and shelter them from heat and cold. They must last a relatively long time and maintain their value. To achieve these goals, the architect must detail them properly, keep abreast of the latest technological developments, and specify proven materials.

There has been much talk about the air quality inside airtight buildings being built in recent years. Research has pointed to the health hazards resulting from breathing the fumes emitted by sealants, adhesives, paints, and synthetic materials used in new construction. This, however, must not be an excuse for making the building envelope less airtight. This section demonstrates the hazards of such a course of action. The problem should be dealt with by improving the products to make them less harmful. In the meantime, better ventilation and filtration systems can lessen the adverse effects of these fumes.

To recap, walls and roofs must be designed in a way that prevents water vapor from condensing and damaging the insulation and the structural components. To do this, air and vapor barriers must be included in wall assemblies. The air barrier prevents the wind from forcing the rain into the interior. The vapor retarder prevents water vapor from exfiltrating and condensing inside the wall assembly. Vapor that does infiltrate through walls and attics must be vented to the outside.

3.4.4 Sources of More Information

American Society of Heating, Refrigeration and Air Conditioning Engineers, Inc. *ASRAE Handbook of Fundamentals* latest edition. 1791 Tullie Circle, N.E., Atlanta, GA 30329. 404-636-8400.

American Society of Testing Materials. 1989. *Water Vapor Transmission Through Building Materials and Systems: Mechanisms and Measurements*. STP 1039. H. R. Trechsel and M. Bomberg, ed. 1916 Race Street, Philadelphia, PA 19103, 215-299-5400.

Latta, J. K. March 1976. *Vapor Barriers: What Are They? Are They Effective?* Canadian Building Digest No. 175, Ottawa, Canada: National Research Council.

Quirouette, R. L. 1985. *The Difference Between a Vapor Barrier and an Air Barrier*. BPN 54. National Research Council, Canada, Division of Building Research.

Shipp, Paul H., and Marchello, Maurice J. 1989. "What You Ought to Know About Air Barriers and Vapor Retarders." *Form & Function*. Construction Technology Laboratory, USG Research Center, United States Gypsum Corporation.

Tobiasson, Wayne. 1989. *Vapor Retarders for Membrane Roofing Systems*. Misc. Paper 2489. Reprinted from the Proceedings of the 9th Conference on Roofing Technology with permission from the National Roofing Contractors Association by the U.S. Army Corps of Engineers, Cold Regions Research and Engineering Laboratory (CRREL), 72 Lyme Road, Hanover, NH 03755-1290.

3.5 THE RAIN-SCREEN PRINCIPLE

3.5.1 General

Water leakage in exterior walls is one of the headaches that are difficult and costly to correct once the wall is built. Exterior walls based on the rain-screen principle are more common in Europe and Canada but are beginning to be constructed in the United States. This system, if understood and applied properly, goes a long way toward minimizing the danger of leaks.

The Norwegians were first to investigate the mechanics of water leakage in wood casement windows and later into walls about forty years ago. They were followed by scientists at the Canadian National Research Council who did research based on a similar approach and introduced the term "rain screen."

3.5.2 Causes of Leaks

For leaks to occur, three essentials are required: water, an opening in the wall, and a force to move the water through the opening.

Figure 3-8 shows the kind of forces that may be instrumental in causing a leak. These forces are:

A. Gravity. This force can be minimized by providing a slope toward the exterior.
B. Kinetic energy. This is the momentum that carries wind-driven raindrops through openings of sufficient size. To resist this force, cover battens, splines, or internal baffles may be used.
C. Surface tension of water. Although this is not a force, it nevertheless contributes to leakage by clinging to the underside of horizontal surfaces, where it can be pushed by the wind through the opening.
D. Capillary action. Water flows through narrow openings by capillary action. Introducing a discontinuity, or gap, in the joint of greater width, as shown in figure 3-9, disrupts the flow.

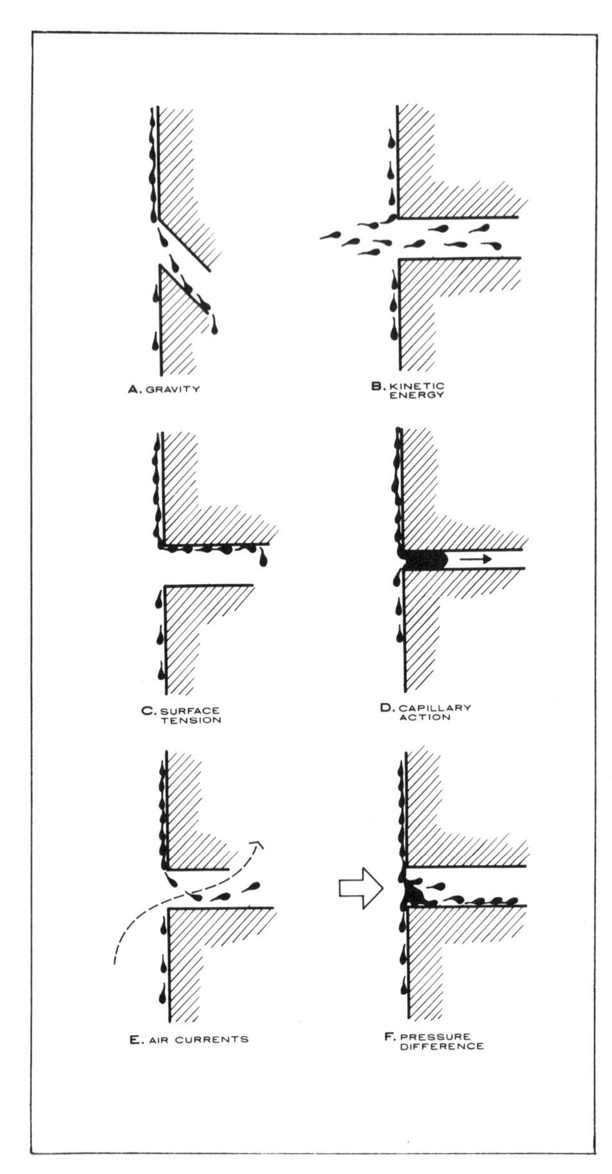

FIGURE 3-8. Forces Acting to Move Water Through an Opening (Reprinted, by permission, from American Architectural Manufacturers Association, *Aluminum Curtain Wall Design Guide*).

E. Air currents. These may result from differences in wind pressures over the wall surface or from convection within wall cavities. Air currents may push or pull water into the wall.

F. Pressure difference. Leakage may also be caused by a difference in pressure between one side of the wall and the other. This pressure drop causes the water to move toward the low-pressure side. This movement may be perpendicular to the wall, from the exterior to the interior, or, in cavity walls, parallel to the wall, from one side of the cavity to the other.

Forces in A, B, C, and D are well recognized and readily controlled. E and F represent wind forces that are the most critical and most difficult to combat. Differential pressure causes most of the leakage at wall joints.

3.5.3 Methods Used to Prevent Leaks

The conventional defense against leaks is to attempt to eliminate all openings by employing a good-quality sealant. While this approach works in theory, it has shortcomings. Joint sealants must be perfect; perfect in formulation and applied by an experienced crew to perfectly prepared substrates under perfect weather conditions (see Sec. 3.7). Oddly enough, high-quality sealants have a good track record. They do, however, have a limited life span and require costly periodic maintenance because they are subject to attack by ultraviolet rays, pollution that includes acid rain, and other factors. The rain-screen approach offers a more permanent alternative if done properly.

3.5.4 How the System Works

The system (Fig. 3-10) is comprised of an outer skin that acts as a rain screen and an inner skin that acts as an air barrier. These two skins are separated by a cavity. The underlying principle is to equalize the air pressure on both sides of the rain screen to overcome the force that moves water in the cavity. Thus, the forces described in E and F above are neutralized and leakage is prevented.

Because wind pressure is higher at the corners and top of buildings, this pressure will create strong currents inside the cavity that travel from these zones to zones of lower pressure.

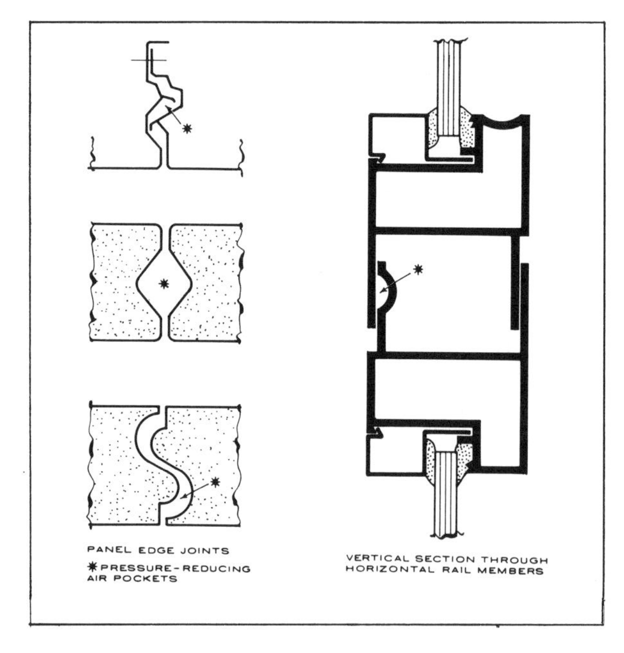

FIGURE 3-9. Typical Capillary Breaks in Joints (Reprinted, by permission, from American Architectural Manufacturers Association, *Aluminum Curtain Wall Design Guide*).

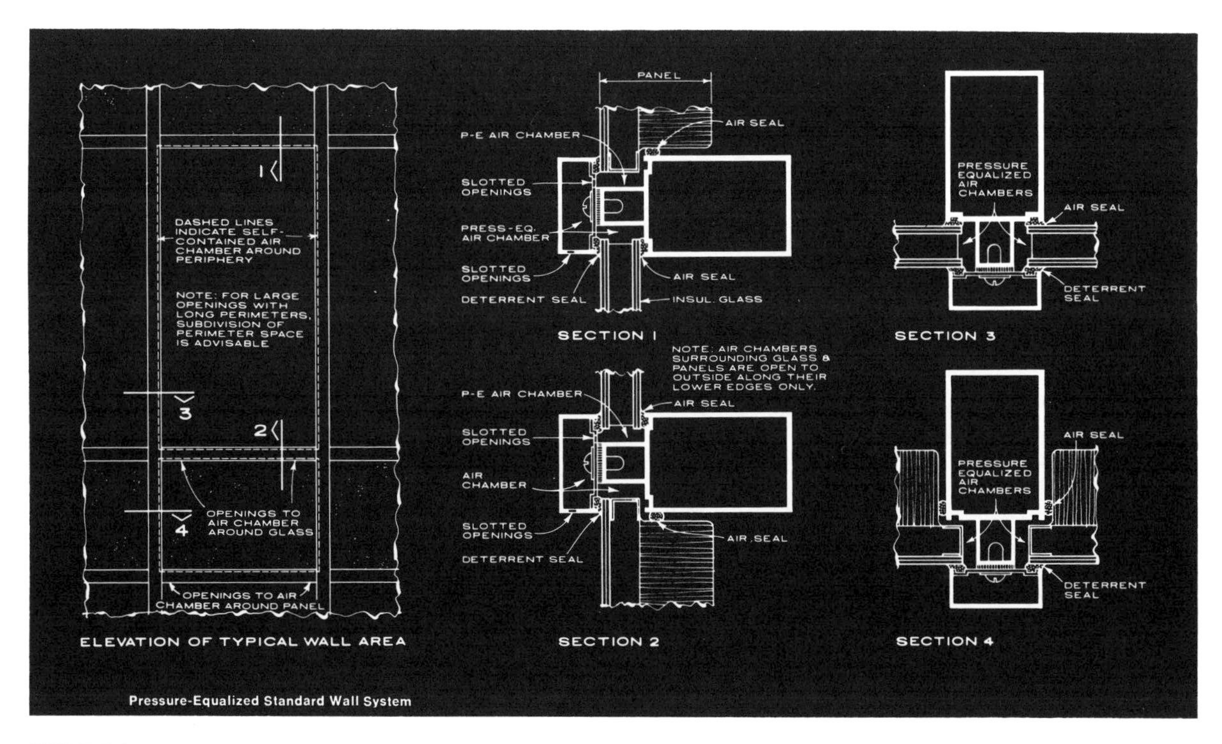

FIGURE 3-10. Pressure-Equalized Standard Wall System (Reprinted, by permission, from American Architectural Manufacturers Association, *Aluminum Curtain Wall Design Guide*).

These air currents force rain into the cavity and into any imperfection in the air barrier. To prevent this, the cavity must be subdivided into spaces confined between partitions and vented to the exterior.

When a wind gust attacks the wall, it is allowed to enter the cavity through vents (protected by baffles) until the pressure inside the cavity becomes equal to the exterior pressure. In this system, pressure equalization happens very rapidly before the rain has a chance to circumvent the baffles placed in its way and enter the cavity. The cavity is also provided with flashing and drainage openings to the exterior so that even if a small amount of water enters, it can drain before causing any damage.

Generally, it is advisable to place these openings in the soffit areas of horizontal joints, where they are protected from heavy wetting and may be shielded by interior baffles if necessary. To prevent water movement by capillary action, these openings should not be less than ¼ inch in width.

3.5.5 Size of Cavities

The rain-screen principle works for both large and small cavities. A cavity can be as small as voids within a mullion (Fig. 3-10) or as large as a whole one- or two-story bay. The size of each cavity compartment should be commensurate with the pressure it is subjected to; the higher the pressure, the smaller the compartment. *The Aluminum Curtain Wall Design Manual* quotes Messrs. Dalgliesh and Garden of Canada's National Research Council as follows:

> By compartmentation of the cavity, the range of pressure differences acting on any cavity compartment may be greatly reduced. It is proposed that until further pertinent information becomes available vertical closures should be provided at each outside corner of a building and at 4-foot intervals for about 20 feet from the corners. Horizontal closures should be used near the top of the wall. It is also considered advisable that both vertical and horizontal closures be positioned up to 30 feet on centers over the total wall area. It should be noted that these cavity closures (or partitions) need not provide a complete air seal but must be sufficient to allow the appropriate pressure difference between cavity compartments to develop. It is possible to follow a similar approach to deal with the problem of achieving pressure equalization in spaces or cavities where abrupt variations occur at projections and recesses.

3.5.6 Types of Construction

The rain-screen principle can be applied to almost any type of construction. It is most successful in metal curtain walls (Fig. 3-11) because the metal air/vapor barrier is impervious. It can also be used in precast walls if the joints are detailed and drained properly and designed according to the principles described in this section.

Wood frame construction using clapboards, although it evolved independently over time, has all the ingredients of a rain-screen system (Fig. 3-12). Brick cavity wall construction, shown in the same figure, can qualify as a rain-screen system if the cavity is subdivided horizontally by shelf angles

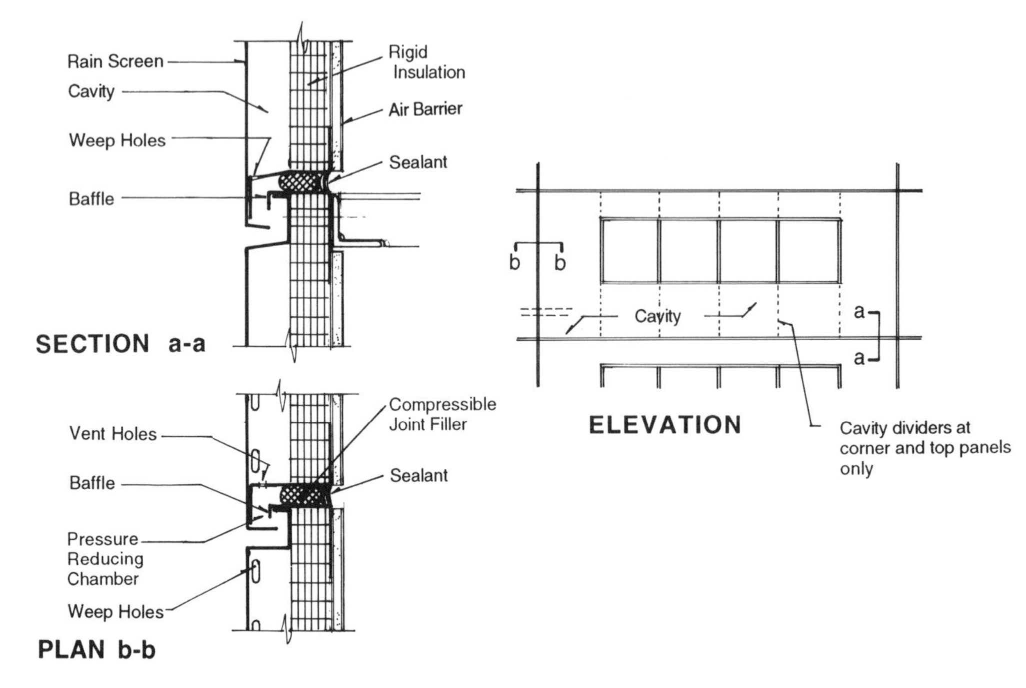

FIGURE 3-11. Example of a Metal Rain Screen.

at every floor or at every other floor, and baffled vertically as recommended in Subsection 3.5.5 above.

Summary

Applying the rain-screen principle to the design of exterior walls has many advantages. It does not rely on a "perfect" line of defense against attack by the forces of man and nature (pollution, acid rain, freeze-thaw, wind, and UV rays) but rather harnesses the forces of nature itself (pressure equalization) to provide that defense. This is a more permanent approach since the first line of defense, namely the rain screen, is virtually without sealant, and the second line of defense (the air/vapor barrier), which requires a sealant, is in a more sheltered location.

The rain-screen principle depends on the rain screen to defend against the bulk of the rain. It is backed by a subdivided cavity vented and drained to the exterior to equalize the wind pressure on both sides of the rain screen, canceling the propeling force pushing the rain and draining any small amount of rain that may enter the cavity. For this

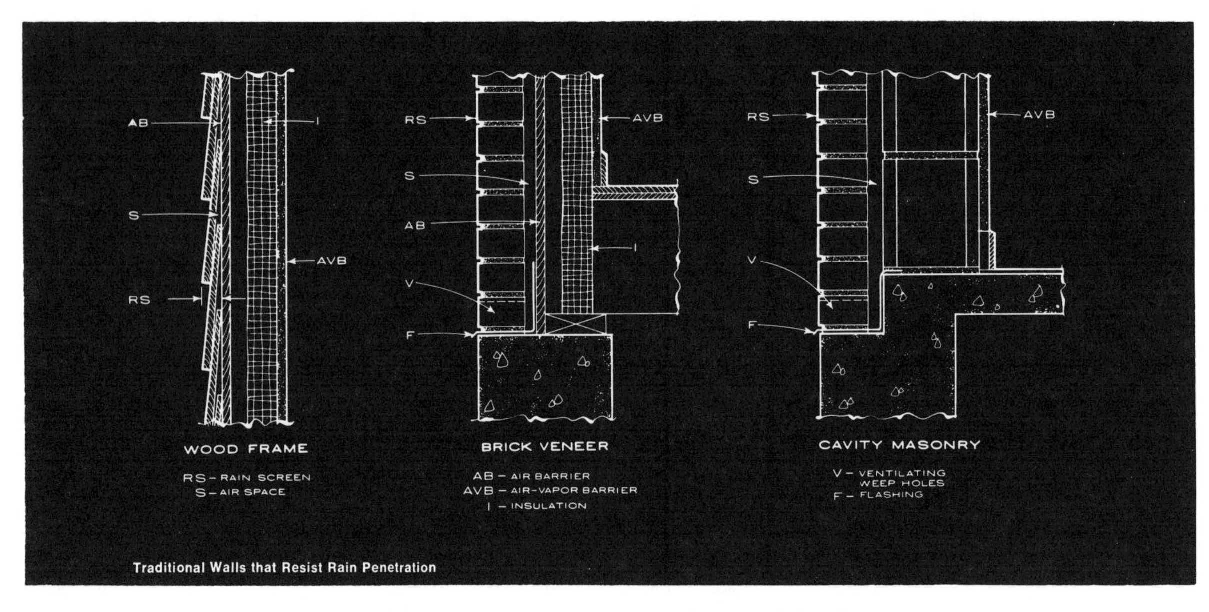

FIGURE 3-12. Traditional Walls That Resist Rain Penetration (Reprinted, by permission, from American Architectural Manufacturers Association, *Aluminum Curtain Wall Design Guide*).

system to work properly, three ingredients are necessary: baffles placed at the joints and/or vents located at the rain screen to prevent rain from having direct access to the cavity, a continuous air barrier to defend against air infiltration at the back of the cavity and cavity subdivision into vented compartments sized in a proportionate way to the intensity of the wind (the higher the wind pressure, the smaller the compartment).

The intent of this section is to introduce the reader to the subject only. The advice of experts in the field should be sought before details are finalized. Rain-screen design is a complicated subject affected by many imponderables, including the ever-changing wind direction and velocity, unforeseen air leakage points, and the right size of vents, to name a few.

3.5.7 Sources of More Information:

Anderson, J. M., and Gill, J. R., ed. 1988. *Rainscreen Cladding*. Stoneham, MA: Butterworth Architecture.

American Architectural Manufacturers Association. 1979. *Aluminum Curtain Wall Design Guide Manual*. 2700 River Road, Des Plaines, IL 60018. 708-699-7310.

*Dalgleish, W. A., and Garden, G. K. 1968. *Influence of Wind Pressures on Joint Performance*, CIB Symposium on Weathertight Joints for Walls, Oslo, Norway, September 1968. NBRI Report 51C. "Weathertight Joints for Walls," No. 26D. (NRC 9873).

Ganguli, U., and Quirouette, R. L. 1987. *Pressure Equalization Performance of a Metal and Glass Curtain Wall*. Appeared in Proceedings 1987 CSCE Centennial Conference, Montreal, Quebec, Vol. 1, pp. 127–44. (IRC Paper No. 1542). NRCC 29024.

Isaksen, Trygve. 1965. *Rain Penetration in Joints, Influence of Dimensions and Shape of Joints in Rain Penetration*, Reprint 119. RILEM/CIB Symposium on Moisture Problems in Buildings. Helsinki: Norwegian Research Institute.

*Latta, J. K. 1973. *Walls, Windows, and Roofs for the Canadian Climate, a Summary of the Current Basis for Selection and Design*. Special Technical Publication Number 1 of the Division of Building Research, National Research Council, Canada. Reprinted December 1979. (NRCC 13487).
13487).

Rousseau, Madelaine Z. 1990. *Facts and Fiction of Rain Screen Walls*. Appeared in Construction Canada 90 03. Vol. 32, No. 2, pp. 40–47. (IRC Paper No. 1666) NRCC 32332.

3.6 FENESTRATION

3.6.1 General

Webster's dictionary defines fenestration as "the arrangement, proportioning, and design of windows and doors in a building." This section focuses on the windows mentioned in that definition.

The first step in defining the fenestration for a project is taken by the designer when he chooses the type of window based on aesthetics. After the window type is selected, the architect considers other factors that have a bearing on the design. Robert D. Foster, director of research and development at Kawneer Company, Inc., provides the following recommendations (letter to the author, November 1991):

1. To be most effective in helping a design professional select and specify a window product, we would like the following information.
   a. Window type—"fixed, projected, casement, etc."
   b. Thermal or non-thermal requirements[2]
   c. Grade of window
      —R: residential
      —C: commercial
      —HC: Heavy commercial
      —AW: architectural grade (new)
   d. Special design features
   e. Other design or performance requirements
2. Wind loads are the principal loads which windows must resist. These load requirements vary throughout the United States.

   The structural loads become a basic design consideration when determined by the maximal allowable glass size.

   Window mullions serve three purposes. They serve as a means to join together two separate window units. They serve as a structural element to resist a given wind load requirement, and they provide for expansion and contraction.

   Window mullions may be an integral part of the window frame members. They may also be a separate structural profile. Either may project internally or externally.

   Typically, window manufacturers will provide wind load charts and corresponding mullion data. For custom applications, the window manufacturers will provide the necessary structural calculations and custom design the mullion accordingly.
3. Always select the highest quality window your budget will allow. Be selective by specifying only three manufacturers that meet both your performance criteria and your budget. Require window manufacturers to furnish certified test reports and a list of past performances. When possible, select standard pretested products and avoid custom designs. Always select reputable manufacturers with long histories in the architectural window industry.

*May be ordered through: Publications Sales, M-20, National Research Council, Canada, Institute of Research in Construction, Ottawa, Canada K1A OR6.

[2]Examples of thermally treated windows are double-glazing in thermally broken frames. Glass may be tinted or reflective.

When possible specify experienced tradesmen for both installation and glazing.

Do not extend windows beyond the manufacturer's recommendations and certification without first consulting the manufacturers.

Do not approve voluntary alternates or substitutions without **thoroughly** researching the manufacturer and its history in the architectural window industry.

When possible, avoid the use of formed trims and accessories. Rather, specify extruded trims.

4. As previously discussed, mullions may be an integral part of the window frame member. This type of structural mullion generally costs less than the separate structural elements which are more costly because of additional material and erection labor. This type of mullion offers more design flexibility.

   On a scale of 1 to 10, with no mullions being 1 and a separate structural mullion being 10, integral mullions can best be classified as a 6. Obviously, in either case the more stiffness required to meet a specific wind load, the more material and the higher the cost.

   Window manufacturer's details will provide wind load tables and mullion options. Additional information and cost comparators may be obtained from the manufacturers' representatives.
5. Glass technology is expanding at a rapid rate. It is recommended that the architect contact the major architectural glass manufacturers for the most current information.

3.6.2 Technical Considerations

Regardless of window type (individual window, strip window, storefront, or curtain wall), wind pressure determines the mullion size to be used. Gravity also affects that choice in cases where an intermediate horizontal mullion separates two wide panes of heavy glass. The larger the area of glass and the heavier the glass type (e.g., insulating glass), the more load is transferred to the framing members, both horizontally and vertically, resulting in heavier, deeper, and more costly members. Where cost is an issue, and in most cases it is, designers must weigh the desire for large, uninterrupted viewing areas against the cost of heavier mullions, thicker and stronger glass (heat strengthened, tempered, or laminated), as well as the energy loss through the larger glass area.

The main considerations in the design process are:

1. Selection of the mullion or frame type to complement the design and resist wind and gravity loads. There are several framing systems to choose from, including aluminum, steel, structural glass, structural silicone, and gasketed systems. Aluminum framing systems are the most prevalent. Aluminum frames are either anodized or painted.
2. Glass type selection depends on both aesthetics and energy conservation. The energy crisis made designers more aware that glazed areas do not offer as much R-value as a well-insulated solid wall. Codes have set standards defining the total R-value a building envelope must provide. Consequently, designs now favor more limited-vision glass areas to conserve energy. The main types of glass to choose from are annealed, heat strengthened, insulating, tempered, and laminated. Each has its distinctive characteristics that favor a particular application (see Sec. 3.6.4 for more detailed information). Aesthetics and energy conservation also require the architect to make a selection from a wide variety of glass treatments which include tints and coatings. Finally, if the glass pane is very large, it must be checked against the maximum size available before the design is finalized, to avoid surprises.
3. Window treatment must be selected (blinds, louvers, shades, overhangs). This selection, including the color, is part of the information to be given to the HVAC engineer because it affects the heating and cooling load calculations. In cold climates, solar gain is desirable, window treatment in those areas is less crucial than in hot parts of the country.
4. Window perimeter protection (type of sealant or pressure equalization), flashing, and method of attachment to the structure must be determined. If the building is a high-rise structure, decisions concerning the window-washing system may affect the frame design by incorporating a window-washing platform track or tie pins in it.
5. Because the lighting load may consume over 50 percent of the energy needed to run a building, every effort should be made to reduce this load. Measures such as providing computer-controlled lighting related to the level of natural lighting, choosing glass with a lighter tint to increase the level of lighting, and proper building orientation should be considered. The help of the mechanical engineer is crucial in this type of research.

Finally, window design should provide for easy replacement of broken glass. This should be an important factor in the frame selection process. If at all possible, a system that allows reglazing from the interior of the building should be chosen in low- and mid-rise structures. High-rise buildings have integral window-washing systems that make it possible to reglaze from the exterior.

3.6.3 Framing Systems

Windows are identified by different names, such as individual or "punched" windows, strip windows, storefronts, and curtain walls. Each of these window types requires a choice from a number of framing systems that include metal, structural silicone, structural glass, and zipper gaskets. The following is a brief description of each.

Metal Systems

Metal systems can be either aluminum or steel. Figure 3-13 shows an example of an aluminum-curtain wall application. Metal systems represent the largest segment of the framing market. The majority is aluminum. Steel frames are required for fire-rated and detention windows. Because steel is stronger than aluminum, larger lites can be accommodated in mullions with a relatively small cross section, and, because thermal expansion is about half that of aluminum, horizontal shearing movement in wet seals is minimized, resulting in longer-lasting sealants. Furthermore, steel's lower coefficient of heat transfer reduces heat loss through the frames. Steel frame manufacturers claim that present-day finishes have overcome the rust problem associated with steel windows in the past. Galvanizing is an option offered at extra cost.

For curtain wall design, a rule of thumb is that if the area of the glazing exceeds 200,000 square feet, a window consultant should be engaged in the design process. Curtain wall systems are designed and engineered according to the rain-screen principle (Sec. 3-5). These systems drain any infiltrated water through the horizontal mullions. The bottom sill sealant must be placed so as not to block the weep holes located at the bottom of the leading edge of the mullion.

Four-Side Structural Silicone

This system (Fig. 3-14) is combined with aluminum framing. It is favored by architects because it presents a clean exterior uninterrupted by framing members. Its advantages are reduced thermal stress in glass caused by mullion-covered edges (since mullion edges do not exist in this system), good thermal insulation due to the absence of metal on the exterior,

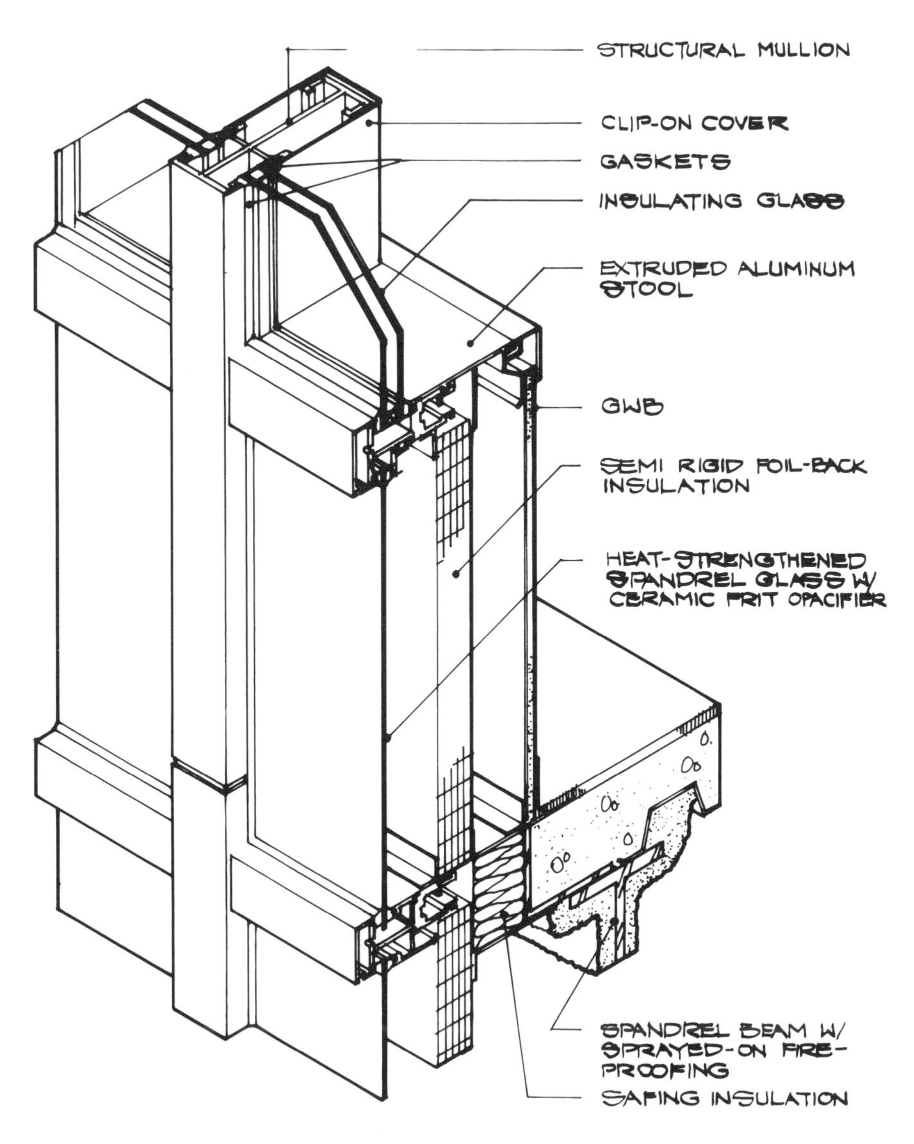

FIGURE 3-13. Example of a Metal Mullion Glazing System.

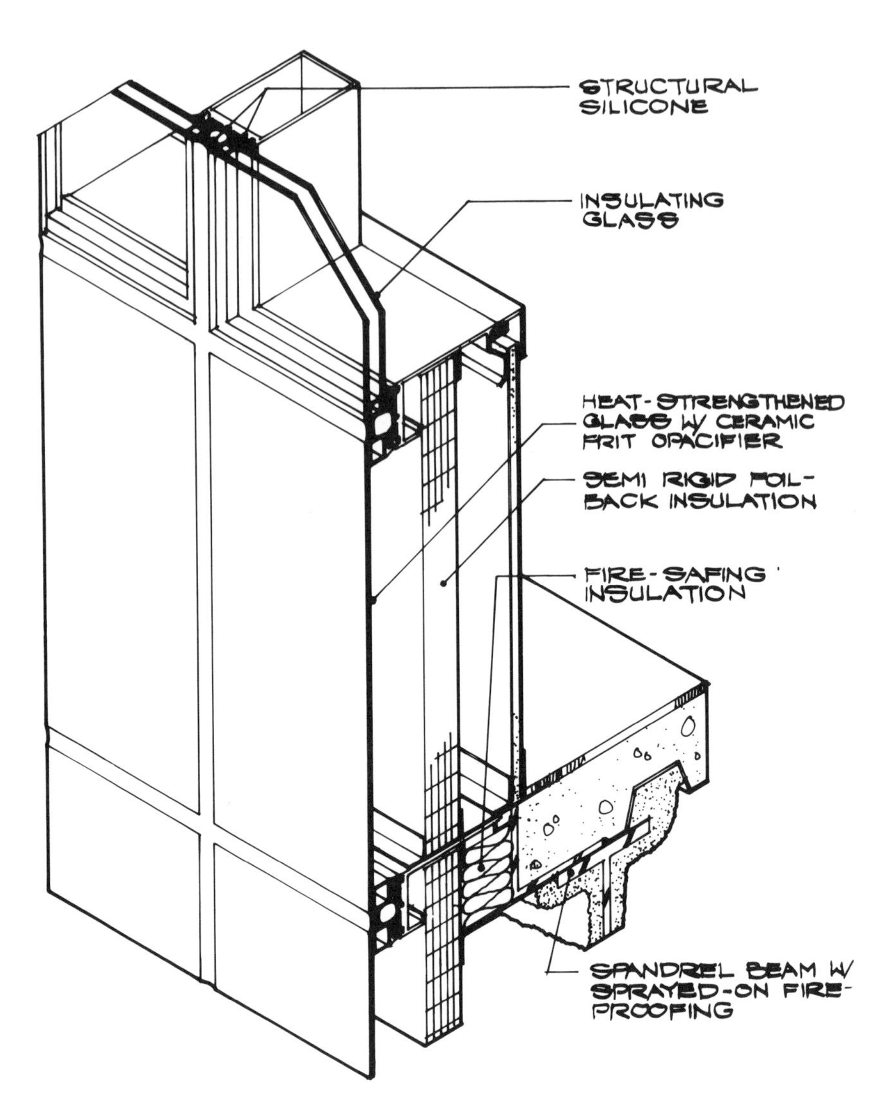

FIGURE 3-14. Example of a Structural Silicone Glazing System.

and the reduced possibility of air and water infiltration. Disadvantages include the danger of losing adhesion, especially in humid environments, unknown life expectancy, and a higher cost.

Structural Glass Systems

This is one of the more expensive systems. It is an all-glass system using thick tempered or heat-strengthened glass assembled with sophisticated hardware (Fig. 3-15). It is used in applications where uninterrupted viewing is important, such as enclosed horse-racing grandstands. It is also used in atria and main entrances of prestigious projects. Where the glass has to span a long distance, glass mullions or fins are employed to provide support. This system is also capable of accommodating insulating glass.

Gasketed Systems

These systems (Fig. 3-16), also known as elastomeric lock-strip or zipper gaskets, were first used in the auto industry. They were introduced in an architectural application more than thirty years ago. Gaskets are usually attached to metal frames, but may be installed in precast concrete grooves or fitted over concrete tongues (Fig. 3-17). Most applications use "H" or spline gaskets. *Sweet's Catalog* lists only one manufacturer for this system. This manufacturer, Stanlock, claims that the system costs less than other systems, is equal to thermally broken systems, does not require periodic maintenance, and is a good noise and vibration barrier. While these claims may be true, one must consider the following:

a. Because wind charts are based on rigid framing systems rather than on gasketed ones, the design should take that into consideration, especially in high wind zones where gasket "rollout" and windowpane release are dangers. The designer must work closely with the manufacturer and check the building code for acceptability.
b. A single-source system is not conducive to competitive

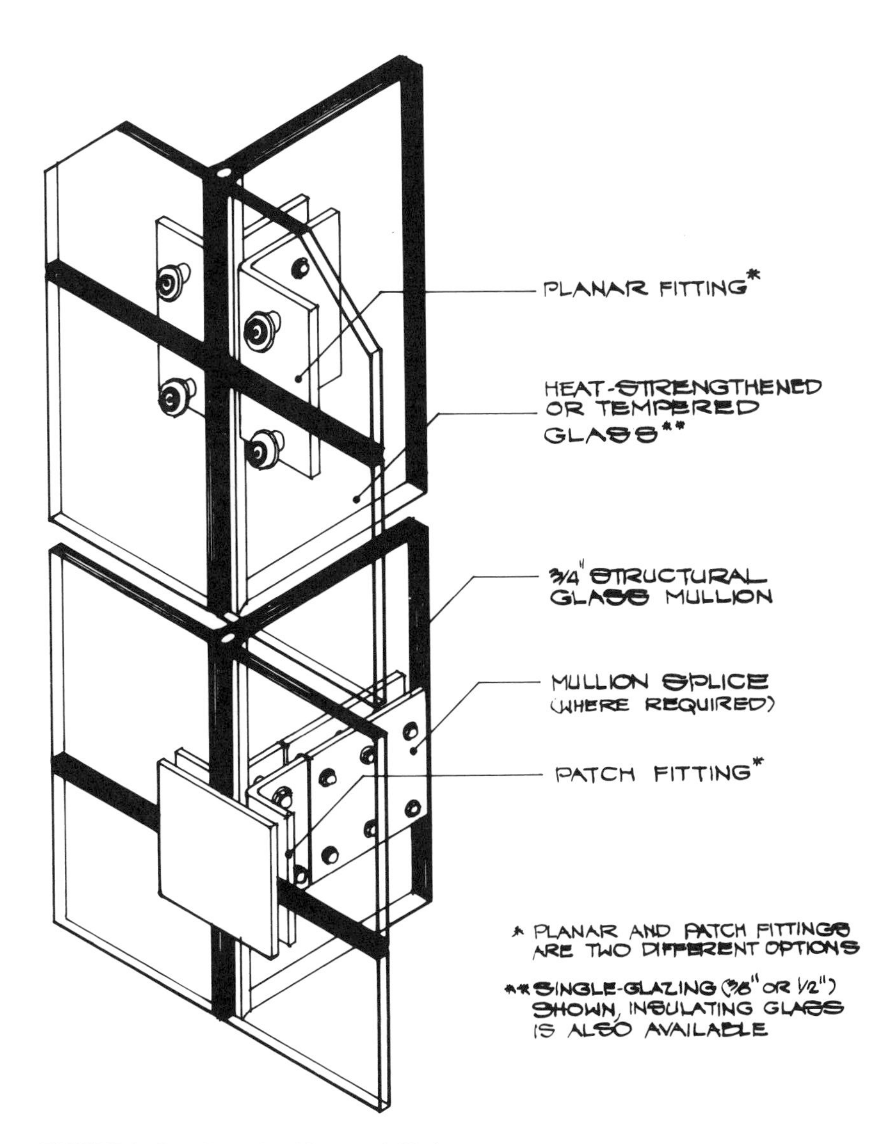

FIGURE 3-15. Example of Structural Glazing.

bidding. A maximum cost guarantee should be secured before bidding or the system should be bid as an alternate to prevent cost overruns.

To select one of the foregoing systems, the architect should consider cost, energy conservation, durability, and maintenance in addition to aesthetics. Framing systems may be dry glazed (using gaskets) or wet glazed (using sealants). The first method has the advantage of replacing broken glass from inside the building and avoiding dependence on field workmanship. However, it is not as watertight as wet glazing. Excessive leaking may result from gasket shrinkage. This possibility can be minimized by specifying gaskets with vulcanized corners. Gaskets may also roll into pockets and cause uneven stress on the glass, or glass may "walk" laterally. To lessen this possibility, specify "shore A65± durometer neoprene" edge blocks.

Wet glazing produces a watertight seal, reduces the possibility of "walking," and protects insulating glass edge seals from water damage if done properly. On the downside, it requires application from the exterior and, like any sealant application, depends on the skill of the applicator, proper surface preparation, and uncertain field conditions.

It is always wise to assume that the seals will fail, so drainage to the exterior as well as flashing to protect the building should be provided. Axonometric drawings are most useful in making the design intent clear to the fabricator and erector of the system. For individual punched windows and strip windows, use frames that provide a return to accommodate the backer rod as well as the sealant (Fig. 3-18).

3.6.4 Glass Types

Different types of glass are used for different applications, project conditions, and loading. The following is a partial listing showing the most common types.

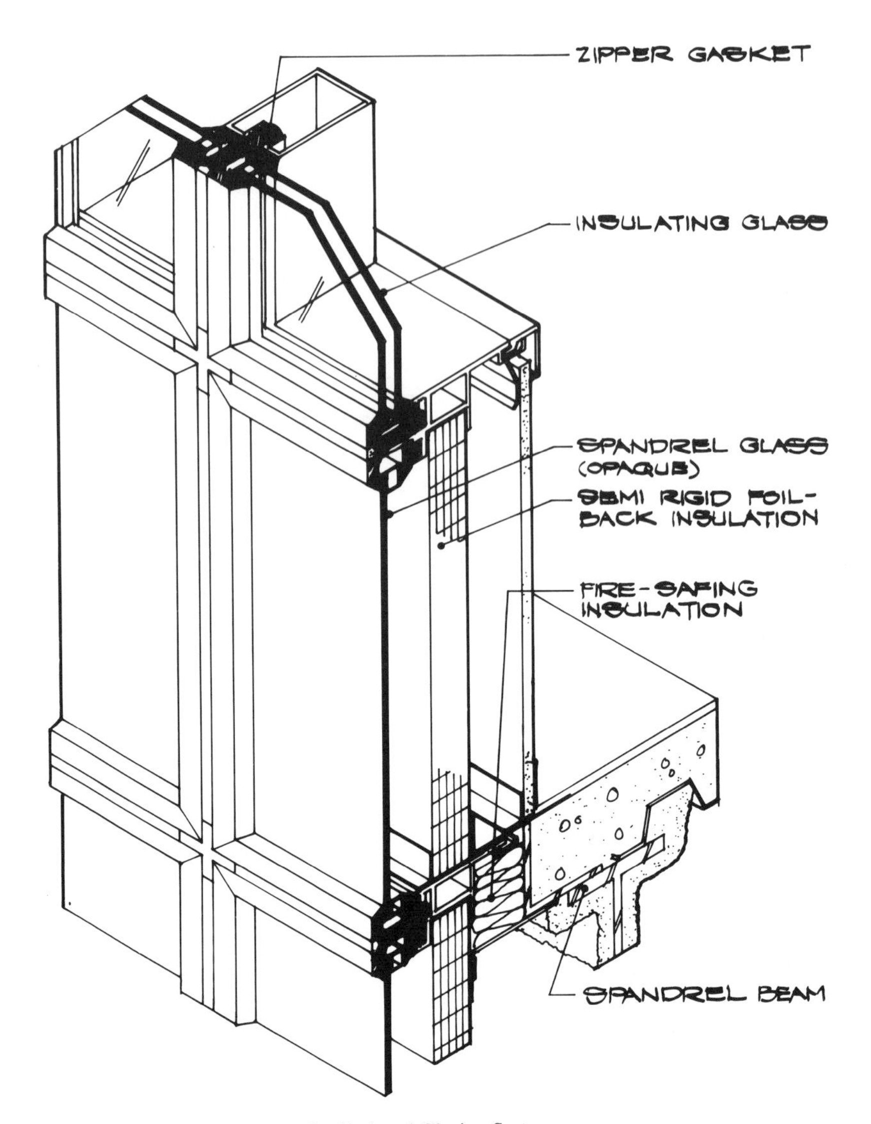

FIGURE 3-16. Example of a Gasketed Glazing System.

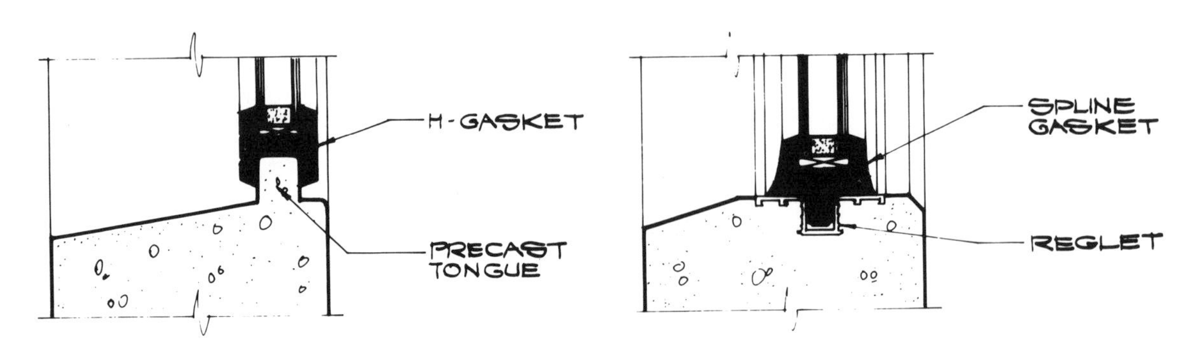

FIGURE 3-17. Gaskets in Precast Concrete Openings.

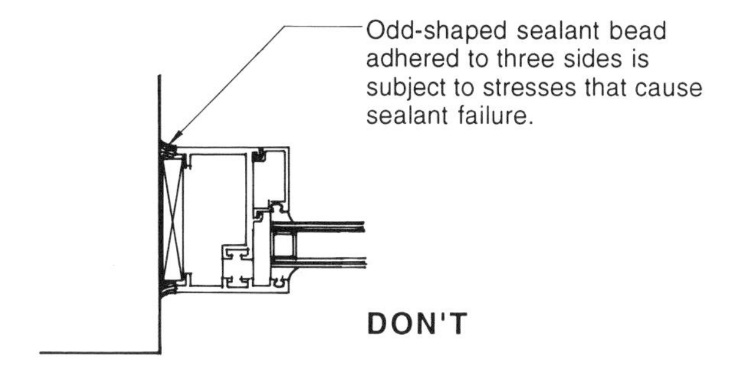

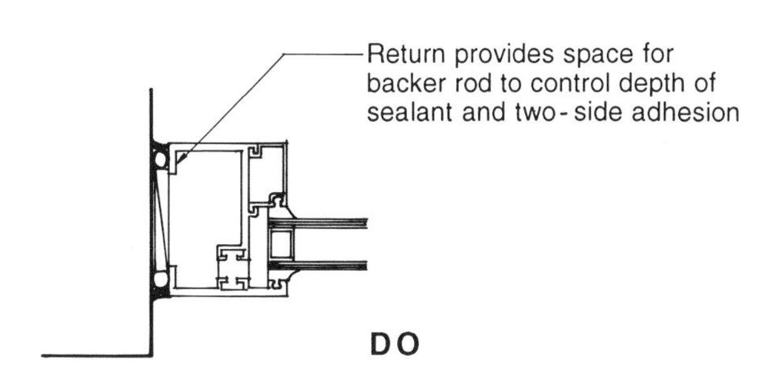

FIGURE 3-18. Window Frame Returns for Proper Sealant Application.

Annealed Float Glass

This is the most common type of window glass. It is manufactured by floating the melted glass over molten tin, hence the name "float" glass. Prior to the early sixties, a different, time-consuming method was used to manufacture glass using polishing equipment to produce plate glass. This method is hardly used anymore for architectural applications. Clear float glass is the least expensive type of glass.

Heat-Strengthened Glass

After float glass is cut to length, glass plate may be reheated and exposed to rapid cooling by air jets to produce heat-strengthened glass. This process causes the surface to shrink and compress while the hotter core is in tension. The resulting product reduces the risk of fracture due to thermal stresses caused by certain conditions such as when part of a window is heated by the sun while a substantial part of the glass is shaded by an overhang. Heat-strengthening also reduces the risk of impact breakage and resists higher wind load than float glass. It is twice as strong as annealed glass and more expensive.

Its most prevalent use is in spandrel areas to resist the heat buildup associated with this application. The accumulation of heat in spandrels is due to the fact that insulation placed behind the spandrel prevents solar heat from dissipating by contact with the cooler air inside the building.

The production process gives the glass a slightly wavy look, which is discernible in all-glass facades. It is recommended that the inner face (#2 surface in single-glazed spandrels and #4 surface in double-glazed ones) be coated with a plastic or ceramic frit opacifier to prevent read-through of the insulation behind it. This may not be necessary if certain reflective glasses are used. Consult with the manufacturer and review actual samples or buildings using the proposed glass.

Insulating Glass

This is a glazing system that incorporates two or more layers of glass separated by an air or gas-filled gap. This arrangement results in an increased R-value, hence the name "insulating."

The most common type is the organically sealed double-glazed unit incorporating a metal spacer and a desiccant to control humidity in the cavity.

For easy reference, glass surfaces are numbered from the exterior to the interior. The exterior surface being Number 1 and the interior surface is Number 4 in a double-glazed unit (Number 6 in a triple-glazed unit). The most preferred surfaces used to deposit reflective coatings are surfaces Number 2 and 3. Because solar heat gain is desirable during the long winter season in cold climates, reflective glazing is not as prevalent in the North as it is in the South.

Wet exterior seals prevent moisture from entering into the glazing pocket to cause deterioration of the seal.

Low E glass (E stands for emissivity) employs a special coating applied directly to the glass or to a plastic film suspended between the glass lites. It cuts down on UV (ultraviolet) transmission and reflects long-wave, low-temperature radiation, improving the effective R-value of the glass. This type of glazing costs more than regular insulating glass.

Safety Glazing

Glazing in this category includes wired glass, tempered or laminated glass, and certain rigid plastics. Using one of these more expensive glazing materials is required by codes for locations where safety is important. These include skylights and sloped glazing (more than 15 degrees from vertical), entrance doors and adjacent panels, storefronts with sills located less than 18 inches above the floor level, and other locations. Check the applicable code(s) for details and consult the glass manufacturer representative if in doubt.

Most codes specify wired glass to be installed in fire-rated doors and partitions and limit the area of glazing. Fire-rated glass without wire is a new product introduced in 1989. It is manufactured by Nippon Electric Glass Company of Japan and costs significantly more than wired glass. It has a maximum fire rating of 90 minutes for openings up to 100 square inches and 60 minutes for openings up to 1,296 square inches, which is the same as wired glass.

Tempered glass is about twice as strong as heat-strengthened glass. It is manufactured according to the same principle used in heat-strengthened glass, except that the air cooling is done more quickly. It has distinctive tong marks at the edges and is considered a safety glass because it shatters into harmless granular fragments upon impact.

Laminated glass, as the name implies, consists of two glass lites fused together with a clear vinyl layer. It offers more impact resistance and when broken, it remains in place. Bulletproof glass is a form of laminated glass with several layers of glass lites.

3.6.5 Glass Treatments

Glass may be tinted or coated, or both. These treatments serve two purposes; one is for aesthetics and the other is to improve performance by reducing heat transfer. This is achieved by reducing the shading coefficient, which is the ratio of solar energy (radiation) that passes through a specific kind of glass relative to the radiation that passes through ⅛-inch clear glass. The lower the value, the more efficient is the glass in resisting heat penetration into the building. This value is used as a yardstick to compare products manufactured by different glass companies. It is also used in HVAC calculations. The shading coefficient is affected by glass thickness, tint (if tinted glass is used), and reflective coating (if used).

John Schleuter of PPG Industries in a letter to the author (December 1991) provided the following insights about glazing:

> Tinting is done by adding colorant to the raw glass batch; on reflective glass spandrels, a plastic opacifier rather than ceramic frit is used; and insulated glass may be used in spandrel panels to provide a closer match to the adjacent vision panels.

To make a glass selection, follow these steps:

1. Select a glass with a U-value and a shading coefficient that meets the value determined by the mechanical engineer. Depending on the prevalent weather conditions, the glazing may be single or double. As mentioned above, color, reflective coating, and glass thickness (the last is determined by wind load) affect the coefficient.
2. Specify all acceptable manufacturers, taking into consideration the difference in color, performance, and spandrel read-through as well as the fact that no two manufacturers produce identical products.
3. If the size and scale of the project warrants, specify a full-scale mock-up of a window and spandrel assembly to be placed on the site to test reflections peculiar to the site. A performance test should also be conducted to determine water and air infiltration of the system, train the erector crew, and correct any flaws before full production. The AAMA (American Architectural Metal Association) publishes an informative booklet on the subject titled *Methods of Test for Metal Curtain Walls*.

Every decision has cost implications. Clear annealed glass is the least expensive, tinted glass could cost twice as much, and reflective glass could cost up to five times as much. A double-glazed unit is also about five times the cost of a single-glazed unit. Laminated glass is even more expensive than double-glazed units. One must consider, however, that glass represents only 15 to 20 percent of the cost of a fenestration system.

3.6.6 Design Calculations

Glass manufacturers' tables, used to determine glass thickness, are divided into two categories, two- and four-sided supported lites. The first means that the system is a strip window without vertical mullions (butt glazing with silicone sealant). The second means that the lite is surrounded by a metal frame.

Window design calculations begin by consulting the building code to define the wind load. Codes divide the area of their jurisdiction into zones defined by wind conditions. For example, coastal, mountain, and flat inland zones have different wind velocities or exposures. The structural engineer or, in some cases, the architect determines the design wind pressures on the facade. These are usually higher at the corners and upper floors. For high-rise projects, a wind-tunnel test may be necessary to determine these pressures because they are usually affected by surrounding structures.

Wind loads are used to determine glass thickness, maximum windowpane size, and the appropriate mullion to resist these loads. The following example is based on the Massachusetts Building Code, fifth edition. It is included here to illustrate the methodology of calculating the wind load. Other codes may use different methods of calculation. Please follow the instructions of the code applicable to each project.

Example

Design the windows at the third floor of a low-rise office building in a Boston suburb given the following information:

1. Height from grade to the center of the windows = 28 feet.
2. The windows are double-glazed, 5′ wide × 6′ high.

Solution

Referring to the code, Boston is located in Zone 3, the suburb is defined as exposure B (Art. 1112.2).

Wind pressure at 0 to 50′ above grade for Zone 3 exposure b = 21 psf (Table 1112.1).

Required design pressure = 21 × 1.2 = 25.2 psf.

Required pressure at salient corners = 21 × 1.7 = 35.7 psf (Table 1112.2).

Relative resistance of double-glazing (assuming identical thickness of lites) $= \frac{25.2}{1.7} = 14.8$ psf.

At salient corners

$= \frac{35.7}{1.7} = 21.0$ psf (Table 2202.1a).

Be sure to check the load with the structural engineer before contacting the manufacturers' reps. The structural engineer's calculations may take other factors into consideration. (It is always advisable to double-check the calculations.) With this information, contact the glass manufacturer's representative to find the appropriate glass thickness. Then refer to the frame manufacturer's catalogs to choose the appropriate mullion from the charts.

Choosing the right combination of tint and coating (if needed) requires consultation with the mechanical engineer and the glass manufacturer to choose the right shading coefficient, transmittance, and reflectance to produce the design U-value.

3.6.7 Sources of More Information:

American Architectural Manufacturers Association. 1979. *Aluminum Curtain Wall Design Guide Manual*. 2700 River Road, Des Plaines, IL 60018. 708-699-7310.

American Architectural Manufacturers Association. 1983. *Methods of Test for Metal Curtain Walls*. 2700 River Road, Des Plaines, IL 60018. 708-699-7310.

American Architectural Manufacturers Association. 1988. *Window Selection Guide*. 2700 River Road, Des Plaines, IL 60018. 708-699-7310.

Johnson, Timothy E. 1991. *Low-E Glazing Design Guide*. Stoneham, MA: Butterworth Architecture.

3.7 SEALANTS

3.7.1 General

One of the more difficult technical chores an architect has to perform is to decide on the right location for control joints on the facade of a building, determine their proper width to accommodate the expansion and contraction of the building panels, and choose from a bewildering array of sealants available to the building industry. In addition to expansion and contraction, sealants are subjected to extreme weather conditions—freeze-thaw cycles, ultraviolet rays, chemical attack such as acid rain, dirt pickup, aging (Table 3-4), and, where located within reach, vandalism.

Sealants are elastic joint materials capable of expanding and contracting with the motion of the joints. In his authoritative book *Chemical Materials for Construction,* Philip Maslow states, "Oil based caulks, on the other hand, have limited capabilities for elongation and compression. As they age and their skin surface thickens, their limited capabilities are reduced still further. These caulking materials are thus limited to areas where little or no movement is expected. Under these circumstances no width-to-depth ratio is recommended since it is irrelevant."[3] Tables 3-5 and 3-6 give detailed information about some of the better-known sealants and caulks available on the market.

3.7.2 Joint Characteristics

Joints should be located at any change in plane or material. Placing them too infrequently requires a proportionate increase in the width of the joint to accommodate the increased expansion/contraction of the adjacent panels. This requires the choice of a sealant that has the ability to resist sagging.

[3]Reprinted, by permission, Maslow, *Chemical Materials for Construction*, 1981, McGraw-Hill, Inc.

TABLE 3-4. Joint Filler Compounds After Aging

Generic Group	*Type*	*Initial Classification*[1]	*Classification After 3 Years Outdoors*	*Maximum Movement Capability*[2]
Linseed oil caulk	100 % solids	Mastic	Elastoplastic or plastic	±2 %
Linseed oil-isobutylene caulk	100 % solids	Mastic	Elastomastic or elastoplastic	±5 %
Butyl caulk	Solvent base	Elastomastic	Elastomastic or elastoplastic	±10 %
Acrylic	Solvent base	Elastomastic	Elastomastic or elastoplastic	±10 %
Acrylic	Emulsion base	Elastomastic	Elastomastic or elastoplastic	±10 %
Neoprene	Solvent base	Elastomastic	Elastoplastic	±10 %
Hypalon	Solvent base	Elastomastic	Elastoplastic	±25 %
Polysulfide 1 part	100 % solids	Elastomastic to Elastomer	Elastomer	±25 %
Polysulfide 2 part	100 % solids	Elastomastic to Elastomer	Elastomer	±25 %
Silicone 1 part	100 % solids	Elastomer	Elastomer	±25 %
Urethane 1 part	100 % solids	Elastomastic to Elastomer	Elastomer	±25 %
Urethane 2 part	100 % solids	Elastomer	Elastomer	±25 %
Polymercaptan 1 part	100 % solids	Elastomastic	Elastomastic to elastomer	±25 %

[1] After 1 month exposure at 75°F and 50 % R.H.
[2] Movement either in extension or compression from the mean.

TABLE 3-5. Comparative Characteristics and Properties of Sealing Compounds

		Butyls		*Acrylics*	
	Oil Base	*Skinning Type*	*Non-Skinning Type*	*Solvent-Release Type*	*Water Release Type*
Chief ingredients	Selected oils, fillers, plasticizers, binders, pigment	Butyl polymers, inert reinforcing pigments, non-volatile plasticizers and polymerizable dryers	Butyl polymers, inert reinforcing pigments, non-volatizing and non-drying plasticizers	Acrylic polymers with limited amounts of fillers & plasticizers	Acrylic polymers with fillers and plasticizers
Primer required	in certain applications	none	none	none	none
Curing process	solvent release, oxidation	solvent release, oxidation	no curing: remains permanently tacky	solvent release	water evaporation
Tack-free time (hrs.)	6	24	remains indefinitely tacky	36	36
[1]Cure time days	continuing	continuing	N/A	14	5
Max. cured elongation	15 %	40 %	N/A	60 %	not available
Recommended max. Joint movement, %	±3 % decreasing with age	±7½ %	N/A	±10 %	±5 %
Max. joint width	1″	¾″	N/A	¾″	⅝″
Resiliency	low	low	low	low	low
Resistance to compression	very low	moderate	low	very low	low
[2]Resistance to extension	very low	low	low	very low	low
Service temp. range °F	−20° to 150°	−20° to 180°	−20° to 180°	−20° to 180°	−20° to 180°
Normal application temp. range	+40° to +120°	+40° to +120°	+40° to +120°	+40° to +120°	+40° to +120°
Weather resistance	poor	fair	fair	very good	not available
Ultra-violet resistance, direct	poor	good	good	very good	not available
Cut, tear, abrasion resistance	N/A	N/A	N/A	N/A	N/A
[3]Life expectancy	5 to 10 years	10 years +	10 years +	20 years +	not available
Hardness Shore A	20–80	20–40	N/A	20–40	30–35
[4]Cost/ft. material (⅜ × ½)	.02–.06	.04–.08	.04–.08	.08–.13	.04–.08
[4]Cost/ft applied (⅜ × ½)	.20–.40	.25–.45	.25–.45	.40–.65	.25–.45
Applicable specifications	FS:TTC-598B	TTS-001657	none	FS:TTS-230C 19-GP-5 (Canadian)	none

[1] *Cure time* as well as pot life are greatly affected by temperature and humidity. Low temperatures and low humidity create longer pot life and longer cure time; conversely, high temperatures and high humidity create shorter pot life and shorter cure time. Typical examples of variations are:

Two-Part Polysulfide

Air Temp.	Pot Life	Initial Cure	Final Cure
50°	7–14 hrs.	72 hrs.	14 days
77°	3–6 hrs.	36 hrs.	7 days
100°	1–3 hrs.	24 hrs.	5 days

[2] *Resistance to extension* is better known in technical terms as modulus. Modulus is defined as the unit stress required to produce a given strain. It is not constant but, rather, changes in values as the amount of elongation changes.

The outcome may be the choice of a sealant that is too stiff to handle the movement. Placing the joints too close together raises the cost and increases the chances of failure due to mistakes in field application. The structural engineer and sealant manufacturers' reps should be involved in determining the location and width of expansion joints.

The sealant should adhere to both sides of the joint only. A bond breaker or a closed-cell polyethylene backer rod must be placed at the back of the joint (Fig. 3-19) to prevent stresses from forming within the sealant and causing its failure. As a general rule, the sealant width should be twice its depth. Where the width is less than ½ inch, both dimensions should be about equal. Most sealant manufacturers include recommendations for depth-to-width ratios for different applications in their literature.

3.7.3 Types of Joints

There are many kinds of joints associated with buildings. Single-stage joints are the most prevalent in the United

TABLE 3-6. Comparative Characteristics and Properties of Sealing Compounds

	Acrylics		*Butyls*	*Butyls*	
	Solvent Release Type	*Water Release Type*	*Skinning Type*	*Non-Skinning Type*	*Hypalons*
Chief ingredients	Acrylic terpolymer with limited amounts of selected fillers	Acrylic polymers with fillers and sometimes plasticizers	Butyl polymers, inert reinforcing pigments, non-volatizing plasticizers and polymerizable drying vehicles	Butyl polymers, inert reinforcing pigments, non-volatizing and non-drying plasticizers	Chlorosulfonated polyethylene and selected fillers
Surface preparation	contact surfaces must be reasonably clean and dry			Contact surface must be	
Primer requirements	none required	as recommended by manufacturer	none required	none required	none required
Curing process	solvent release and very slow chemical cure at exposed surfaces; interior remains soft	water evaporation; relatively uniform consistency throughout	oxidation forms skin on exposed surfaces, progresses slowly inward	no curing; remains permanently tacky	solvent release; develops uniform consistency throughout
Tack-free time (40–80 % rh, 50–80°F)	1 to 7 days	1 to 3 days	1 to 6 hours	indefinite	6 to 24 hours
Hardness, Sh. A @ 75°F Aged 1–6 mos.	0–20	15–40	10–20	5–15	10–20
Aged 5 years	45–55	not available	30–50	10–20	30–40
Service temperature range, °F	−30 to 200	not available	−10 to 200	−20 to 200	−10 to 200
Resiliency	little or none	low	low	little	low to moderate
Resistance to extension	very low	low	low to moderate	low	moderate
Resistance to compression	very low	low	moderate	low	moderate
Recommended maximum joint movement, %	±10 to 25	±10 to 15 (estimated)	±5 to 10	±10 to 20	±10 to 15 (estimated)
Dirt pick-up (after tack-free)	low	low to modate	high	very high; not recommended for exposed locations	moderate
Puncture and tear resistance	not applicable	poor to fair	fair	not applicable	fair
Ultra-violet resistance Direct:	good to excellent	not available	fair to good	fair, but not recommended for use in exposed locations	good
Through glass:	good to excellent		fair to good		good
Color range	unlimited	not available	limited range	limited range	unlimited
Staining of masonry	none	not available	may occur	should not be used on masonry	none
Degree of toxicity in handling & applying	none	none	none	none	none
Life expectancy**	20 years +	not available	10 years +	15 years +	10 to 20 yrs.
Applicable specifications	FS:TTS-230C or 19-GP-5 (Canadian)	none	FS:TTC598B TTS-001657	NAAMM Specs for Non-Skinning Bulk Compounds	FS:TTS-230C

** Estimated for appropriate top-grade compound; may vary widely
* With proper use of primer and proper application of sealant

States. They depend upon the sealant to resist rain, wind pressure, thermal expansion/contraction, ultraviolet rays, etc. They demand perfection in execution, which is difficult to achieve in the field, and require periodic repairs and maintenance.

Two-stage joints are more common in Europe and Canada. They are used in the rain-screen design approach (Sec. 3.5). The exterior seal is usually a rubber gasket or tube acting as the rain screen. In the middle of the joint is a pressure-equalizing chamber and the interior seal is the air barrier (Fig. 3-20). The air barrier seal may be applied from the front if the joint is wide enough, otherwise, it must be accessible from inside the building for both application and maintenance.

Exterior expansion joints are another type of joint used to subdivide large buildings vertically into separate structures

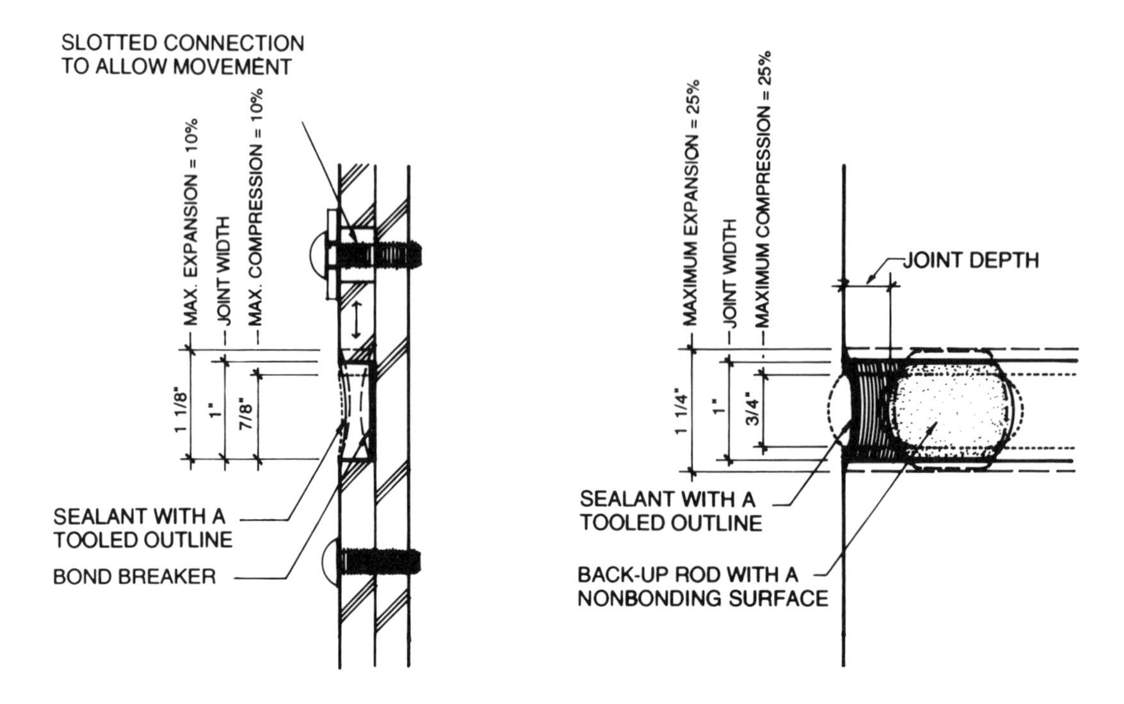

FIGURE 3-19. Joint Details.

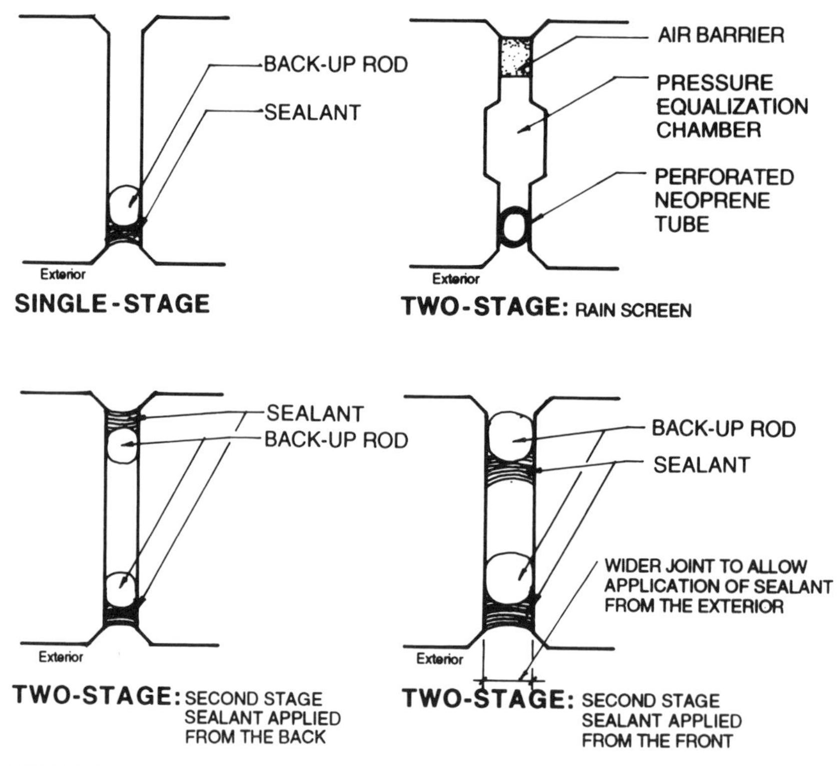

FIGURE 3-20. Single- and Two-Stage Joints.

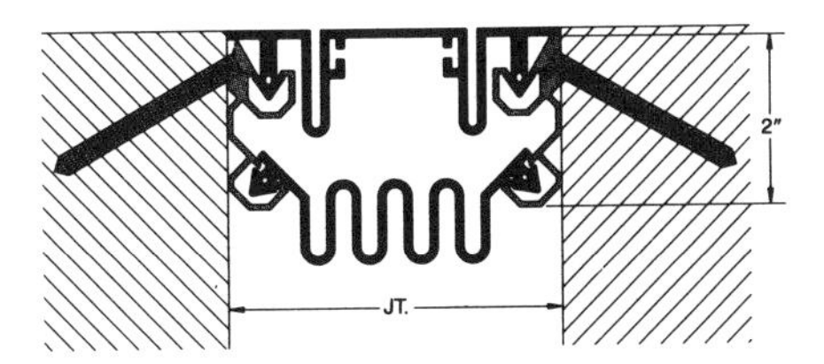

FIGURE 3-21. Building Expansion Joint (MM Systems Corporation, 1991).

to allow the building frame to expand or contract. These joints may be sealed with gaskets (Fig. 3-21) or a high-performance sealant.

3.7.4 Causes of Joint Failure

Many things can go wrong with a joint. The following are a few causes of joint failure:

1. Improperly prepared substrate. This can be due to the presence of a form release agent on the surface resulting in loss of adhesion, or the substrate may have been improperly prepared using a chemically incompatible primer.
2. The field conditions may be unsuitable for sealant application. This may be due to cold weather or wet surfaces caused by dew, recent rain, or frost.
3. Negligent or unskilled applicators who may omit the bond breaker at the back of the joint or place the sealant to a depth other than that recommended by the manufacturer. The applicator may also mix a two-component sealer improperly or omit tooling the surface to force the sealant against the sides for better adhesion.

These are a few of the things that can go wrong. Fortunately, most applicators employ skilled personnel and with good supervision bad results can be avoided.

3.7.5 Sources of More Information

Callender, John Hancock, ed. Latest edition, *Time-Saver Standards*. New York: McGraw-Hill, Inc.

Maslow, Philip. 1982. *Chemical Materials for Construction*. New York: McGraw-Hill, Inc.

Precast/Prestressed Concrete Institute (PCI). 1989. *Architectural Precast Concrete*, 2nd ed. Edited by Sidney Freedman and Alan R. Kenney. Chicago: PCI.

Ramsey, Charles G., and Sleeper, Harold R. latest edition. *Architectural Graphic Standards*. New York: John Wiley & Sons, Inc.

3.8 ROOFING

3.8.1 General

Choosing the appropriate roofing system for a project is not a simple task. The number of systems on the market today requires careful research. This section is intended to provide general information about low-slope roofing, the most prevalent kind of roofing. There are three main systems—built-up roofing (BUR), single-ply roofing including modified bitumen, and metal standing seam. Other systems such as spray-applied polyurethane foams (PUF), cold mastic, and high-slope metal roofs are beyond the scope of this book.

One cannot overemphasize the important of this subject. Premature roofing failure is one of the most common headaches connected with building construction. Many things can go wrong with a roof. In loose-laid ballasted single-ply systems, the membrane may be blown right off the roof if the ballast weight is inadequate. This can happen if the stone is delivered wet and becomes lighter than the minimum weight specified when it dries out on the roof. The seams may separate, causing major leaks, if the contractor deviates from procedures recommended by the manufacturer. The fasteners in mechanically attached systems may back out and punch through the membrane or corrode and leave the membrane without adequate anchorage to resist wind uplift. Other roofing systems have their problems. In general, three factors are at the root of all roof problems: choosing a bad product, poor roof design and detailing, and poor execution.

Architects must choose the roof system carefully and study the terms of the warranty. A long-term warranty that offers to replace materials only is much worse than a shorter-term warranty that offers total replacement including the cost of labor. Proper drainage and careful detailing of flashings, expansion joints, and penetrations can make the difference between a long-lasting roof and a leaky one. Finally a backup drainage system, such as scuppers or overflow drains, is required in case the roof drains are stopped up. One must remember that water weighs 5.2 pounds per square foot for each inch of depth. An accumulation of 4 inches exceeds the 20 pound per square foot live load specified by most codes for the supporting structure. Although a vapor retarder is sometimes required to be placed under the insulation in a roof assembly (see Sec. 3.4), the decision to install it will have to be weighed against the fact that if a leak develops, the vapor retarder will trap the water and nobody will notice until it is too late and the roof is damaged beyond repair. Without the retarder, the water leak will be noticed immediately and measures will be taken to repair it.

Most architects specify a roof that has a long track record backed by a nationally known manufacturer. Every once in a while, circumstances may force them to specify a relatively new product. The procedure in this case is to call up the manufacturer and get all the information, including warranty information and an address list of installed roofs that includes the names and phone numbers of contacts to be called and checked before specifying the product.

If the architect does his homework properly, chooses the appropriate system for the project backed by a good warranty, writes a specification that screens out cheap substitution, and does proper field observation, chances are the roof will not develop any problems during the life of the warranty. Problems that occur soon after occupancy can result in litigation and damage to reputations. This is the reason many architects and developers sometime turn to roofing consultants to aid in this difficult task.

3.8.2 Types of Roofing

As mentioned in the beginning of this section, there are three main types of low-slope roofing. The following is a brief description of each.

Built-up Roofing (BUR)

BUR has been the most prevalent roofing for over one hundred forty years. It continued to be preeminent until the energy crisis of the seventies. Increasing the insulation to conserve energy prevented heat from escaping through the roof. The result was that the membrane became exposed to temperature extremes that caused the roofing to split. This problem occurred in the North during the first thaw when melting snow combined with rain aggravated the leak problem. To correct this shortcoming, fiberglass felts with better tensile strength were substituted for organic felts and the plies were sometimes increased from two or three to four plies. In most cases, it costs less than single-ply roofing and probably lasts longer if—and that's a big if—everything is done properly, a good system is chosen, the details are correctly done, and the roofing is applied by a competent crew.

BUR is considered an adhered system, although the asphalt or coal tar used as moppings between the felts does not act as a glue but as the actual roof membrane. Because of the high cost of labor in the North, this system, which is labor-intensive, is losing ground to single-ply systems. It is, however, still going strong in other parts of the country, especially in the Southwest, where two-ply asphalt-impregnated systems are prevalent. The membrane is topped with a ballast to protect the final mopping from ultraviolet degradation. Ballast also improves fire-resistance and lessens the impact of temperature extremes. The minimum slope recommended for this type of roofing is ¼ inch per foot.

The following is a description of the components used in BUR:

Bitumens

Two types of bitumen are used—asphalt and coal tar. The first is tied to the price of oil and the second is a by-product of the steel industry.

Felts

Ninety percent of felts in use today are fiberglass. Impregnated glass fiber felts allow vapors to be vented during application. Type IV is the most prevalent type in new construction. A base sheet is the first ply placed on the roof. It acts as a vapor retarder to prevent the intrusion of moisture into the membrane. Moisture is the main cause of ridges and wrinkle cracking that result in roof leaks. A cap sheet is the top roof ply designed to receive the bituminous flood coat.

Base Flashing

This component is placed on top of the plies at the perimeter of the roof and around roof penetrations. The base flashing must be tough because it has to withstand most of the stress caused by wind uplift and differential movement caused by different abutting construction. This component may be constructed from a heavy saturated felt laminated to asphalt-impregnated fabric or fiberglass-reinforced felt. It may also be a roofing material such as rubber, neoprene, PVC, or polymer-modified bitumen. Base flashings are mechanically attached at the top of the vertical run, which is located at a minimum 8 inches above the roof and adhered to the substrate with hot asphalt (Fig. 3-22).

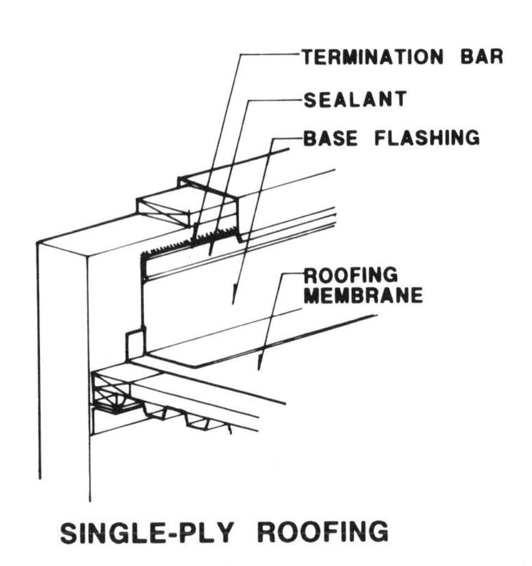

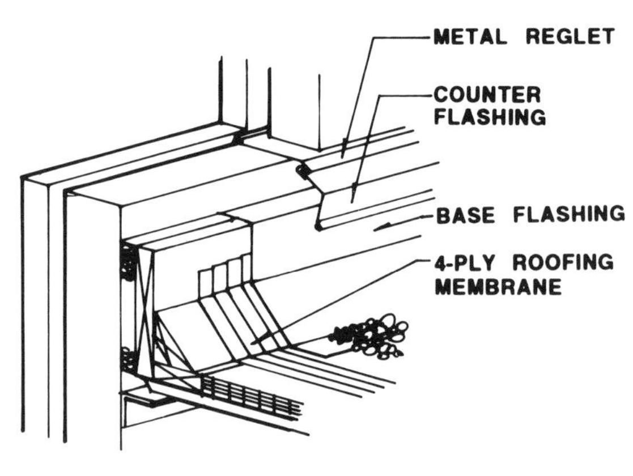

FIGURE 3-22. Base Flashing.

Counter Flashing

Also known as cap flashing, counter flashing is placed above and in front of the base flashing to prevent water from penetrating behind the base flashing and into the building.

Cants

There are two kinds of cants—treated wood and fiber cants. Treated wood stabilizes curbs and, unlike fiber cants, will not rot if exposed to moisture. Cants ease the junction between horizontal and vertical planes to prevent the membrane from cracking. Cants are usually not required for single-ply roofs.

Roof Drains

Drains should be placed near midspan because this is the area where deflection occurs. Placing them next to a column may leave them high and dry and contribute to water ponding on the roof, one of the causes of membrane deterioration.

Ballast

Ten-pound-per-square-foot gravel is the most prevalent kind. Where ballast is not required to anchor the roof, some developers prefer to substitute a fibered aluminum coating because, unlike gravel, a coating makes it easier to spot leaks.

Recommendations for Built-up Roofing:

A. Use coal-tar pitch instead of asphalt in locations with extreme temperature variations.
B. Impress upon the mechanical engineer the importance of grouping roof penetrating items in as few locations as possible (Fig. 3-25). Penetrations should be located not less than 2′ 0″ from the nearest obstruction.
C. Place expansion joints on curbs. Cheaper joints installed on the surface of the roof cause major leaks if damaged.
D. If the roof is constructed of lightweight insulating concrete, specify a galvanized slotted deck to vent water vapor emitted from the concrete.
E. Insulations must be anchored firmly in the bitumen before it congeals. In large-area roofs, only as much insulation as can be covered by the membrane in the same day should be used.
F. Design roof slopes to drain away from penthouses.
G. Attach base flashing to curbs, independent of walls at the roof perimeter to avoid differential movement. Perimeter curbs may not be necessary if the roof is carried on bearing walls.
H. Fascias and copings must be attached with continuous cleats to prevent them from being pealed back during high winds.
I. Place a compressible joint filler in building expansion joints to prevent condensation from forming on the underside of the metal cap flashing.

Single-Ply Roofing

Single-ply refers to roofing membranes known by the initials of their chemical formulation. For instance, Ethylene Propylene Diene Monomer is mercifully referred to as an EPDM membrane. There are many kinds of single-ply roofing membranes including CPE (Chlorinated Polyethylene), CSPE (Chlorosulfonated Polyethylene), CPE/CSPE composite, PIB (Polyisobutylene), PVC (Polyvinyl Chloride), Modified Bitumen, and other products. Three widely used types for public buildings are EPDM, PVC, and Modified Bitumen. Each type has strengths and weaknesses and each varies in sheet widths, method of application, and detailing.

There are four methods of attaching a membrane to the roof—fully adhered, partially adhered, mechanically fastened, and loose-laid ballasted. Care must be exercised if a ballasted system is chosen. In high wind locations, winds in excess of 80 mph have blown stones off roofs and damaged adjacent property and injured people. In addition, one must consider the added weight. Ballasted systems weigh almost thirty times as much as other single-ply systems. On the plus side, some factory seamed systems which are used in conjunction with ballast can be delivered to the roof in sheets up to 100′ × 150′. This avoids the hazard of leaks caused by poor seam workmanship in the field.

Fully adhered and partially adhered systems are no panacea, either. In high winds, there is always the risk of the membrane pulling the facer off the insulation board. The most effective method in roofs exposed to high wind is to mechanically fasten the membrane to the roof, using the specified number of corrosion-resistant fasteners and a reinforced membrane to prevent the membrane from tearing around the fastener. Both U.L. and Factory Mutual certify systems according to wind uplift. In addition, Factory Mutual tests mechanically fastened systems for fastener corrosion. The most popular fasteners are stainless steel, carbon steel, multicoated screws, and carbon steel embedded spikes for concrete decks.

Single-ply is a relative newcomer to roofing. It was introduced about twenty-eight years ago and yet, according to a survey conducted in 1990 by the National Roofing Contractors Association (NRCA), single-ply accounts for 36.9 percent of all new roofs in the United States. Modified bitumen accounted for an additional 13.2 percent. This represents a great gain over BUR, which accounted for only 30 percent.

Recommendations for Single-Ply Roofing:

In choosing a single-ply roofing membrane, the architect should evaluate the following characteristics:

A. Membranes vary in thickness from 30 to 160 mils. The thicker the membrane, the more it resists puncture caused by fastener pull-out, roof traffic, and stresses caused by a compressible substrate.

B. Tensile strengths vary from 1,100 psi to 3,000 psi. The higher the strength, the better the resistance to stresses caused by building movement, wind uplift, and thermal loading.
C. Depending on whether the membrane is reinforced or not, elongation varies from 27 percent to 500 percent. This characteristic is useful in figuring membrane resistance to rupture caused by an unstable substrate and building movement. In general, single-ply membranes are very good at resisting these forces compared to BUR.
D. Tear resistance indicates the ability of the material to resist initiation and propagation of a tear. This quality is especially important in evaluating mechanically fastened membranes.
E. Water absorption. Membranes with a high absorption rate are not suitable for roofs that have areas where water may pond and at protected membrane roofs (PMR).
F. Type of seam. Seams are the weakest link in single-ply systems. Although solvent welding is almost three times faster than hot air welding, it takes longer to stick in cold weather and evaporates too fast in hot weather. The architect must exercise good judgment in making the selection.
G. Color. White and light-colored membranes are more energy efficient. A ¼-inch-per-foot minimum slope is required to prevent dirt from accumulating on the surface. White membranes show aging more readily than black membranes.

Metal Roofing

Low-slope standing seam metal roofing is a relatively recent development. It was introduced by the pre-engineered building industry and is currently used as roofing for many types of projects. This system differs from conventional standing seam roofing because it requires a minimum slope of only ¼ inch per foot instead of 3 inches required by the conventional metal systems. This is quite an advantage when covering a large roof. For instance, a 50-foot slope would require only 12½ inches as opposed to the conventional roof's 12′-6″. The savings in structural support, building volume, and roof area are substantial.

The characteristics of the newer system are also different. The seams, which range from 2 to 4 inches in height, are mechanically connected or "zippered" and sealed on the roof, and the metal roof is designed to float to allow for expansion and contraction of the metal. This is done by attaching the roofing with slotted clips that move with the roof while maintaining a strong connection to the substructure capable of meeting the UL.I 90 (90 pounds per square foot) wind uplift requirement. The standard span is 5′-0″ between purlins for most systems. Care must be taken in detailing to make sure that no detail hampers the roof from movement. Roof panels which may have a run of 200 feet are typically constructed from 40-foot-long sheets. Splices must be staggered and supported, but *not* fastened to the structure. Roof penetrations must not act as anchors but be designed to allow the roof to move. Flashings must also move independently from sidewalls or parapets. The movement can be as much as 2 inches in a 200-foot-long panel, which is usually fastened at the low end and allowed to move at the ridge. For longer slopes, an overlapping step-up detail is required. Consult roof manufacturers before finalizing the details.

Most metal roofs provide long-term weather tightness and favorable life-cycle costs. This, of course, means that the initial cost is higher than membrane roofing. It is lightweight, weighing 2 pounds per square foot on average compared to 5–9 pounds for BUR. This, and the fact that it is not adversely affected by added insulation, makes it a good candidate for reroofing over existing BUR roofs, which require added slope and insulation.

On the downside, metal roofing has a limited ability to handle complicated slopes and excessive roof penetrations. It requires more skill to construct, is noisier during rain and hailstorms, and is not as comfortable to walk on as membrane roofing. In addition, condensation may occur at the clips unless the system is designed with thermal breaks.

The metals used most on low-slope roofs are aluminum, galvanized aluminum-coated steel, and aluminum-zinc-coated steel. The last named is produced by Bethlehem Steel under the trademark Galvalume.

Conventional standing seam is constructed from several metals including:

- Stainless steel. This roofing is often annealed or rough-rolled to reduce the brightness of the metal. Corrosion may cause pinholes in the sheets.
- Terne. This is an alloy of lead and tin bonded to mild or stainless steel. It is recommended that the surface be primed with iron oxide primer mixed with linseed oil, then painted. This requires repainting every ten years.
- Copper. One of the more expensive and longer-lasting roofing materials. Surface corrosion forms a patina that protects the metal. This roofing must be detailed carefully to prevent runoff from staining adjacent surfaces.
- Cor-ten. This is a trademark of United States Steel. Roofs using Cor-ten steel are generally thicker and heavier. Like copper, it develops a protective layer that has a tendency to stain adjacent surfaces if not detailed properly.
- Aluminum. Roofing constructed from aluminum is more expansion and contraction prone. If exposed, it forms a protective gray oxide layer. It can be anodized, painted, or dipped in hot zinc for corrosion resistance.

Paintable roofs can be finished with polyester enamels, which come with a ten-year warranty, or the more expensive silicone-modified polyester covered by a twenty-year warranty, or the longer-lasting fluorocarbon coatings, with a twenty-year film integrity and fade guarantee. Heat-laminated acrylic films are used mostly over aluminum and galvanized steel and are usually dependable.

The most authoritative book on detailing conventional roofs is *Standard Practice in Architectural Sheet Metal Work* by the Sheet Metal and Air Conditioning Contractors National Association, usually referred to as the SMACNA book. It gives information on what gauges to use, size of gutters, downspouts, expansion joints, and standard details for most roofing systems and other useful information.

3.8.3 Protected Membrane Roofing (PMR)

A protected membrane roofing assembly, sometimes referred to as "upside-down roof," is a method of construction that can be applied to many types of roofing membrane. The design (Fig. 3-7) calls for placing the ballasted insulation on top of the membrane. This protects the membrane from ultraviolet rays and the excessive heat differential that ages membranes in conventionally constructed roofs. Like everything else in roof construction, PMR is not without potential problems. Because the membrane is hidden from view, it is very difficult to locate the source of a leak without removing a substantial amount of ballast and insulation. This may be facilitated somewhat if polystyrene boards backed with a concrete layer (integral ballast) is used as ballast.

A well-sloped membrane with low water absorption is required for this system because the membrane stays damp longer than in conventional construction. For this reason, organic felts should not be used. The insulation must also be chosen to withstand long periods of continuous immersion without loss of R-value. In addition, in stone-ballasted systems, a protective mat must be placed between the insulation and the ballast to prevent debris from blocking water passages to the drain.

3.8.4 Warranties

Warranties for single-ply or any type of roofing are like an insurance policy. They are as good as the company that offers them. Richard Coursey, J.P. Stevens' Elastomerics technical products manager, and Bruce R. Wilby, Stevens' business manager, offer the following suggestions concerning warranties to architects specifying a roofing system:[4]

1. Ascertain what kind of coverage is provided. Check the materials, labor, workmanship, and various combinations that are covered in the warranty.
2. Check for well-defined and reasonable exclusions. Some of the exclusions that may be meaningless without further definition are environmental fallout, windstorms, roof ponding, substrate movement, building occupancy/use changes, and water penetration through and around walls. Though some warranties cover winds up to "hurricane force," a measurable 70 miles per hour, most cover up to "gale force," which ranges from 70 mph down to as low as 30 mph.
3. Understand who backs the warranty. It is important to know whether the manufacturer, the distributor/supplier, or a third party backs the warranty, as well as the financial stability of the backer. "Ask if there's a dedicated warranty program—how it works, whom you call, and whether there's a proven track record," recommends Coursey.

   "A warranty from a manufacturer may be more secure than one from a contractor," says Wilby. "In any case, if the contractor goes out of business, the warrantee may 'deep pocket' back to the manufacturer. Remember, there is no foolproof roofing system anywhere; they all have problems. An owner should be sure to ask, 'Does this company ever walk away from a roof problem?' "
4. Check out how the warranty is issued. Are there inspections before it is issued? Can it be altered? By whom? Under what circumstances?
5. Weigh whether less is more. "If the warranty sounds too good to be true, it usually is," says Wilby. "Generally speaking, the longer the warranty, the less that is actually covered; the length may be a 'dress-up' process. A better-written warranty is preferable to a longer one, for everybody concerned, because if a long list of exclusions keeps a manufacturer from being liable, a dissatisfied client may attempt to sue the architect or the builder."
6. Stress maintenance to the owner. "Everyone's job is made easier if there is an early detection of a problem, before a leak destroys the insulation and it has to be removed and replaced," says Coursey. Stevens' maintenance guidelines, for example, suggest that the roof owner have a working knowledge of the roof, limit access to it, inspect drains and equipment fastened to the roof at least every three months, and inform the manufacturer immediately if a problem arises.

[4]Reprinted by permission from *Architecture*, January 1988.

3.8.5 Roof Plans

Although some single-ply systems claim that their membrane can be constructed dead flat, it is advisable to create a minimum slope of 1/4 inch per foot to direct rainwater toward the drains. This can be done in two ways—by building a horizontal roof deck and creating the slopes by installing tapered insulation or by tilting the deck. The latter method is usually cheaper, using a constant thickness of insulation over most of the roof area. Only the possibility of adding floors to the structure at a future date would justify using the tapered insulation method. Figures 3-23 and 3-24 show an example using each method. Please note the difference in insulation thickness, which is 12″ at the perimeter of the flat roof and only 2″ at the perimeter of the sloped roof, except at the crickets.

The examples show the location of drains adjacent to the columns. This can only be done if the deflection in the beams, which normally occurs at midspan, is negligible, or if the slope is steep enough to offset the effect of the low spot at midspan and prevent the water from ponding. The architect should ask the structural engineer about the maximum deflection in roof beams before determining the slope and locating the roof drains.

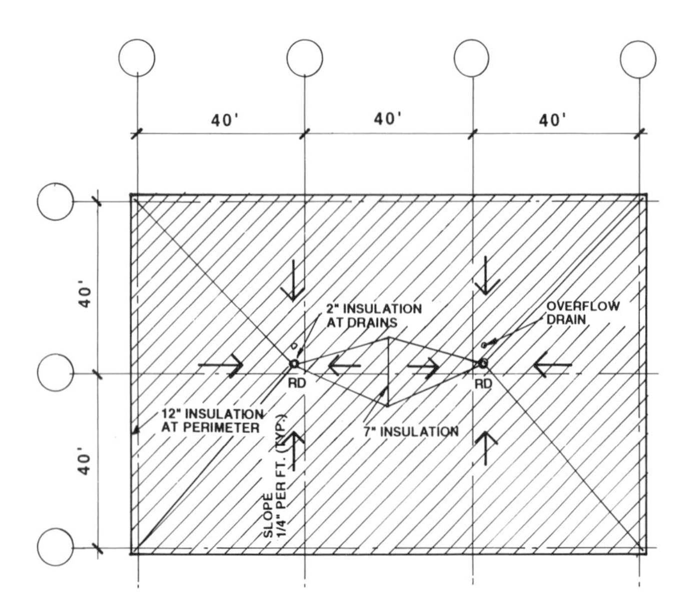

FIGURE 3-23. Tapered Insulation on a Flat Structure (area of insulation is shown hatched).

3.8.6 Detailing

Roof details depend on the type of roofing, the complexity of the building, the roof edge condition, the mechanical equipment, and roof penetrations. In general, if any of the following items are present, they must be detailed—flashing and counterflashing, roof drains and overflow drains or scuppers, equipment and screen supports, pipe and vent penetrations, parapet coping, gravel stop, curbs, window-washing davit supports and cable anchorage (high-rise), and expansion joints. Metal roofs include gutters and downspouts, eave and ridge details, and crickets located upstream of a major obstruction in addition to side flashing details.

Roof plans must show the locations of roof and overflow drains or scuppers, slopes with arrows indicating the direction and rate of slope, cants (if present), traffic pads to service mechanical equipment and drains, skylights (if included), roof hatches, penthouses, ladders, and major mechanical equipment supports and screens around them. It is not necessary to show every small penetration caused by plumbing vents and the like as long as a typical detail for pipe penetration is shown. It is, however, worthwhile to work with the MEP consultant to try to combine these penetrations into as few locations as possible (Fig. 3-25). Detailing may be generic, e.g., applicable to most products of the particular type of roofing used, or based on one product, in which case the following note should be added to the detail sheet: "These details are based on details for the *(fill in the name of the manufacturer)* system. They show intent only and are not intended to preclude any other system that meets the specifications," or similar wording to that effect.

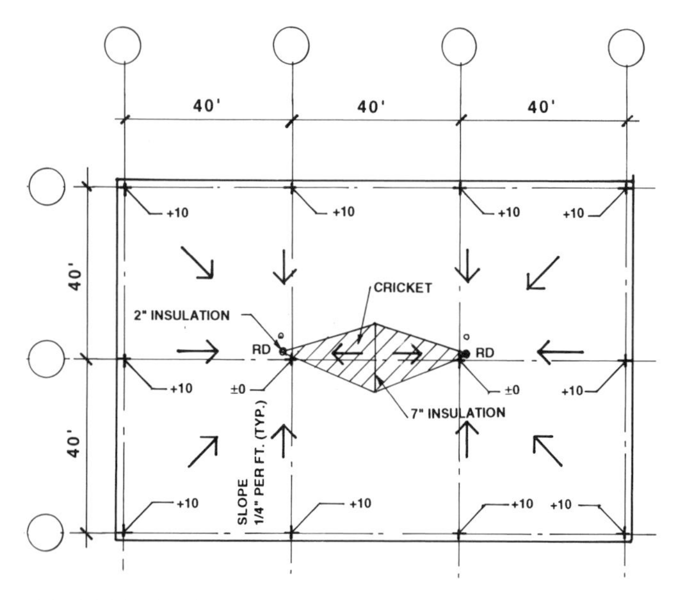

FIGURE 3-24. Tapered Insulation on a Sloped Structure (shown hatched).

3.8.7 Roof Drainage

To locate roof drains, the architect must take many things into consideration. Drains must be located so that they do not interfere with the roof supports. They should be compatible with the type of roof and be spaced evenly. Scuppers or overflow drains must also be installed to prevent subjecting the roof to excessive water loads should one of the drains be clogged with debris. If scuppers are used, the distance between the drain and the scupper must not exceed certain limits to prevent roof overloading. For instance, it the roof is designed to carry a maximum of 3½″ of accumulated water, and the slope is ¼″ per foot or 1″ per 4 feet, the distance between the drain and the scupper should not exceed 14′ (3 ½″ × 4′ = 14′) and the distance between drains should not exceed 28′ assuming a high point located midway between the drains. Because overflow drains are installed to protrude approximately 3″ above the roof, they free the architect to place the drains without the restrictions required for the scuppers.

Where overflow drains and scuppers are not desirable, such as in public plazas with upside-down roofing topped by a precast pedestal system (Fig. 3-7), a number of roof drains in addition to those required by calculations may substitute for overflow drains.

One factor to take into consideration is that if the roof is bordered by a high facade, the rain hitting that facade flows down to the roof. This factor is a must in determining the number and size of the drains. As soon as the roof structure and insulation thickness are determined, supply the plumbing engineer with a sketch of a section through the roof and a layout showing the suggested location of the drains and the

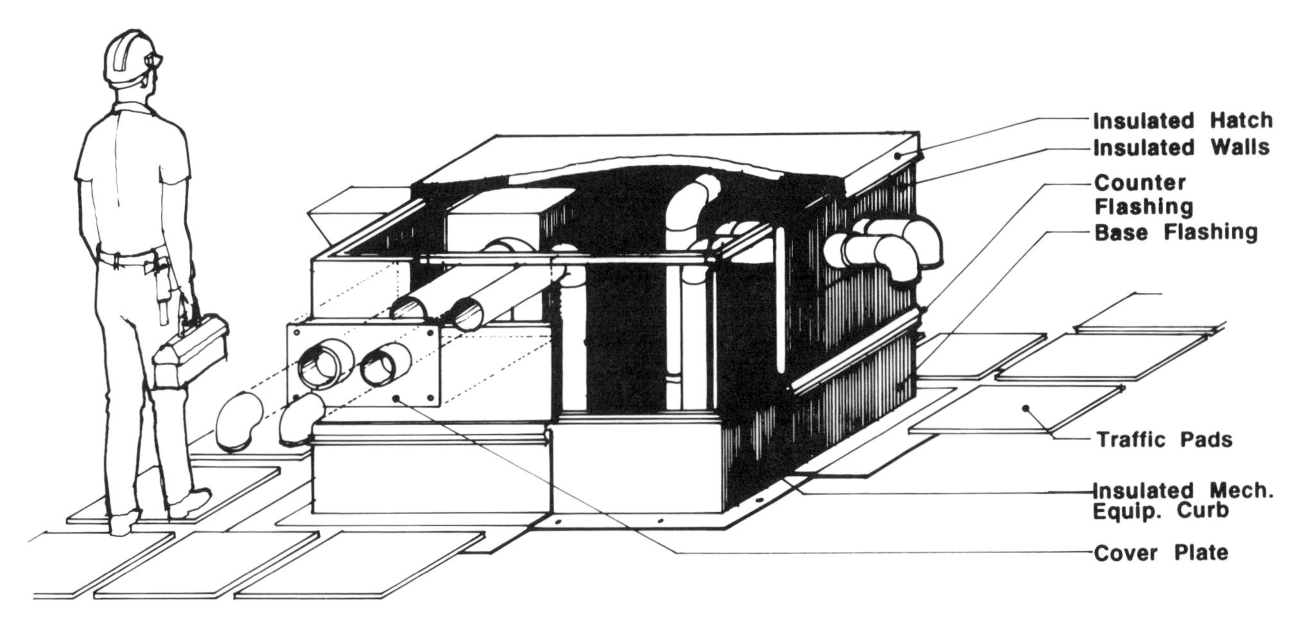

FIGURE 3-25. Grouping of MEP Roof Penetrations.

back-up system (overflow drains or scuppers) and ask him to provide a drain "cut sheet" and any modifications or improvements he may want to suggest. If a plumbing engineer is unavailable, refer to plumbing catalogs such as *JOSAM*, which include an easy-to-use method to size drains and horizontal and vertical leaders. Vertical leaders (roof drainage pipes) are usually connected to horizontal lines that lead to several roof drains. These vertical leaders must be located early in the project to be taken into consideration when planning the space. They may be located in mechanical shafts or in furred-out chases combined with column enclosures usually referred to as wet columns (Fig. 3-26).

Horizontal leaders and the thickness of the roof deck or slab, the roof low point, the main sprinkler pipes, depth of girders (including sprayed fireproofing), the ceiling/light fixture depth, and the mechanical ducts are all factors in determining the clearance between the roof and the ceiling.

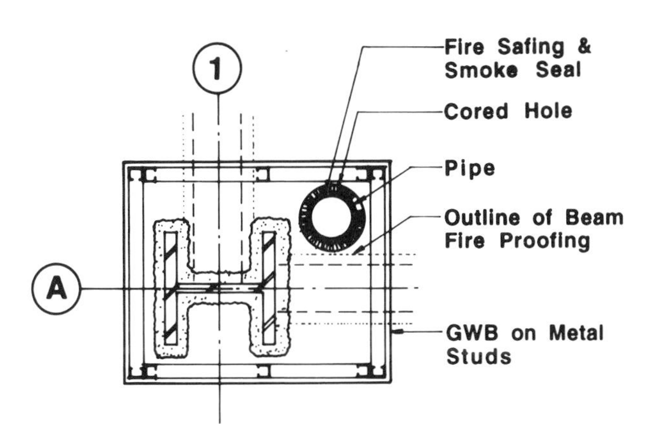

FIGURE 3-26. Example of a "Wet" Column.

3.8.8 Terminology

The following are a few terms used in roofing and their definitions:

Alligatoring: Cracks that appear in a coating or mastic.

Batten: A strip of steel or aluminum used to mechanically fasten the membrane to the deck to resist wind uplift.

Blistering: A bubble in BUR caused by gases. One theory states that they are caused by fluorocarbon gas emitted by rigid insulation.

Cold Process Roofing: One or more plies of coated felts bonded with cold-applied asphalt roof cement.

Fish Mouth: A tear in an edge wrinkle, usually in BUR.

Flashing: A reinforcing ply located at roof edges, walls, and curbs in membrane roofing or at valleys and hips in metal roofing.

"Hot Stuff": Term used by roofers for hot bitumen.

Leader: A vertical pipe similar to a downspout, but placed inside the building.

Mil: 0.001 inch.

Pitch Pocket: A flanged metal container filled with bitumen and constructed around roof-penetrating elements, such as equipment supports, to waterproof them.

Scrim: Membrane reinforcement made of polyester or glass fiber.

Slip Sheet: A sheet placed between two layers to prevent adhesion.

Thermoplastic Membrane: A membrane that can be softened repeatedly by heating.

Thermoset Membrane: A membrane that, once it cures, cannot be softened by heat or chemical means.

Vulcanized Membrane: Same as thermoset.

As can be readily seen, roofing is a vast subject encompassing many disparate and highly complicated systems evolving continuously as new products and systems are introduced into the market. The architect must educate himself, use consultants if the budget permits, analyze the roofing, choose the most appropriate system, calculate initial and life-cycle cost, select the insulation, calculate the R-value of the system, choose the appropriate warranty, detail the roof properly, and write a spec to guarantee that no inferior substitutions occur. The intent of this section is to introduce the reader to the subject. He should continue to pursue the latest information available.

3.8.9 Sources of More Information:

Addleson, Lyal. 1990. *Building Failures, A Guide to Diagnosis, Remedy and Prevention,* 2nd ed. Stoneham, MA: Butterworth Architecture.

Asphalt Roofing Manufacturers Association (ARMA), 6288 Montrose Road, Rockville, MD 20852, 301-231-9050.

Callender, John Hancock, ed. *Time-Saver Standards*, latest edition. New York: McGraw-Hill, Inc.

Exteriors (journal), P.O. Box 6318, Duluth, MN 55806

Factory Mutual Research Organization (FM), 1151 Boston-Providence Turnpike, Norwood, MA 02062, 617-762-4300

Griffin, C. W. 1982. *Manual of BUR Systems*, 2nd. ed. New York: McGraw-Hill, Inc.

McCampbell, B. Harrison. 1992. *Problems in Roofing Design.* Stoneham, MA: Butterworth Architecture.

National Roofing Contractors Association. *NRCA Roofing and Waterproofing Manual*, latest edition. 10255 W. Higgins Road, Suite 600, Rosemont, IL 60018. 708-299-9070.

Ramsey, Charles G., and Sleeper, Harold R. *Architectural Graphic Standards*, latest edition. New York: John Wiley & Sons, Inc.

Roof Consultants Institute, 7424 Chapel Hill Road, Raleigh, NC 27607. 919-859-0742.

The Roofing Industry Educational Institute, 7006 South Alton Way #B, Englewood, Co 80112

Sheet Metal and Air Conditioning Contractors' National Association (SMACNA). *Standard Practice in Architectural Sheet Metal Work.* 4201 Lafayette Center Drive, Chantilly, VA. 703-803-2980.

The Single-Ply Roofing Institute (SPRI), 104 Wilmont Road, Suite 201, Deerfield, IL 60015. 780-940-8800.

Underwriters Laboratories, Inc. (U.L.), 333 Pfingsten Road, Northbrook, IL 60062, 708-272-8800.

Watson, John, 1983. *Commercial Roofing Systems.* Reston Publishing Co.

3.9 AREA CALCULATIONS

3.9.1 General

The gross area of a building can be defined as the sum of the floor areas within the outer surface of the exterior walls. Net areas are the spaces that are actually used by the building inhabitants. These are broad descriptions that may differ somewhat from the specific guidelines used by the AIA, BOMA (Building Owners and Managers Association International), and other organizations.

The gross area is used to make a rough cost estimate of the project, to be presented to the client at the end of the schematic design phase. For example, if the median cost of construction for a school building is $150 per square foot and the gross area of the building is 100,000 square feet, a rough estimate for the project can be arrived at by multiplying 150 by 100,000 sq. ft. to state that the project would cost roughly $15,000,000. Of course the architect must use good judgment before defining the cost per foot. Cost can be affected by many factors, including inflation, how active the contracting business is at the time of bidding, as well as the size of the project. The bigger the project, the more competitive the bids get.

Net and gross areas are also used to calculate the efficiency of a building. Building efficiency is arrived at by dividing the net area by the gross area and multiplying by 100. Efficiency varies from roughly 60 percent to 95 percent, depending on the design and building type. As an example, research lab buildings usually require a disproportionate area for plumbing chases that result in a lower efficiency. Site limitations may also affect efficiency.

A small building is usually less efficient than a large building because the percentage of the area occupied by service areas such a stairs, elevators, etc., in relation to the usable spaces is, in most cases, higher in the small building. Other factors also affect efficiency. By making this calculation, the architect finds out if the plan needs to be modified to become more efficient. This can usually be done by getting rid of wasted spaces, choosing a different mechanical system that houses all the equipment in a mechanical penthouse instead of air-handling units on each floor, reducing corridor widths, and other measures.

The volume of a building may also affect efficiency. For instance, if the building contains an enclosed atrium several stories high, part of this unoccupied space is added to the gross area at each floor, adversely impacting the efficiency of the building. This does not mean that atriums are undesirable and should not be used. On the contrary, an atrium can enhance the total appeal of the building and make it more marketable, bringing in higher rents. Computing the volume is just another yardstick to be used in evaluating the building.

Two of the most common methods for calculating areas are described in the following subsections.

3.9.2 AIA Document D101

Figures 3-27 and 3-28 explain area calculations as recommended by the American Institute of Architects and are reproduced by permission from the AIA.

3.9.3 The BOMA International Method

The Building Owners and Managers Association International (BOMA) employs the "Standard Method for Measuring Floor Area in Office Buildings," which is based on ANSI Z65.1–1980. It is used to allow comparison of values based on a generally accepted method of measurement. The following text is reproduced by permission from BOMA and describes standard methods of measuring office building Usable, Rentable, Store, and Construction Area.

Usable Area

This method measures the actual occupiable area of a floor or an office suite and is of prime interest to a tenant in evaluating the space offered by a landlord and in allocating the space required to house personnel and furniture. The amount of Usable Area on a multi-tenant floor can vary over the life of a building as corridors expand and contract and as floors are remodeled. Usable Area on a floor can be converted to Rentable Area by the use of a conversion factor.

THE AMERICAN INSTITUTE OF ARCHITECTS

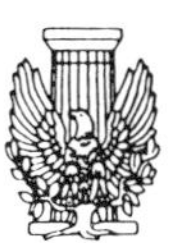

AIA Document D101

THE ARCHITECTURAL AREA AND VOLUME OF BUILDINGS

Establishing Definitions for the Architectural Area and Architectural Volume of Buildings

ARCHITECTURAL AREA OF BUILDINGS

The ARCHITECTURAL AREA of a building is the sum of the areas of the several floors of the building, including basements, mezzanine and intermediate floored tiers and penthouses of headroom height, measured from the exterior faces of exterior walls or from the centerline of walls separating buildings. Discretion is advised in calculating areas of interstitial space.

- Covered walkways, open roofed-over areas that are paved, porches and similar spaces shall have the architectural area multiplied by an area factor of 0.50.
- The architectural area does not include such features as pipe trenches, exterior terraces or steps, chimneys, roof overhangs, etc.

ARCHITECTURAL VOLUME OF BUILDINGS

The ARCHITECTURAL VOLUME (cube or cubage) of a building is the sum of the products of the areas defined above (using the area of a single story for multistory portions having the same area on each floor) and the height from the underside of the lowest floor construction system to the average height of the surface of the finished roof above for the various parts of the building.

AIA DOCUMENT D101 • ARCHITECTURAL AREA AND VOLUME OF BUILDINGS • JAN. 1980 EDITION • ©1980
AIA® • THE AMERICAN INSTITUTE OF ARCHITECTS, 1735 NEW YORK AVE., N.W., WASHINGTON, D.C. 20006 D101—1980 1

FIGURE 3-27. AIA Area and Volume Calculations.

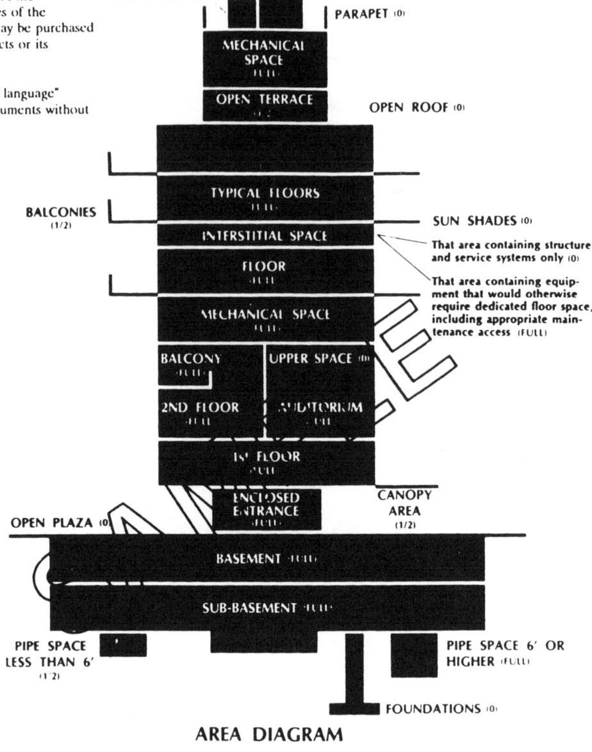

AREA DIAGRAM

AIA STANDARD NET ASSIGNABLE AREA

The STANDARD NET ASSIGNABLE AREA is that portion of the area which is available for assignment or rental to an occupant, including every type of space usable by the occupant.

The Standard Net Assignable Area should be measured from the predominant inside finish of permanent outer walls to the office side of corridors or permanent partitions and from center line of adjacent assigned spaces. Where there are interior spaces surrounded by corridors, measurement shall be from the inside face of enclosing walls. Included should be space subdivisions for occupant use; i.e., offices, file rooms, office storage rooms, etc. Deductions should not be made for columns and projections necessary to the building or for partitions subdividing space.

NOTE: There are two variations of the Net Area which may be useful:

(1) Single Occupant Net Assignable Area includes the space from the inside finish of exterior walls to the office side of permanent partitions and including toilets. The Single Occupant Net Assignable Area is different from the Standard Net Assignable Area by the addition of corridors and toilets, but excludes elevator shafts, elevator lobbies, stair enclosures, mechanical equipment rooms and electrical closets.

(2) Store Net Assignable Areas are from the exterior front face of the building or store front line through to the exterior rear face of the building and from center line to center line of walls between adjacent spaces.

FIGURE 3-28. AIA Area and Volume Calculations.

Rentable Area

This method measures the tenant's pro rata portion of the entire office floor, excluding elements of the building that penetrate through the floor to areas below. The Rentable Area of a floor is fixed for the life of a building and is not affected by changes in corridor sizes or configuration. This method is therefore recommended for measuring the total income producing area of a building and for use in computing the tenant's pro rata share of a building for purposes of rent escalation. Lenders, architects, and appraisers will use Rentable Area in analyzing the economic potential of a building.

It is recommended that on multi-tenant floors the landlord compute both the Rentable and Usable Area for any specific office suite.

Store Area

This method measures the ground floor rentable area of an office building for occupancy as store space.

Construction Area

This method of measurement is to be used primarily to determine building cost or value and is not used for leasing purposes except where an entire building is leased to a single tenant.

Definitions

1. "Finished Surface" shall mean a wall, ceiling or floor surface, including glass, as prepared for tenant use, excluding the thickness of any special surfacing materials such as paneling, furring strips, and carpet.
2. "Dominant Portion" shall mean that portion of the inside finished surface of the permanent outer building wall which is 50% or more of the vertical floor-to-ceiling dimension measured at the dominant portion. If there is no dominant portion, or if the dominant portion is not vertical, the measurement for area shall be to the inside finished surface of the permanent outer building wall where it intersects the finished floor.
3. "Major Vertical Penetrations" shall mean stairs, elevator shafts, flues, pipe shafts, vertical ducts, and the like, and their enclosing walls, which serve more than one floor of the building, but shall not include stairs, dumb-waiters, lifts, and the like exclusively serving a tenant occupying offices on more than one floor.
4. "Office" shall mean the premises leased to a tenant for which a measurement is to be computed.

Conversion Formula

$$\frac{\text{Rentable Area}}{\text{Usable Area}} = \text{Rentable/Usable Ratio ("R/U Ratio")}$$

$$\text{Usable} \times \text{R/U Ratio} = \text{Rentable Area}$$

$$\frac{\text{Rentable Area}}{\text{R/U Ratio}} = \text{Usable Area}$$

The following example (not part of the BOMA text) illustrates the usefulness of the conversion formula:

Example:
The rentable and usable areas of a typical floor in an office building were calculated at 10,000 sq. ft. and 8,000 sq. ft., respectively. A prospective tenant requires a usable area of 1,000 sq. ft. Find the tenant rentable area.

Solution:

$$\text{R/U Ratio} = \frac{\text{Rentable Floor Area}}{\text{Usable Floor Area}} = \frac{10{,}000}{8{,}000} = 1.25$$

$$\begin{aligned}\text{Rentable area for the tenant} &= 1{,}000 \text{ sq. ft.} \times 1.25 \\ &= 1{,}250 \text{ sq. ft.}\end{aligned}$$

Usable Area (Fig. 3-29a)

The Usable Area of an office shall be computed by measuring to the finished surface of the office side of the corridor and

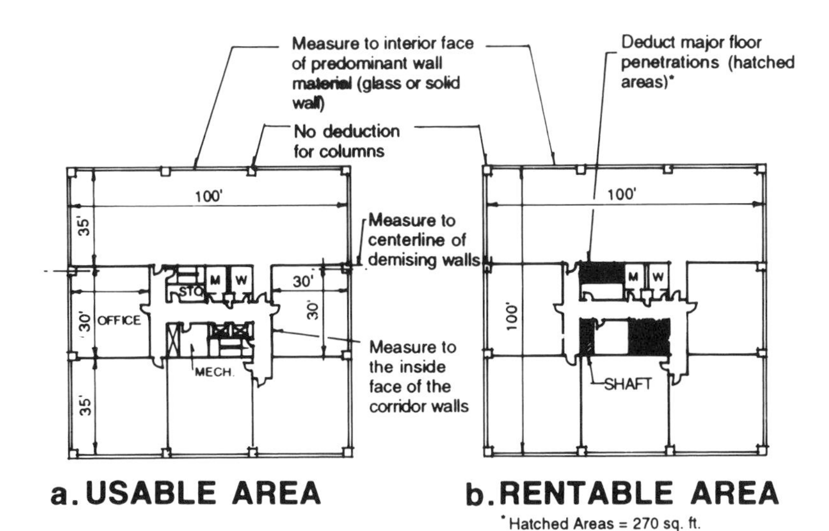

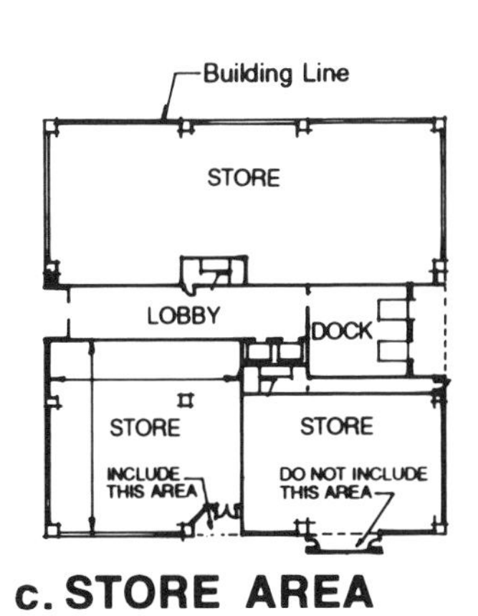

Example: Find the rentable area for the office in figure 3-29

Solution: Rentable floor area = Floor Area (100 x 100) - Floor Penetrations (270) = 9,730 sq. ft.
Total Usable Floor Area = 2(30 x 30) + 2 (100 x 35) = 8,800 sq. ft.
R/U Ratio = 9,730 ÷ 8,800 = 1.11
Rentable Area for the office = Office Area (30 x 30) x R/U Ratio (1.11) = 999 sq. ft.

FIGURE 3-29. Area Calculations (BOMA Method).

other permanent walls, to the center of partitions that separate the office from adjoining Usable Areas, and to the inside finished surface of the dominant portion of the permanent outer building walls.

No deductions shall be made for columns and projections necessary to the building.

The Usable Area of a floor shall be equal to the sum of all Usable Areas on that floor.

Rentable Area (Fig. 3-29b)

The Rentable Area of a floor shall be computed by measuring to the inside finished surface of the dominant portion of the permanent outer building walls, excluding any major vertical penetrations of the floor.

No deductions shall be made for columns and projections necessary to the building.

The Rentable Area of an office on the floor shall be computed by multiplying the Usable Area of that office by the quotient of the division of the Rentable Area of the floor by the Usable Area of the floor resulting in the "R/U Ratio" described herein.

Construction Area

The Construction Area of a floor shall be computed by measuring to the outside finished surface of permanent outer building walls. The Construction Area of a building shall be the sum of the building, including basements, mechanical equipment floors, penthouses, and the like.

Store Area (Fig. 3-29c)

The number of square feet in a ground floor Store Area shall be computed by measuring from the building line in the case of street frontages, and from the inner surface of other outer building walls and from the inner surface of corridors and other permanent partitions and to the center of partitions that separate the premises from adjoining rentable areas.

No deductions shall be made for vestibules inside the building line or for columns or projections necessary to the building.

No addition should be made for bay windows extending outside the building line.

3.9.4 Sources of More Information

The American Institute of Architects, 1735 New York Ave., Washington, D.C. 20006.

Building Owners and Managers Association, 1221 Massachusetts Ave., NW, Washington, D.C. 20005. 202-371-0181.

3.10 MASONRY

3.10.1 General

Masonry is a general term that is used to describe walls built with brick, stone, and concrete masonry units (CMU) also known as concrete block walls. Building walls out of masonry is one of the oldest and most durable methods of construction. New materials and construction techniques have revolutionized the industry and speeded up the process. These new methods, however, have introduced new problems that resulted in costly failures. This section is devoted to two of the most commonly used assemblies, namely brick veneer walls with steel stud backing and CMU construction.

3.10.2 Brick Veneer Walls

Veneer walls have the following advantages:

1. Lightweight construction. This feature effects savings on the building frame and foundations and makes it suitable for residential and multistory construction.
2. Early enclosure of the building. The sheathed stud walls are constructed to provide shelter so that the interior work may proceed while the slower brick veneer is being built.
3. Good insulation value. The space between the studs is suitable for placing insulation batts with high R-values.
4. A multilayered defense against the weather. The brick veneer provides the first line of defense against rain and wind. The cavity drains any water that may penetrate beyond the veneer, the sheathing protects against water access to the interior of the building, the insulation protects against the heat and cold, and the vapor retarder/air barrier (polyethylene and GWB) protects the insulation from vapor diffusion that may damage its ability to insulate. The air barrier also protects the interior from uncomfortable drafts and wind-driven moisture (see Sec. 3.4.3).
5. Appearance. Brick gives a solid appearance to the building. It offers a variety of colors and textures to choose from and almost unlimited potential for design creativity. It also makes the building relate to adjacent older structures if it is situated in an older neighborhood.
6. Relatively low cost of construction. Compared to solid masonry, precast, or high-tech construction methods, brick veneer has a distinct cost advantage.

Unfortunately, there are some pitfalls associated with this method of construction that one must guard against. The following is a partial listing of these problems and the current approach to avoid them:

1. Deflection of studs. Because the stud backing deflects under wind pressure, it may cause the veneer to crack. To lessen the chance of cracking, the Brick Institute of America (BIA) recommends that the studs be sized to allow for a maximum deflection of L/600 to L/720. In addition, a minimum thickness of ½-inch sheathing and GWB are required to cover the entire surface on both sides of the studs. Table 3-7 shows one manufacturer's stud sizes based on a maximum deflection of L/600 for different wind pressures and stud spacing. These tables

TABLE 3-7. Example of Limiting Heights for Stud Walls Used as Backup for Brick Veneer (L/600 Max. Deflection)

SJ Studs: Curtain Wall Limiting Heights — **Brick Exterior**

Wind load	Stud spacing (in o.c.)	362 SJ (3⅝")				40 SJ (4")			
		20 gauge	18 gauge	16 gauge	14 gauge	20 gauge	18 gauge	16 gauge	14 gauge
15 psf	12	12'5"	13'0"	13'8"	14'3"	13'0"	13'8"	14'4"	15'0"
	16	11'11"	12'5"	12'11"	13'5"	12'4"	12'11"	13'6"	14'1"
	24	10'5"	10'10"	11'3"	11'9"	10'10"	11'4"	11'10"	12'4"
20 psf	12	11'4"	11'10"	12'5"	12'11"	11'10"	12'5"	13'1"	13'8"
	16	10'10"	11'3"	11'9"	12'2"	11'3"	11'9"	12'3"	12'10"
	24	9'5"	9'10"	10'3"	10'8"	9'10"	10'3"	10'9"	11'2"
25 psf	12	10'6"	11'0"	11'6"	12'0"	10'11"	11'6"	12'1"	12'8"
	16	10'0"	10'6"	10'11"	11'4"	10'5"	10'11"	11'5"	11'11"
	24	8'9"	9'2"	9'6"	9'11"	9'1"	9'6"	10'0"	10'5"
30 psf	12	9'10"	10'4"	10'10"	11'4"	10'4"	10'10"	11'5"	11'11"
	16	9'5"	9'10"	10'3"	10'8"	9'10"	10'3"	10'9"	11'2"
	24	8'3"	8'7"	8'11"	9'4"	8'7"	9'0"	9'5"	9'9"
35 psf	12	9'5"	9'10"	10'3"	10'9"	9'9"	10'4"	10'10"	11'4"
	16	9'0"	9'4"	9'9"	10'1"	9'4"	9'9"	10'2"	10'8"
	24	7'10"	8'2"	8'6"	8'10"	8'2"	8'6"	8'11"	9'3"
45 psf	12	9'0"	9'5"	9'10"	10'3"	9'4"	9'10"	10'4"	10'10"
	16	8'7"	8'11"	9'4"	9'8"	8'11"	9'4"	9'9"	10'2"
	24	7'6"	7'10"	8'2"	8'5"	7'10"	8'2"	8'6"	8'11"

Wind load	Stud spacing (in o.c.)	60 SJ (6")				725 SJ (7¼")			80 SJ (8")		
		20 gauge	18 gauge	16 gauge	14 gauge	18 gauge	16 gauge	14 gauge	18 gauge	16 gauge	14 gauge
15 psf	12	16'4"	17'6"	18'8"	19'9"	20'0"	21'5"	22'8"	21'6"	23'0"	24'6"
	16	15'4"	16'4"	17'4"	18'3"	18'6"	19'9"	20'11"	19'10"	21'3"	22'6"
	24	13'5"	14'3"	15'1"	15'11"	16'2"	17'3"	18'3"	17'4"	18'6"	19'8"
20 psf	12	14'10"	15'11"	16'11"	17'11"	18'2"	19'5"	20'7"	19'6"	20'11"	22'3"
	16	13'11"	14'10"	15'9"	16'7"	16'10"	17'11"	19'0"	18'1"	19'3"	20'5"
	24	12'2"	12'11"	13'9"	14'6"	14'8"	15'8"	16'7"	15'9"	16'10"	17'10"
25 psf	12	13'10"	14'9"	15'9"	16'8"	16'10"	18'0"	19'1"	18'1"	19'5"	20'8"
	16	12'11"	13'9"	14'7"	15'5"	15'7"	16'8"	17'7"	16'9"	17'11"	19'0"
	24	11'4"	12'0"	12'9"	13'5"	13'8"	14'7"	15'5"	14'8"	15'8"	16'7"
30 psf	12	13'0"	13'11"	14'10"	15'8"	15'10"	17'0"	18'0"	17'1"	18'3"	19'5"
	16	12'2"	12'11"	13'9"	14'6"	14'8"	15'8"	16'7"	15'9"	16'10"	17'10"
	24	10'8"	11'4"	12'0"	12'8"	12'10"	13'8"	14'6"	13'9"	14'8"	15'7"
35 psf	12	12'4"	13'2"	14'1"	14'11"	15'1"	16'1"	17'1"	16'2"	17'4"	18'5"
	16	11'7"	12'4"	13'1"	13'9"	14'0"	14'11"	15'9"	15'0"	16'0"	17'0"
	24	9'2"	10'9"	11'5"	12'0"	12'2"	13'0"	13'9"	13'1"	14'0"	14'10"
45 psf	12	11'10"	12'8"	13'5"	14'3"	14'5"	15'5"	16'4"	15'6"	16'7"	17'8"
	16	11'1"	11'9"	12'6"	13'2"	13'4"	14'3"	15'1"	14'4"	15'4"	16'3"
	24	8'0"	10'3"	10'11"	11'6"	11'8"	12'5"	13'2"	12'6"	13'4"	14'2"

Limiting heights are for SJ members with 40 ksi yield strength (Fy) and based on lateral bracing provided by mechanically fastened gypsum board or sheathing each side. Stress based on the properties of the stud alone with a 33% increase for wind loading; Deflection limitations of L/600 based on composite wall assembly (gypsum sheathing and brick veneer exterior and drywall or plaster interior) with noncomposite addition of brick veneer stiffness. See Design Considerations, page 34.

(Reprinted, by permission, from Unimast, Inc. 1991)

are provided for preliminary design only. The specifications must stipulate that stud size, gauge, spacing, and bracing must be based on wind loads conforming to code requirements. As a general rule, studs should be a minimum 18 gauge zinc-coated steel, conforming to ASTM A525, grade G-90. Shop drawings must be checked by the structural engineer.

2. Corrosion. Moisture leaking through the wall may cause leaching of the salts present in the mortar to corrode the metal ties and separate the veneer from the backing. To minimize this possibility, ties should be a minimum 3/16 inch in diameter, similar in design to those shown in Fig. 3-30. The top two are the most commonly used. They should be spaced 24-inch on center maximum as shown in Fig. 3-31 and embedded a 2-inch minimum into the joint. They should be hot-dipped galvanized in accordance with ASTM A153, Class B-3, and fastened to the studs with screws which are cadmium-plated, hot-dipped galvanized, or composite zinc- and polymer-coated.

   Shelf angles (Fig. 3-32) must also be protected by flashing and painting. Galvanizing is required in areas with high humidity or severe weather conditions.

   Studs may also corrode if water vapor penetrates through gaps in the vapor retarder. Taking care during construction to seal all potential vapor passages is very important to prevent this from happening (see Sec. 3.4.2).

3. Cracking. Brick expands and contracts due to temperature changes. Minor dimensional changes are also caused by changes in moisture content. If control joints are placed too far apart, these movements will cause the wall to crack. Vertical joints (Fig. 3-33) should be placed near corners, between low and high wings, and at suitable intervals. Some experts recommend a spacing of 20–25 ft. o.c. The BIA recommends using the following formula to determine joint width:

$$W = [0.0005 + 0.000004\,(T\ \text{max.} - T\ \text{min.})]\,L$$

where W = the total expected movement of the brick

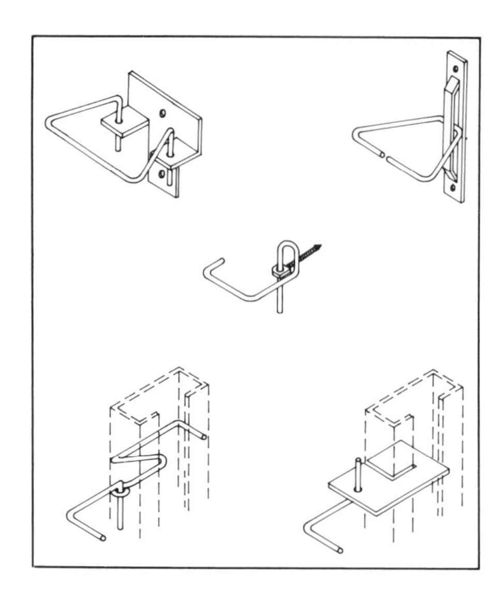

FIGURE 3-30. Cavity Wall Ties (Reprinted, by permission, from Brick Institute of America, *Technical Notes on Brick Construction*, 28B Revised II).

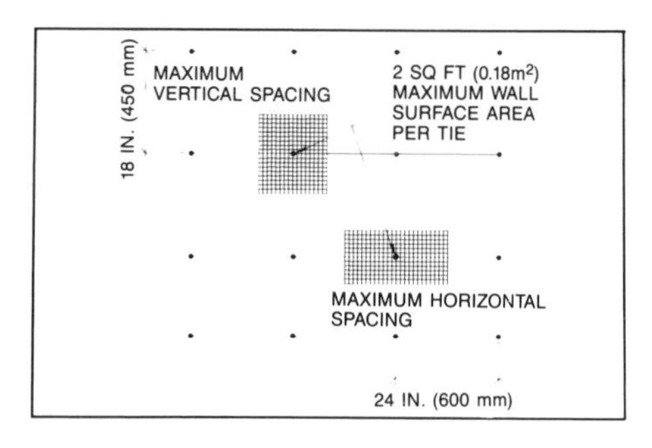

FIGURE 3-31. Spacing of Metal Ties (Reprinted, by permission, from Brick Institute of America, *Technical Notes on Brick Construction*, 28B Revised II).

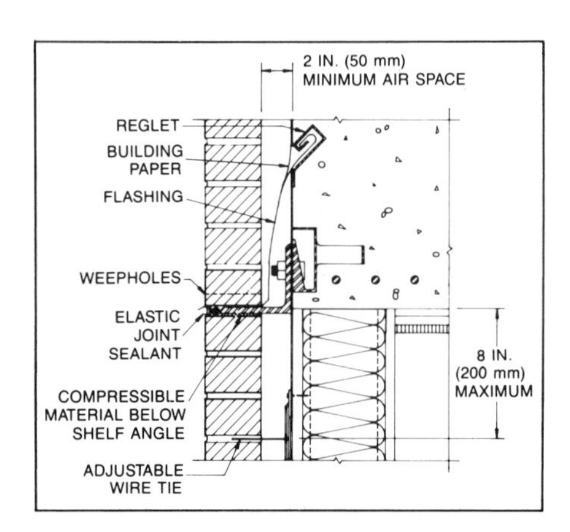

FIGURE 3-32. Typical Shelf Angle Detail (Reprinted, by permission, from Brick Institute of America, *Technical Notes on Brick Construction*, 28B Revised II).

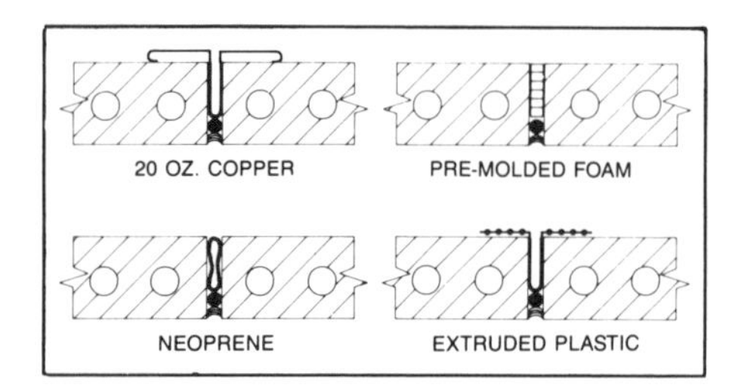

FIGURE 3-33. Typical Expansion Joint Details (Reprinted, by permission, from Brick Institute of America, *Technical Notes on Brick Construction*, 28B Revised II).

veneer in inches. Since most exterior sealant materials are only about 50 percent efficient, the joint width should equal 2 × W.

T max = Maximum mean temperatures of the brick veneer, degrees Fahrenheit.

T min = Minimum mean temperatures of the brick veneer, degrees Fahrenheit.

L = Distance between joints.

It should be pointed out that joint width must not be less than one half inch and that joint movement must not be impeded by other elements of the assembly such as shelf angles. These must be interrupted at each joint. Shelf angles must be designed for a maximum deflection of L/600.

4. Poor drainage. Veneer walls are designed with flashing and weep holes at the base of the wall and above shelf angles (Fig. 3-34). If mortar falls in the cavity during construction and blocks the weeps, the wall will not drain properly and water damage will occur at this location. To avoid this, some masons place a board on the ties to receive any mortar that may drop in the cavity. Before placing the next tier of ties, they lift the board and repeat the operation. This method, however, may weaken the

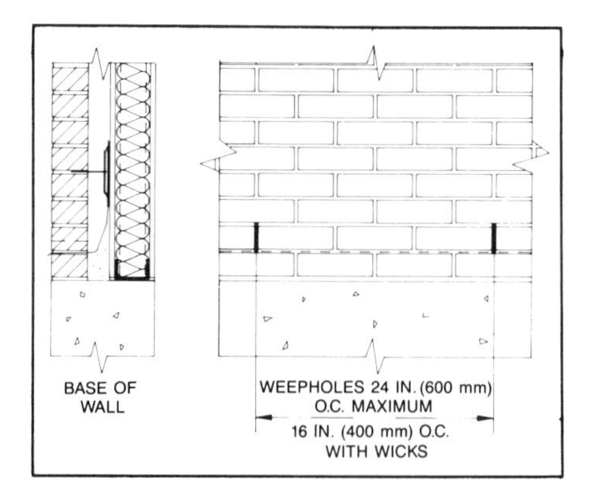

FIGURE 3-34. Flashing and Weep Holes (Reprinted, by permission, from Brick Institute of America, *Technical Notes on Brick Construction*, 28B Revised II).

wall if the board nudges the bricks and breaks the bond between the brick and the mortar during the lifting procedure. Most masons depend upon their skill to prevent mortar from dropping in the cavity. To prevent unskilled masons, who usually think they are skilled, from clogging the works, some designers indicate a 3-inch layer of gravel graded between ⅜″ and ½″ in diameter at the bottom of the cavity to ensure water drainage through passages in the gravel, even if mortar falls into the cavity.

5. Leakage. Water leakage can be caused by any of a number of factors. It may occur through expansion/contraction cracks (see #3 above) or through mortar joints, if they are improperly tooled, or through cracks caused by structural inadequacy. Leakage may also be caused by improper flashing or sealant failure in joints.

   The following measures are recommended to prevent water from penetrating into the building: a. Choose a good-quality sealant such as polysulphide to seal all expansion joints and use a closed-cell backer rod placed at the right depth (Fig. 3-19). b. Specify that the mason must tool the joint to a concave or V shape to consolidate the surface and make it denser to prevent water from seeping in. c. Detail complicated flashing configurations by drawing an axonometric drawing of the assembly, especially to show end dams above openings (Fig. 3-35a). d. If at all possible, avoid using brick sills and copings because they have too many joints that may crack in these exposed locations. Use precast, stone, or metal instead.

6. Efflorescence. This phenomenon is caused by water passing through the mortar, dissolving the salts present in the mix, and depositing them on the face of the wall in the form of white scum.

   To prevent efflorescence from forming, the top of the wall should be protected during construction by waterproof sheeting extending 2 feet down both sides of the wall and securely weighted down. Another measure is to specify "non-staining" portland cement and lime that contain less than .0006 free alkali. A water saturation test conducted after the panel is finished should be observed after three days for signs of efflorescence. The test panel should be built two weeks before the start of masonry work.

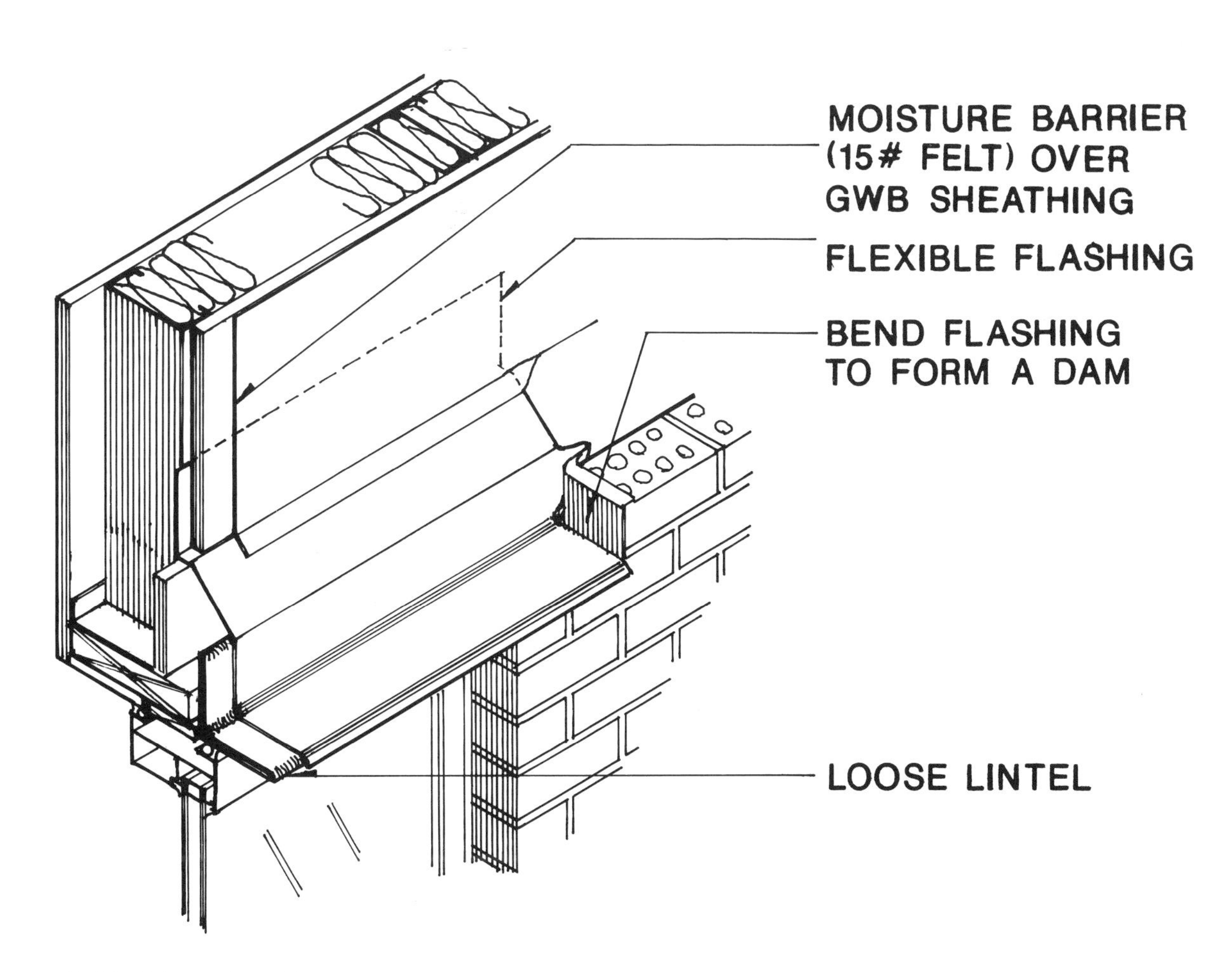

FIGURE 3-35a. Loose Lintel.

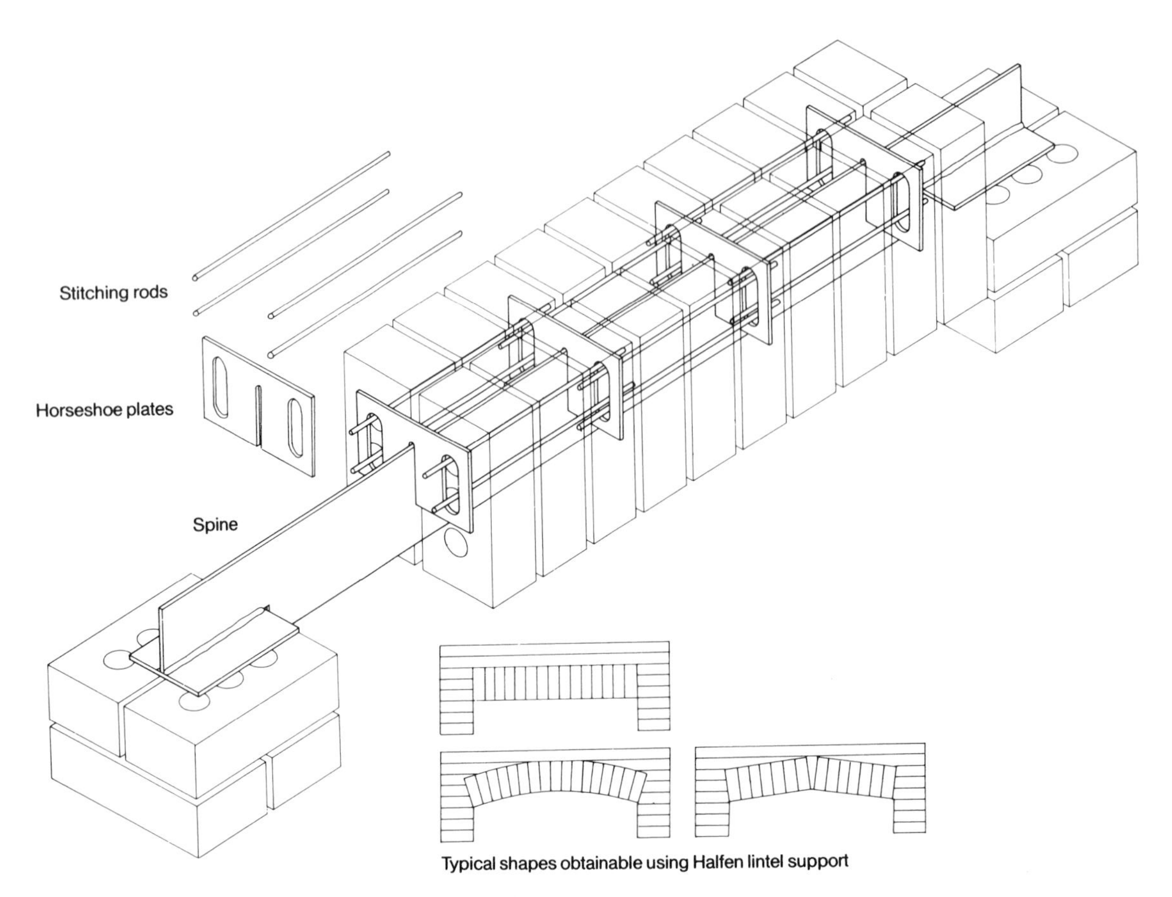

FIGURE 3-35b. The Halfen Lintel System. (Reprinted, by permission, Halfen Fixing Systems, Ltd., 1982, and Halfen, Inc., 1990).

3.10.3 Loose Lintels

Loose lintels (Fig. 3-35a) are steel angles, other steel shapes, or stone beams spanning over an opening and supported on the jamb walls. Steel lintels must be hot-dipped galvanized if the building is located in humid or corrosive atmospheres. All steel lintels must be protected by flashing and painted with a durable paint. It is recommended that the minimum size for a steel angle be 3½″ × 3″ × 5⁄16″ placed with the 3½″ horizontal. At least two thirds of the brick width should bear on the angle. The structural engineer usually includes a lintel schedule in his set of drawings.

An alternative to the familiar steel angle loose lintel has recently been introduced in the United States by Halfen Anchoring Systems of Charlotte, NC (Fig. 3-35b). This system, unlike typical loose lintels, is totally hidden from view and is suitable for openings up to 10′ wide and a thickness of at least one full brick. The Halfen Anchoring System costs more and should be considered for projects where the exposed lintel flanges would be aesthetically objectionable. Consult the manufacturer for information regarding wider openings and other anchoring systems for arches and corbels.

3.10.4 Guidelines

The following is a listing of additional guidelines to take into consideration when detailing brick walls:

1. Always keep in mind during the design process that concrete shrinks over time while brick expands. If precast elements are incorporated in the facade or if the structure is a concrete frame, isolate the two materials both vertically and horizontally to avoid cracking and leakage. Vertical soft joints are easy to achieve. Isolating the two materials horizontally can be a challenge. For concrete frame buildings, some designers allow space for a horizontal expansion joint below the shelf angle equal to the floor-to-floor height in inches × .0018. This formula provides adequate space for the combined shrinkage of the concrete and expansion of the brick.
2. Parapets, if required, should be backed by reinforced brick masonry. The BIA states that stud backing is not recommended in this location.
3. Attach windows to either the veener or the backup wall, never to both.

4. Place ties as close as possible above shelf angles and lintels. Tie the wall to the structure as well as to the studs. Also specify closer tie spacing if the cavity is wider than 3″.
5. Relate plan dimensions to brick size to avoid excessive cutting of bricks in the field. For modular bricks, dimensions should be a multiple of 4″. Use a brick scale to draw wall sections.
6. If special brick shapes are used to accommodate special angles and curves, details must be drawn to dimension each special shape. Consult with the brick representative to determine the most economical dimensions.

   Standard dimensions for the most commonly used brick are shown in Fig. 3-36.
7. Use type N (which stands for Normal) mortar for most applications. Use type S (which stands for Strong) mortar for walls extending two or more stories without the support of shelf angles or in locations exposed to wind forces in excess of 25 psf. Type S is also used to build arches. Check the code before specifying the type of mortar.
8. A crude but effective method to determine whether brick needs wetting at the site is to draw a circle on a brick using a quarter as a guide. Then place twenty drops of water inside the circle using a medicine dropper. If the water disappears in less then a minute and a half, the brick needs wetting before placement in the wall. If brick is hot or of a kind that is too absorbent, it will draw water from the mortar and result in poor adhesion and a weak wall.

3.10.5 Concrete Masonry Unit (CMU) Walls

Concrete masonry units, also referred to as concrete blocks or CMU, are versatile products used to build both exterior bearing or nonbearing walls as well as partitions. A recently developed block, the BI-X Block (Fig. 3-37), is designed to accommodate electrical wiring and piping without the need for furring. For exterior use, CMU comes in many textures and colors, including glazed, fluted, and embossed.

Insulation may be poured in the cells (perlite, vermiculite), preformed to fit in the cells (polystyrene), placed on the interior face, and covered with a vapor retarder and GWB or on the exterior face as part of an Exterior Insulation and Finish System (EIFS) such as Dryvit.

CMU and other heavy masonry walls possess thermal mass. Unlike lightweight construction such as stud walls, masonry has a tendency to store solar energy during the day and radiate it during the night. This phenomenon should be considered during the design phase if CMU is to be used as part of the exterior envelope of the project.

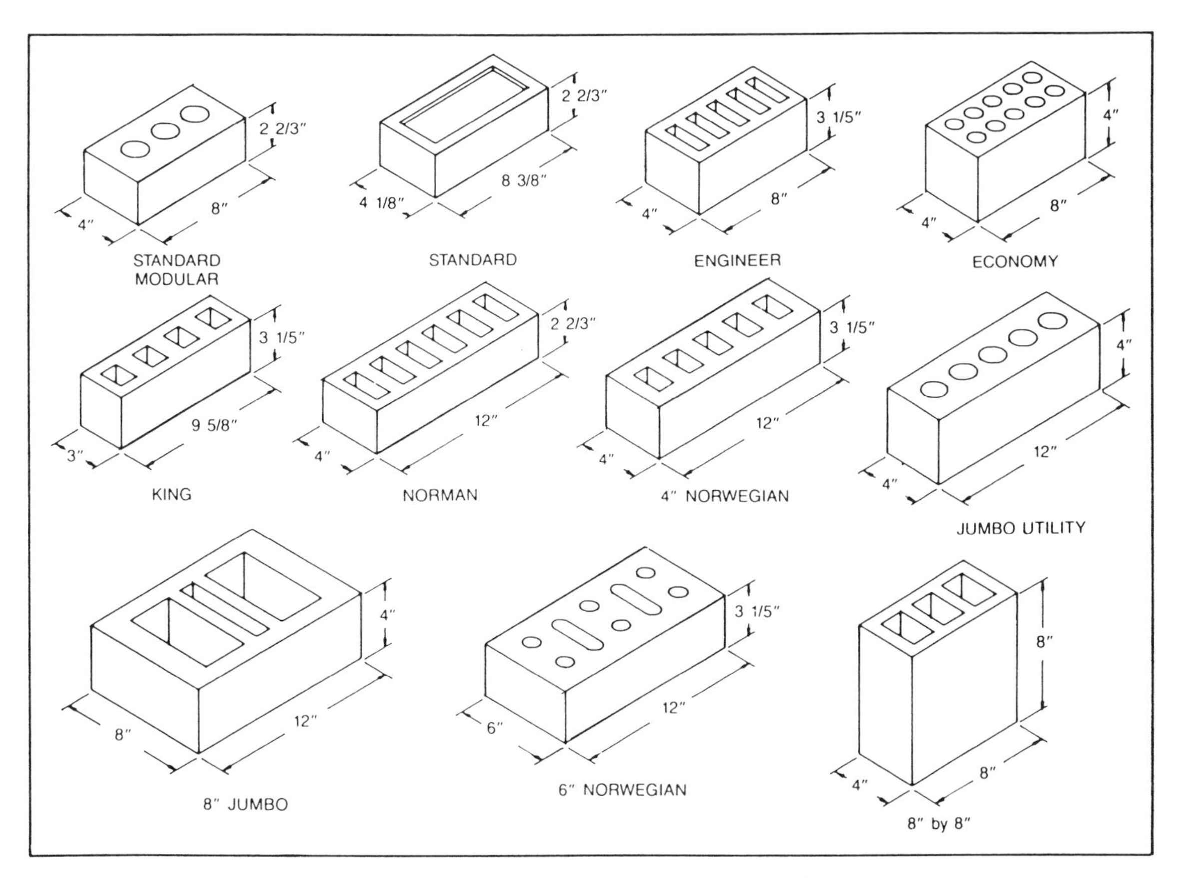

FIGURE 3-36. Brick Sizes (Nominal Dimensions) (Reprinted, by permission, from The Brick Institute of America, *Principles of Brick Masonry*).

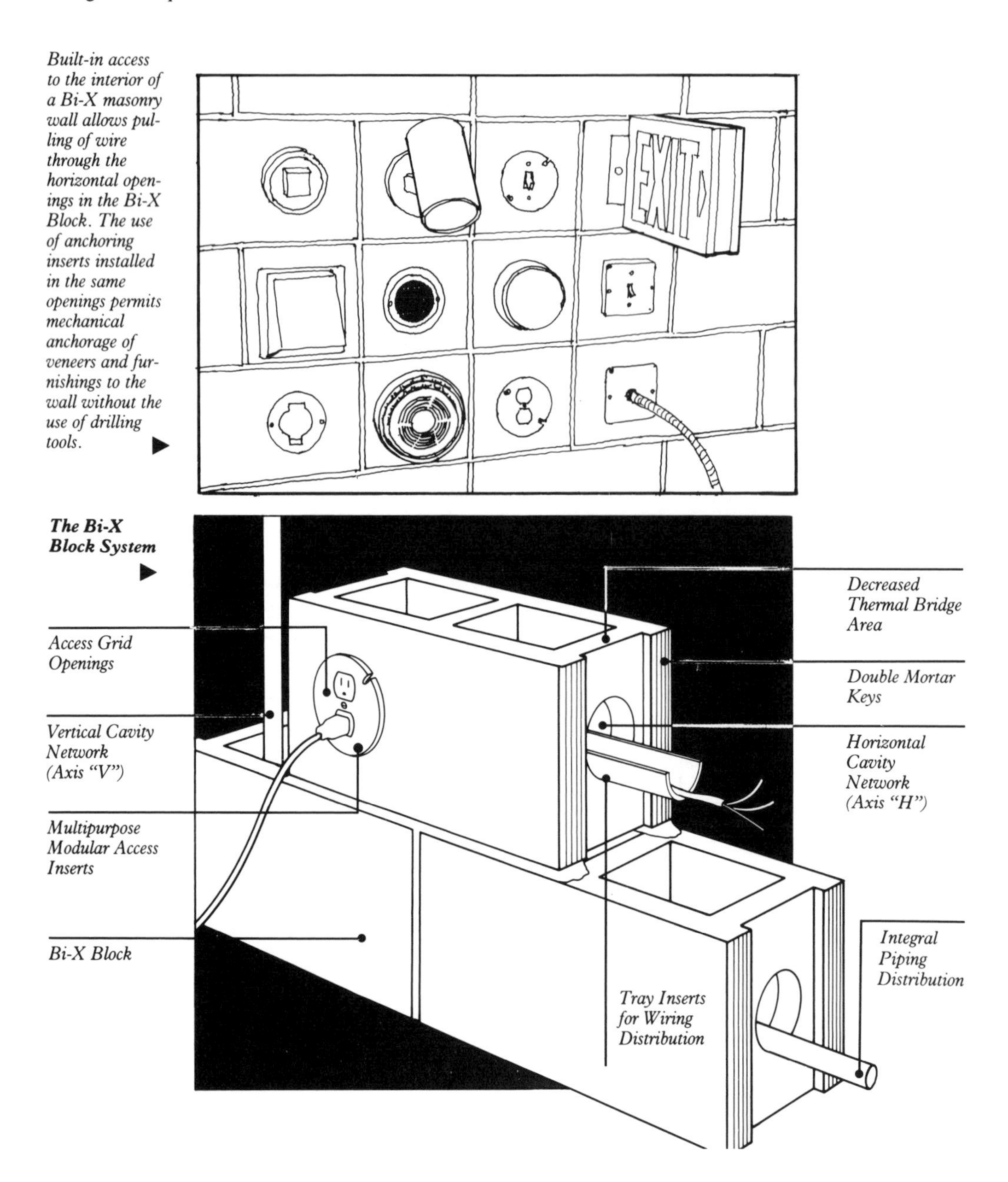

FIGURE 3-37. BI-X Block System (Reprinted, by permission, from Elmiger, *Architectural & Engineering Concrete Masonry Details for Building Construction*).

CMU walls may be reinforced and the cells filled with grout to form retaining walls. They may be used as filler blocks in floor slabs (Fig. 3-38a), sound-absorbing blocks in industrial plants (Fig. 3-38b), as backup for brick veneer, and as lintels (Fig. 3-39).

3.10.6 Guidelines

1. For dimensioning, it is advisable to use a multiple of 4″ to minimize cutting in the field. Refer to Table 3-8 for standard CMU dimensions.
2. Locate control joints at 20–25 feet to accommodate horizontal movement. If the design calls for joints at intervals of up to 40 feet, use type I moisture-controlled units instead of the more common and cheaper type II.
3. Ladder- or truss-type reinforcement is customarily placed in the mortar every two courses. Do not use truss-type reinforcement if CMU is used as a backup for brick veneer. In the presence of moisture, brick expands while CMU shrinks. Trying to tie the two with braced reinforcement such as truss type may create internal stresses in the wall that are detrimental to its structural integrity. Use ladder type instead.
4. Most codes specify the required ratio of wall span to thickness. Horizontal spans are defined as the length of wall between supports such as pilasters and intersecting

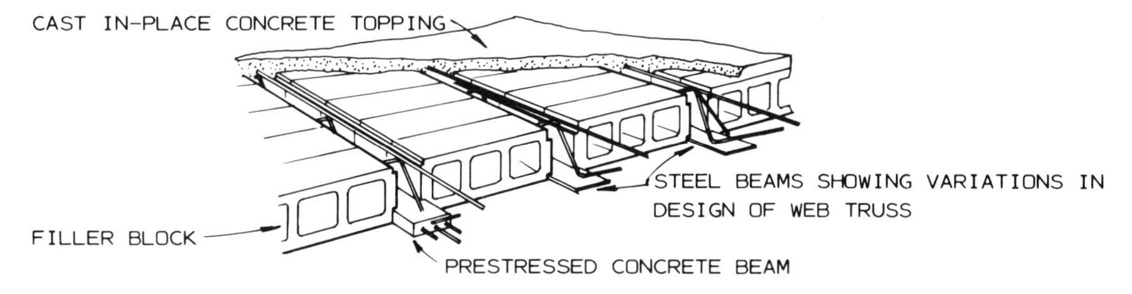

FIGURE 3-38a. Floor Slab Filler Blocks (Reprinted, by permission, from Elmiger, *Architectural & Engineering Concrete Masonry Details for Building Construction*).

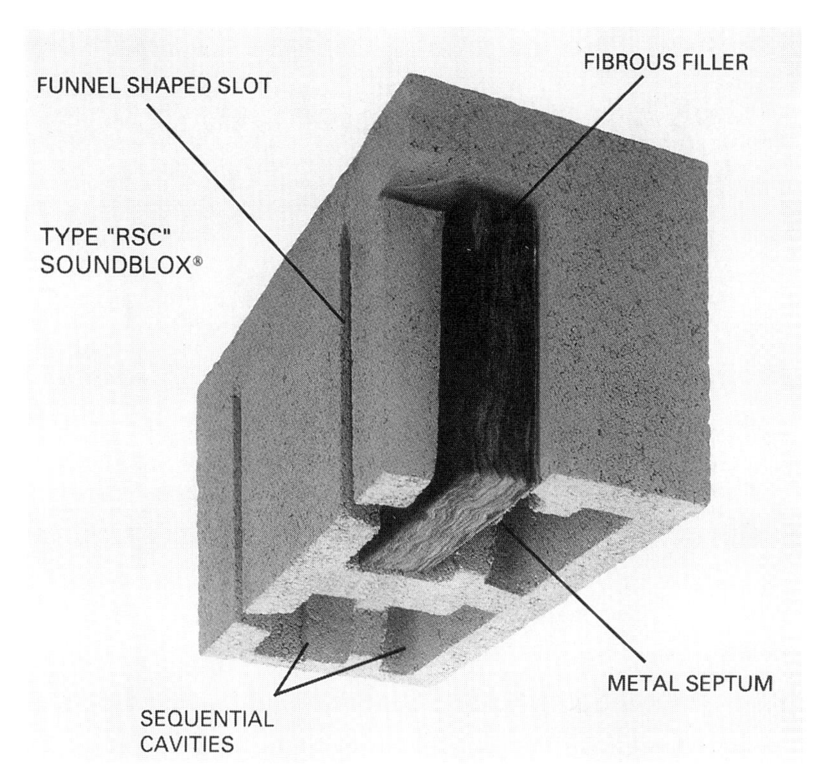

FIGURE 3-38b. Sound-Absorbing Block (*The Proudfoot Company*, 1991).

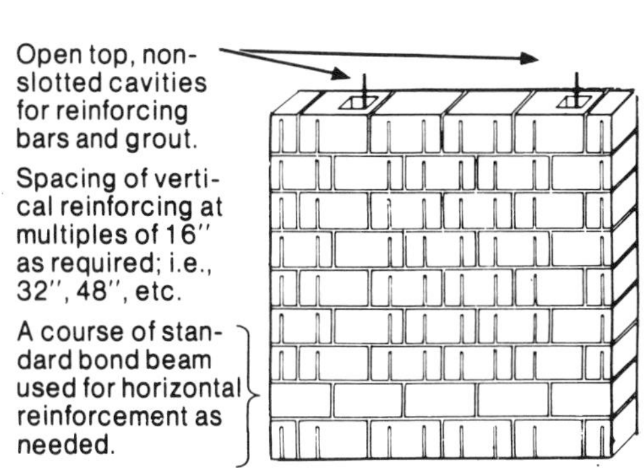

walls. Vertical spans require anchorage to beams and slabs. The National Concrete Masonry Association recommends the thicknesses shown in Table 3-9 for unreinforced walls.

5. Hollow metal door frames are usually installed by anchoring them to the floor before the CMU wall is built. T-shaped sheet metal anchors are placed in the frame and the frame is filled with grout as the wall construction progresses (see Sec. 4.4). Door height and frame face thickness must relate to the coursing of the wall (Fig. 3-40 and 3-41).
6. Refer to Table 3-10 for lintel sizes designed for CMU walls. It is customary for the structural engineer to include a lintel schedule in his/her set of drawings. Make sure that all openings are represented in the schedule.
7. Refer to Table 3-11 for STC ratings for typical CMU walls.

3.10.7 Acknowledgments

Except for items 1, 4, 5, and 7 of the Guidelines, information about brick masonry is based primarily on information and

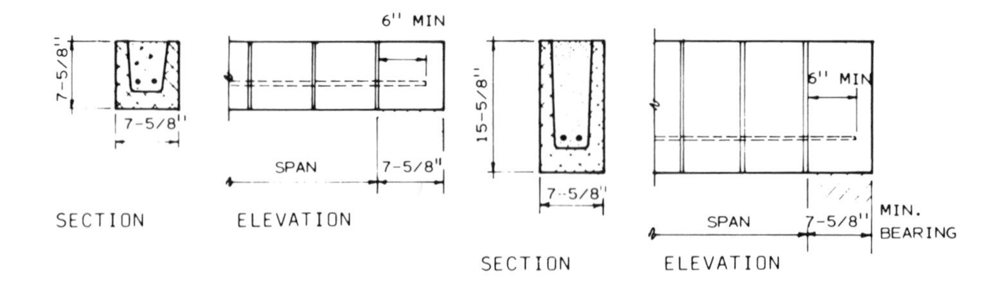

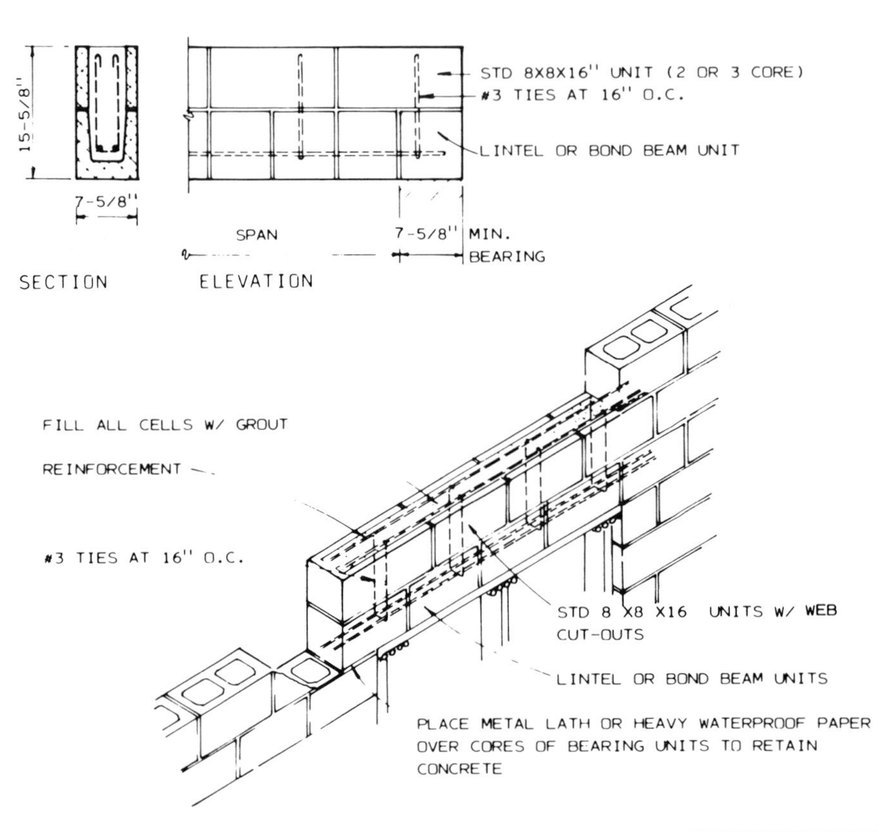

FIGURE 3-39. Lintel Block Details.

TABLE 3-8. Standard Dimensions of Concrete Block

NOMINAL DIMENSIONS OF TYPICAL CONCRETE BLOCK SHAPES

HEIGHT(H)	LENGTH(L)	WIDTH(W)
4	8	2
8	12	4
	16	6
		8
		10
		12

NOTES:

1. Actual dimension is 3/8'' less than nominal shown.
2. All shapes shown are available in all dimensions given in chart except for width (W) which may be otherwise noted. *Available in special sizes (does not refer to table shown).
3. Some manufacturers make either a two or three core unit exclusively; others make some sizes and shapes in both types and others in only one type.
4. Because the number of shapes and sizes for concrete masonry screen units is virtually unlimited, it is advisable for the designer to check on availability of any specific shape during early planning.

(Reprinted, by permission, from Elmiger, *Architectural & Engineering Concrete Masonry Details for Building Construction*)

TABLE 3-9. Lateral Support Requirements for Unreinforced Masonry

MAXIMUM RATIO OF UNSUPPORTED HEIGHT OR LENGTH TO NOMINAL THICKNESS OF WALLS

Type of Masonry	Ratio $\frac{L}{T}$
Solid masonry-bearing walls	20
Hollow unit masonry-bearing walls	18
Cavity walls	18*
Non-bearing walls	36*

* Thickness equal to sum of the nominal thickness of the inner and outer widths.

** Based on actual thickness of partition including plaster

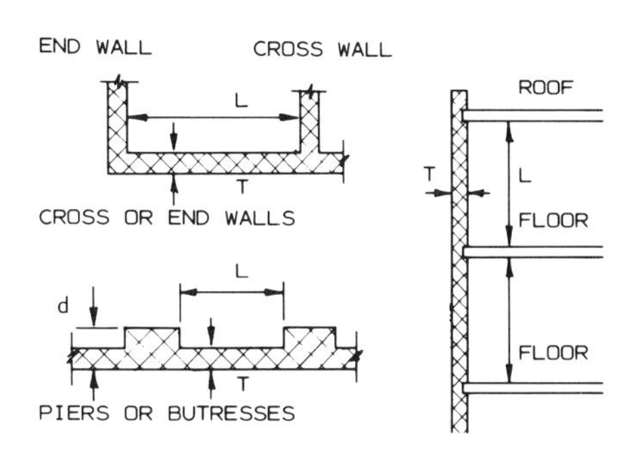

NOTE: Refer to Local Building Code for local code requirements for masonry.

(Reprinted, by permission, from Elmiger, *Architectural & Engineering Concrete Masonry Details for Building Construction*)

illustrations included in *Technical Notes on Brick Construction*, 28B Revised II, February 1987, and other publications issued by the Brick Institute of America, 11490 Commerce Park Drive, Reston, VA.

Subsections 3.10.5 and 3.10.6 are based primarily on information and illustrations taken from *Architectural and Engineering Concrete Masonry Details for Building Construction* by A. Elmiger, published by the National Concrete Masonry Association, P.O Box 781, Herndon, VA 22070

3.10.8 Sources of More Information

American Society of Testing Material. 1990. *Masonry Components to Assemblages*. STP 1063. Edited by J. Matthys. 1916 Race St., Philadelphia, PA 19103.

Beall, Christine. 1988. *Masonry Design and Detailing*. New York: McGraw-Hill, Inc.

Brick Institute of America. 1989. *Principles of Brick Masonry*. 11490 Commerce Park Drive, Reston, VA 22091

Brick Institute of America. Latest ed., *Technical Notes on Brick Construction*. 11490 Commerce Park Drive, Reston, VA 22091.

Callender, John Hancock, ed. *Time-Saver Standards*, latest edition. New York: McGraw-Hill, Inc.

Elmiger, A. 1976. *Architectural & Engineering Concrete Masonry Details for Building Construction*. Herndon, VA: The National Concrete Masonry Association.

Halfen Incorporated, P.O. Box 410203, Charlotte, NC 28241. 1-800-323-6896.

Ibstock Building Products Ltd., ed. 1989. *Brickwork Arch Detailing*. Stoneham, MA: Butterworth Architecture.

National Concrete Masonry Association, 2362 Horse Pen Road, Herndon, VA 22071. 703-713-1900.

ELEVATION

FIGURE 3-40. Butt-Type Door Frame (Reprinted, by permission, from Elmiger, *Architectural & Engineering Concrete Masonry Details for Building Construction*).

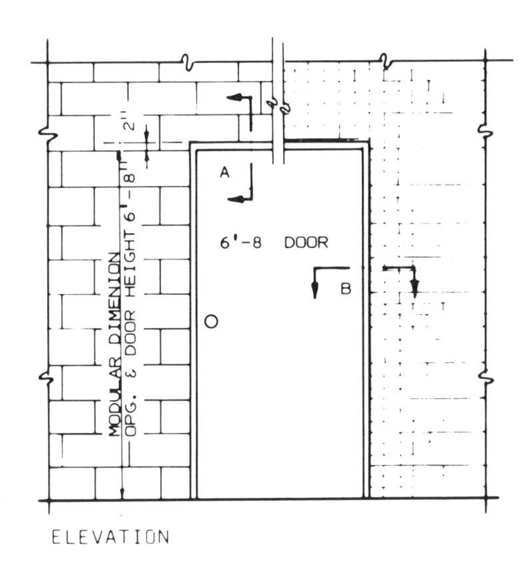

FIGURE 3-41. Wraparound Door Frame (Reprinted, by permission, from Elmiger, *Architectural & Engineering Concrete Masonry Details for Building Construction*).

TABLE 3-10. Lintel Sizes and Reinforcement

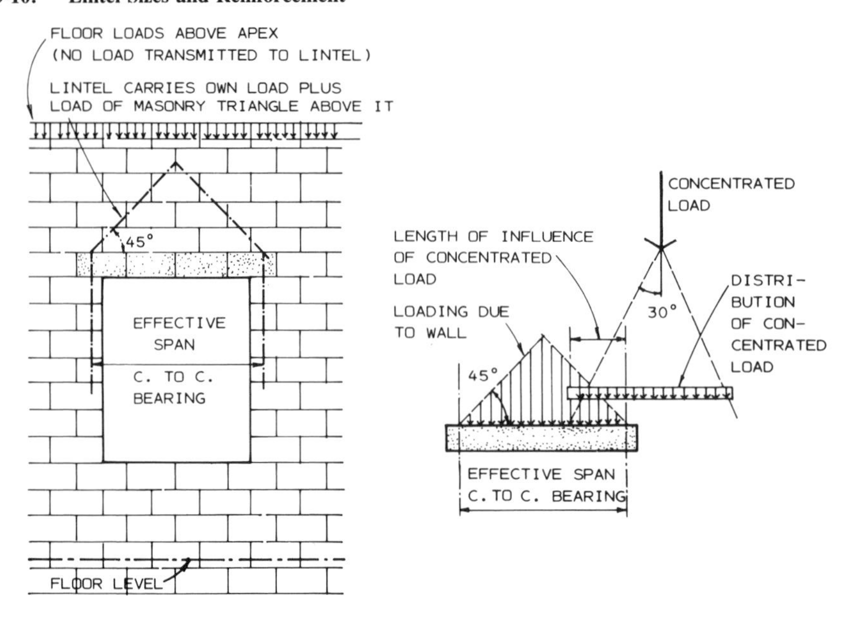

REQUIRED REINFORCING FOR SIMPLY SUPPORTED REINFORCED CONCRETE MASONRY LINTELS

Type of Loading*	Lintel Section Nominal Size (In.)	Required Reinforcing Clear Span							
		3'-4"	4'-0"	4'-8"	5'-4"	6'-0"	6'-8"	7'-4"	8'-0"
Wall Loads	6x8	1-#3	1-#4	1-#4	2-#4	2-#5			
	6x16					1-#4	1-#4	1-#4	1-#4
Floor & Roof Loads	6x16	1-#4	1-#4	2-#3	1-#5	2-#4	2-#4	2-#5	2-#5
Wall Loads	8x8	1-#3	2-#3	2-#3	2-#4	2-#4	2-#5	2-#6	
	8x16							2-#5	2-#5
Floor & Roof Loads	8x8	2-#4							
	8x16	2-#3	2-#3	2-#3	2-#4	2-#4	2-#4	2-#5	2-#5

*Includes weight of lintel

NOTES:
1. Wall loads assumed 300 lbs. per lineal foot.
2. Floor and Roof loads including wall loads assumed 1000 lbs. per lineal foot.
3. 8" lintels assumed to weigh 50 lbs./ft.
4. 16" lintels assumed to weigh 100 lbs./ft.

(Reprinted, by permission, from Elmiger, *Architectural & Engineering Concrete Masonry Details for Building Construction*)

Ramsey, Charles G., and Sleeper, Harold R. Latest edition, *Architectural Graphic Standards*. New York: John Wiley & Sons, Inc.

3.11 PRECAST CONCRETE

3.11.1 General

Precast concrete construction is both durable and versatile. It can be designed as a complete system comprising bearing exterior walls, interior columns, floor slabs, and roof. Alternatively, it can be utilized as a curtain or bearing wall attached to a steel or concrete structure. Precast panels may also be used as both the form and the veneer for concrete columns, window-shading devices, screens, and pavers.

If precast is to be used in the project, the architect should meet with several precast manufacturers as early as possible, preferably during the design development phase, to make sure that the drawings conform to industry standards, are detailed to fit the budget, and to consider other technical issues. These meetings should continue periodically during the construction drawings phase whenever a technical issue needs to be resolved.

The architect must define all precast pieces as clearly as possible. He must draw elevations, sections, and details of each piece rather than leave this task to be defined by shop drawings. This way, all bidders will base their bids on the same assumptions, avoiding future disappointments, surprises, disputes, and ultimately, costly change orders and

TABLE 3-11. STC Ratings for Typical CMU Walls

STC TESTS ON 4 INCH THICK CONCRETE MASONRY WALLS

Wall Description	Wall Weight psf	STC
No surface treatment:		
hollow units, lightweight	18	40
hollow units, normal weight	26.5	41
Surface sealed with:		
paint, both sides; hollow units, lightweight	22	43
paint, both sides; hollow units, normal weight	29	44
paint, both sides; hollow units, cores filled with sand	34	43
1/2" plaster, both sides; hollow units, lightweight	30	48
1/2" plaster, both sides; hollow units, normal weight	42	50
1/2" gyp. board, both sides; hollow units, lightweight	26	47
1/2" gyp. board, both sides; hollow units, normal weight	32	48

STC TESTS ON 6 INCH THICK CONCRETE MASONRY WALLS

Wall Description	Wall Weight psf	STC
No surface treatment: hollow lightweight units	21	44
Surface sealed with:		
paint, both sides; hollow lightweight units	28	46
paint, both sides; hollow normal weight units	39	48
1/2" plaster both sides; hollow lightweight units	31	46
plaster both sides solid normal weight units	54	52
5/8" gyp. board, both sides, hollow units	35	49
1/2" gyp. board on one side, paint on other side, hollow units	27	53

STC TESTS ON 8 INCH THICK CONCRETE MASONRY WALLS

Wall Description	Wall Weight psf	STC
No surface treatment:		
hollow lightweight units	30	45
hollow lightweight units	39	49
hollow lightweight units, cores filled with loose-fill insulation	40	51
hollow normal weight units	53	52
hollow lightweight units, fully grouted and reinforced with #5 @ 40"	73	48
composite wall, 4" brick and 4" hollow lightweight block	58	51
Surface sealed with:		
paint, both sides, hollow lightweight units	30	46
paint, both sides, hollow lightweight units	34	48
paint, both sides, hollow lightweight, full grout & reinforced with #5 @ 40"	73	55
plaster on one side, hollow units	38	52
plaster, both sides, solid block	67	56
plaster, both sides, fully grouted hollow block, reinforced with #5 @ 40"	79	56
plaster, block side — composite wall, 4" brick, 4" hollow block	61	53
1/2" gyp. board, both sides, hollow lightweight units	40	56

TABLE 3-11. STC Ratings for Typical CMU Walls

STC TESTS ON 8 INCH THICK CONCRETE MASONRY WALLS

Wall Description	Wall Weight psf	STC
1/2" gyp. board, both sides, full grout, reinforced with #5 @ 40"	77	60
1/2" gyp. board one side, paint on other, hollow units	43	50
1/2" gyp. board on block side — composite wall, 4" brick 4" hollow block	56	60

STC TESTS ON 10" & 12" THICK CONCRETE MASONRY WALLS

Wall Description	Wall Weight psf	STC
No surface treatment:		
10" cavity wall, 4" concrete brick, 2" air space, 4" hollow lightweight block	56	54
12" solid block	121	55
Surface sealed:		
10" hollow block plastered both sides	49	55
10" solid block plastered both sides	81	58
10" cavity wall, 4" brick, 2" air space, 4" hollow block; plaster on block	59	57
10" cavity wall, 4" brick, 2" cair space, 4" hollow block; 1/2" gyp. board on block	58	59
12" hollow block, painted one side	53	51
12" hollow block, painted one side, 1/2" gyp. board other side	55	57
12" hollow block, painted both sides	53	50
12" hollow block, plaster on one side	56	50
12" hollow block, plaster one side, paint on other	56	50
12" hollow block, plaster both sides	59	50
12" solid block, 5/8: gyp. board on one side	124	58

(Reprinted, by permission, from Elmiger, *Architectural & Engineering Concrete Masonry Details for Building Construction*)

delays. This is especially true for complicated designs. An axonometric drawing may also be needed to clarify intricate relationships between panels, especially those located at the corner of the building. Needless to say, if the design is based on straightforward flat panels with few openings and very little embellishment, panel-type drawings may be unnecessary.

The manufacturers' facilities should be checked to make sure that they are adequate for uniform batching and mixing, that the aggregates to be used are suitable for the desired aesthetic effect, and are available in adequate quantities to finish the work. Furthermore, the architect must verify that the work will be done under quality-control measures by experienced personnel.

The specifications, in addition to defining the quality, construction tolerances, finishes, etc., also define the responsibility for erection. This may be left up to the general contractor or be included in the precaster's work or, in some cases, assigned to a subcontractor specializing in that type of work. The specs should stipulate that the precast work be assigned to the precaster within a short, defined time period after award of the general contract. This allows enough time for the precaster to cast the units using the minimum number of forms. Duplicating forms to meet short deadlines adds substantially to the cost.

3.11.2 Responsibilities

While the architect is responsible for the design of the project as a whole, there are areas that require expertise based on specialized knowledge in the field of precasting. The following is a description of responsibilities that each of the participants undertakes before submitting the drawings for bidding:

The Architect

A. Provides drawings that define the shapes and sizes of units as well as locates joints both functional and false.
B. Selects tolerances and clearances, and specifies performance characteristics.
C. Selects color, texture, and aggregates to produce a certain appearance.

The Structural Engineer

A. Designs the structure to resist gravity, wind, and, in some cases, seismic forces.
B. Designates connection points and defines the maximum loads allowed at these points so that common information can be given to all bidders.
C. Evaluates the effects of structural movement on the performance of the cladding.
D. Sets deflection limitations.

The Precaster

A. Designs the panels for special loads, including all plant- and erection-handling loads.
B. Details bearing, tie-back, and alignment connections.
C. Selects the concrete mix, and provides for the weatherproofing and durability of the panels.
D. Prepares erection drawings, which include reinforcement and hardware based on extensive and specialized experience. Manufacturers choose details suitable for their plants' production and erection techniques. These drawings must be checked thoroughly by both the architect and engineer.

The contract documents should make it clear whether the project contract, specifications, or drawings prevail in case of conflict. The Precast/Prestressed Concrete Institute (PCI) recommends that they prevail in the order stated in the preceding sentence.

A pre-bid conference should be held at least three weeks prior to bidding. At this conference, precasters are invited to submit samples, technical literature, information about the materials they propose to use, outline their plans for the work, discuss the samples, define material sources, and guarantee adequacy of reserves. They should also provide production schedules to be reviewed and evaluated by the architect. Each sample is identified by the precaster's name, dated, given a number, and identified with the project name.

Approved precasters and sample numbers are included in an addendum (or an approved list) given in writing to the general contractor. The architect should treat pre-bid submittals in confidence and allow sufficient time for submitting samples and approvals.

3.11.3 Handling Considerations

The design of precast panels is affected by several factors other than aesthetics. Maximum unit size is governed by production repetition, ease of handling, shipping equipment, crane capacity, and loads imposed on support systems. All these considerations have an impact on construction cost. If each panel has a different design requiring a different mold, the cost of these molds adds considerably to the overall cost of the cladding. If handling requires bigger cranes or if the connection design mandates that the crane stay on the job for a longer period of time, the equipment leasing cost goes up. The architect must take all these issues into consideration during the early stages of the design process.

Another issue which is sometimes overlooked is the size and weight of each panel. The usual payload for a semi truck is 20 tons. If the panel weighs 11 or 12 tons, it is uneconomical to ship because only one can be shipped per truck. Reducing the weight to 10 tons allows two to be shipped at once. To figure the weight of the unit, calculate the cubic feet and multiply by 150 pounds. This weight includes reinforcement and hardware. The average trailer can handle a load within a volume of $8' \times 8' \times 45'$. If the height of the panel is required to be more than 8 feet, a lowboy (step deck) trailer may be used. These trailers can transport heights between 10 and 12 feet without a special permit.

To avoid excessive costs, efforts should be made to allow the panels to be handled in one motion from unloading to positioning. Construction phasing should take this into consideration. The location of a low building wing may prevent erection equipment from operating efficiently and may, in some cases, require double handling of the load (Fig. 3-42).

3.11.4 Panel Thickness and Attachment

Panel thickness is determined by the precaster to accommodate handling, hardware, the type of concrete mix, and reinforcement. Thickness varies depending on the size of the panel. The Precast/Prestressed Concrete Institute states that reasonable slenderness ratios (thickness over unsupported lengths) for flat panels should be 1/20 to 1/50 for regular

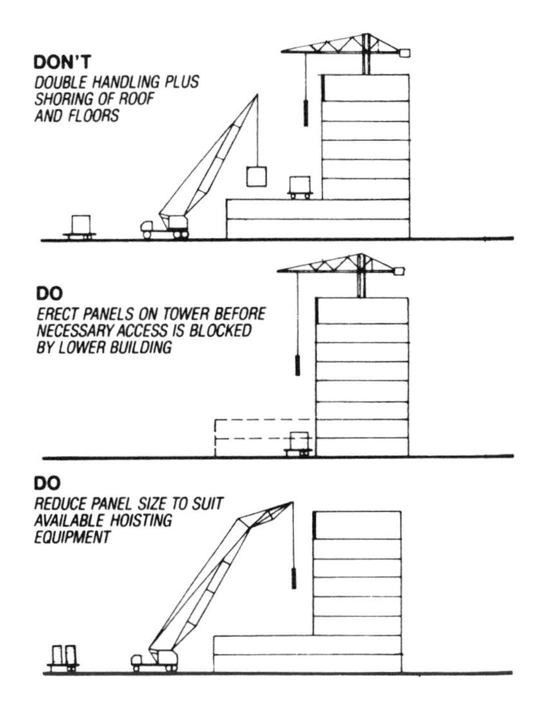

FIGURE 3-42. Access for Trucks and Hoisting Equipment (Reprinted, by permission, from Precast/Prestressed Concrete Institute, *Architectural Precast Concrete*).

panels and 1/30 to 1/60 for prestressed panels. If higher ratios are desired, a more careful structural analysis must be performed.

A minimum thickness of 3 inches is required to achieve ¾ inch minimum cover over conventional reinforcement. The designer must subtract groove depths and reduction in cover due to the use of retarders to determine this minimum dimension. The practical minimum panel thickness for flat panels must be not less than 4 inches. Panel-thickness guidelines are shown in Table 3-12. This rule does not apply to cast stone, which is a type of precast listed under the masonry section in the specifications. Cast stone may have a minimum thickness of 2 1/12 inches and is used for sills, copings, and other applications requiring relatively small precast pieces.

Figure 3-43 shows different methods of attaching panels to the structure. Where dimension "x" is too small to provide adequate resistance to rotation, lateral bracing (shown dashed) must be provided. The condition in "e" may require temporary bracing, which results in a complicated erection procedure.

If at all possible, panels should be supported on columns to avoid the possibility of deflection problems associated with beam-supported panels (Fig. 3-44). The distance between panel bearings should not exceed 30 feet in order to avoid shrinkage problems.

Finally, connections between panels should be designed to allow easy adjustment in all directions. They must also be accessible for welding, bolting, and other activities such as

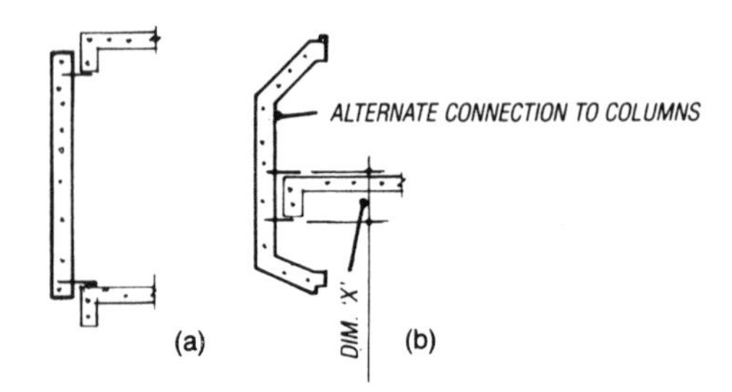

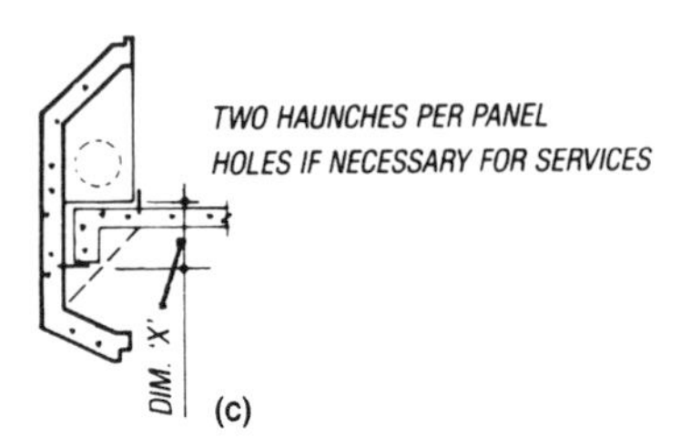

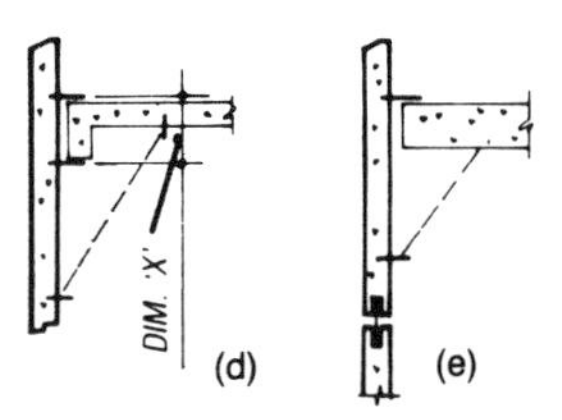

FIGURE 3-43. Panel Support Concepts (Reprinted, by permission, from Precast/Prestressed Concrete Institute, *Architectural Precast Concrete*).

TABLE 3-12 Guidelines for Panel Thickness for Overall Panel Stiffness Consistent with Suggested Normal Panel Bowing and Warping Tolerances*

Panel Dimensions**	8'	10'	12'	16'	20'	24'	28'	32'
4'	3"	4"	4"	5"	5"	6"	6"	7"
6'	3"	4"	4"	5"	6"	6"	6"	7"
8'	4"	5"	5"	6"	6"	7"	7"	8"
10'	5"	5	6"	6"	7"	7"	8"	8"

* This table should not be used for panel thickness selection.

** This table represents a relationship between overall flat panel dimensions and thickness below which suggested bowing and warpage tolerances should be reviewed and possibly increased. For ribbed panels, the equivalent thickness should be the overall thickness of such ribs if continuous from one end of the panel to the other.

(Reprinted, by permission, from Precast/Prestressed Concrete Institute, *Architectural Precast Concrete*)

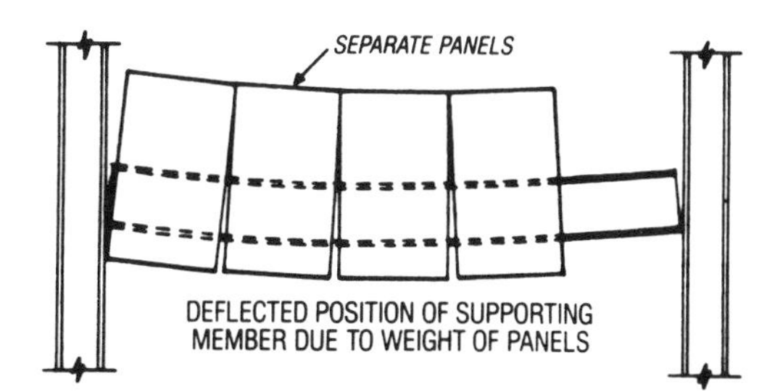

FIGURE 3-44. Deformation of Panels on Flexible Beam (Reprinted, by permission, from Precast/Prestressed Concrete Institute, *Architectural Precast Concrete*).

post-tensioning or grouting. Operations that require overhead work should be avoided especially for welding.

3.11.5 Joints

Joints must be designed to accommodate movement caused by changes in temperature, sealant capability, moisture content, and imposed loads. Joints are an important part of the architectural design of the facade. Because larger panels provide savings in erection costs, require a reduced amount of sealant, have better dimensional control, and are attached with fewer connections, it is recommended that the actual panel size be dictated by these factors. Any additional joints required by aesthetics should be accommodated by grooves known as false joints or rustications simulating the appearance of real joints. False joint depth should be kept as shallow as possible because the effective thickness of the panel is measured from the joint depression to the back of the panel.

Joints must also be designed to accommodate variations in dimensions caused by field conditions. A ¾-inch joint with a tolerance of ± ¼ inch is common. Corner panels should have enough clearance to allow alignment with both of the adjacent panels. The structural engineer should be involved in determining joint widths.

The type of sealant specified is an important factor in determining joint width. Sealants with a greater movement capability do not require as wide a joint as sealants with lesser capability. Refer to Section 3.7 for more information about sealants. The specifications should state that if the sealant has to be applied when the temperature is outside a range of 40° to 90°F, a joint wider than the one specified may have to be employed, or some similar wording. Because the construction schedule is usually determined by factors outside the control of the architect, this clause is necessary. The contractor schedules the construction sequence. If construction is scheduled during the freezing season, the precaster will construct his mold based on a modified joint width to allow for the wider joint. This clause makes provision for this eventuality.

The Precast/Prestressed Concrete Institute (PCI) recommends the following rule of thumb clearances: at least ½ inch between precast concrete members; a minimum of 1 inch between precast members and cast-in-place concrete (with 1½ inch preferred); a minimum of 1 inch between precast members and a steel frame; 1½ to 2 inches in tall, irregular structures; and 1½ inches minimum between column covers and columns, with 3 inches preferred.

Single-stage joint design is the most prevalent kind of joint used in the U.S. at the present time. This kind of joint requires periodic inspection and maintenance. For a description of a single- and two-stage joints and the rain-screen principle, refer to Sections 3.5 and 3.7.

3.11.6 Guidelines

The following guidelines should be observed:

1. If the use of sealers is considered to improve weathering, facilitate cleaning, reduce efflorescence, or for other reasons, the choice must be made very carefully. Some sealers may give the panel a permanently wet look, attract hydrocarbons, or interfere with the adhesion of sealants. In addition, sealers may act as a vapor retardant, trapping water vapor within the structure (see Sec. 3.4). If an appropriate sealer is used, low-pressure airless spray equipment should be used to apply two coats uniformly.
2. Drips should be placed no closer than 1½ inch to the edge of the panel.
3. Chamfers or radii should be used at all panel edges to reduce the potential for damage and lessen the perception of misalignments.
4. The use of white cement and special aggregates affects the cost of the panel. These ingredients enhance the aesthetics and should be considered unless the budget is extremely tight. Galvanizing the reinforcement, on the other hand, adds substantially to the cost. It is usually not required if adequate concrete cover is specified and quality is controlled.
5. Air entraining for face mixes is advisable where the panels are exposed to freeze-thaw cycles.
6. Insulate exterior columns to minimize temperature differentials between these columns and the rest of the enclosed frame.
7. Flat panels meeting at a building corner should be mitered to form a corner quirk. Figure 3-45 shows recommended miter dimensions.
8. The concrete mix used to manufacture stone-veneered panels should closely approximate the shrinkage and thermal expansion coefficient of the stone. The responsibility for stone coordination should be written into the specs so that its cost can be bid. Prestressing is effective in controlling bowing, especially if brick or tile is used as facing.
9. Using the least number of molds should be set as a goal for the designer. A master mold, also referred to as an envelope or box mold, may be used repeatedly. It is

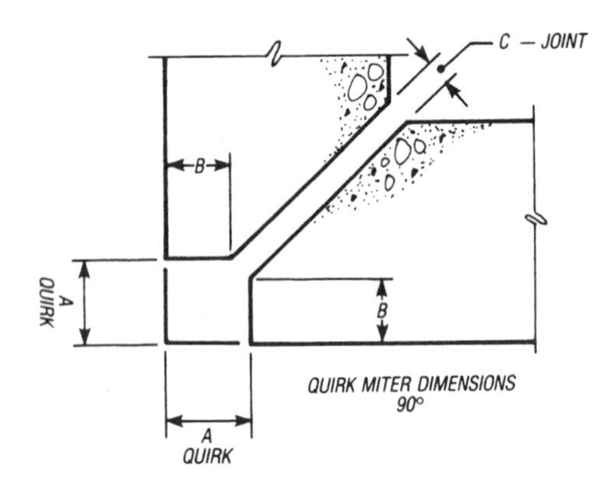

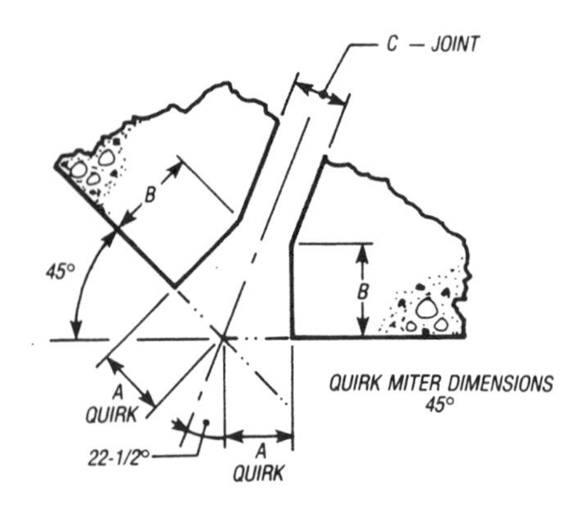

A	B	C
1 1/8	3/4	1/2
1 1/4	7/8	1/2
1 1/2	1 1/8	1/2
1 3/4	1 3/8	1/2
2	1 5/8	1/2

A	B	C
1 1/4	3/4	3/4
1 1/2	1	3/4
1 3/4	1 1/4	3/4
2	1 1/2	3/4
1 1/2	13/16	1
1 3/4	1 1/16	1
2	1 5/16	1

A	B	C
5/8	13/16	1/2
3/4	1 1/8	1/2
7/8	1 7/16	1/2

A	B	C
3/4	13/16	3/4
7/8	1 1/16	3/4
7/8	13/16	1
1	1 1/16	1

FIGURE 3-45. Typical Quirk Details and Dimensions (Reprinted, by permission, from Precast/Prestressed Concrete Institute, *Architectural Precast Concrete*).

relatively easy to alter a mold if variations can be done within its boundaries using bulkheads or blockouts. As mentioned before, enough time should be allowed for the precaster to use this costly mold. If this is not done, many identical molds will have to be built to meet the shorter deadline, causing a substantial increase in the cost. One mold is good for 60–75 castings.

For ease of stripping the panel from the mold, cant the sides a minimum 1 inch per foot to form a draft. A more generous cant is required for delicate pieces.

D.A. Sheppard, a consulting structural engineer, supplied the following recommendations at a seminar conducted by the PCI. He mentioned that these measures are an effective deterrent to inferior work on projects where the architect or engineer does not have control over who can or cannot bid for the project.

Summary and Recommendations for Effective Quality Control of Plant Cast Precast Concrete Projects (reprinted by permission)

1. Fabricators should have a minimum of five years continuous experience in the manufacture of plant-fabricated architectural precast concrete products, and be producer members of either the Architectural Precast Association or the Precast/Prestressed Concrete Institute. Project specifications should also require precast concrete producers to have written quality control procedures with full time quality control employees, or pay for the cost of full time plant inspection by an independent testing laboratory approved by the Engineer of Record.
2. Precast plants should maintain files of material tests and production tests as a part of quality control program documentation.
3. All welded reinforcing bars will be ASTM A706 grade; no tack welding of prefabricated reinforcing bar cages will be permitted.
4. Concrete mix designs should be submitted for approval of the Engineer of Record prior to fabrication, so quality control inspectors can effectively monitor batch plant or transit mix batching operations.
5. Concrete batching operations should have a regular procedure and method for correcting water addition quantities for aggregate surface moisture. The use of superplasticizers should be mandatory to assure low water-cement ratios and workability. Slumps should be less than 2″ before addition of superplasticizers.
6. Product tolerances should be in accordance with PCI MNL-116 and MNL-117; erection tolerances and tolerances with interfacing materials and systems should be in accordance with MNL-127, the *PCI Design Handbook*, and ACI 117.
7. Plant and field patching and crack repair procedures shall be submitted for the architect's and engineer's information prior to commencing fabrication and erection, respectively.
8. Elastomeric bearing pads shall meet AASHTO standards, substantiated by a manufacturer's certificate of compliance, and physical testing.

9. Field handling and bracing shall be shown on erection drawings and specific bracing drawings, prepared by a licensed civil engineer, in accordance with OSHA requirements.

Acknowledgment

Most of the material included in this section is based on information included in *Architectural Precast Concrete*, a publication issued by the Precast/Prestressed Concrete Institute in 1989.

3.11.7 Sources of More Information:

Ambrose, James. 1991. *Simplified Design of Concrete Structures*, 6th ed. New York: John Wiley & Sons, Inc.

American Society of Testing Material (ASTM). 1991. *Exterior Wall Systems*. STP 1034. 1916 Race St., Philadelphia. PA 19103. 215-299-5400.

Precast/Prestressed Concrete Institute (PCI). 1989. *Architectural Precast Concrete*, 2nd ed. Edited by Sidney Freedman and Alan R. Keeney. Chicago: PCI.

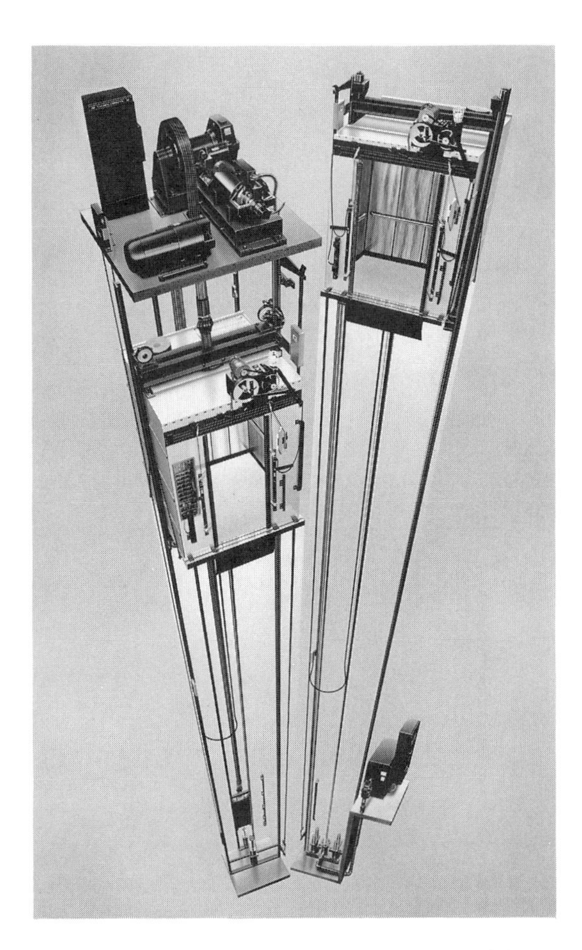

FIGURE 3-46. Types of Elevators (Courtesy of Dover Elevators).

3.12 PASSENGER ELEVATORS

3.12.1 General

The vertical transportation of people and materials to otherwise unobtainable heights was extremely hazardous until Elisha Graves Otis invented the elevator safety in 1852. It consisted of a wagon spring and a rack with teeth on the guide rails. Since that date, elevators have evolved into a complex system utilizing advanced electronics and computer technology. Elevators had a profound effect on the shape of our cities and the density of developments. The current sensitivity to the needs of the physically handicapped has accelerated the use of elevators even in two-story structures. To qualify as handicapped-accessible, the elevator must be equipped with a gong to sound at each floor for the hearing impaired. There must be raised lettering at the control panel, at call buttons, and on the door jamb to indicate the floor designation at each level for the visually impaired. In addition, the door and cab must be sized to accommodate a wheelchair. In most codes, high-rise buildings also require the following provisions: a cab that is able to accommodate a stretcher (this requires certain dimensions for both the cab and the door); venting of the shaft; provisions to override the elevator cab controls in case of fire, to make the elevators accessible to fire fighters from the fire access floor (this usually means that all the elevators can be recalled to the fire access floor—(usually the ground floor); a limit on the number of elevators occupying a single hoistway (usually three); a specified number of elevators are to be connected to the emergency power supply; and the elevator lobby may be required to be separated by a rated wall and doors. Check the building code for specific requirements.

3.12.2 Elevator Systems

There are two general systems—traction and hydraulic (Fig. 3-46). Traction elevators are further divided into geared and gearless elevators. Geared elevators are usually used in medium-rise buildings with elevator speeds up to 350 feet per minute (fpm) and load capacities up to 30,000 pounds. In geared elevators, the machines are connected to the drive sheave by means of a gearbox to provide high gear-reduction ratios (Fig. 3-47). Gearless elevators can be used in buildings of any height and for speeds up to 1,800 fpm. The machine is called gearless because the motor drives the sheave directly without the need for reduction gears (Fig. 3-48).

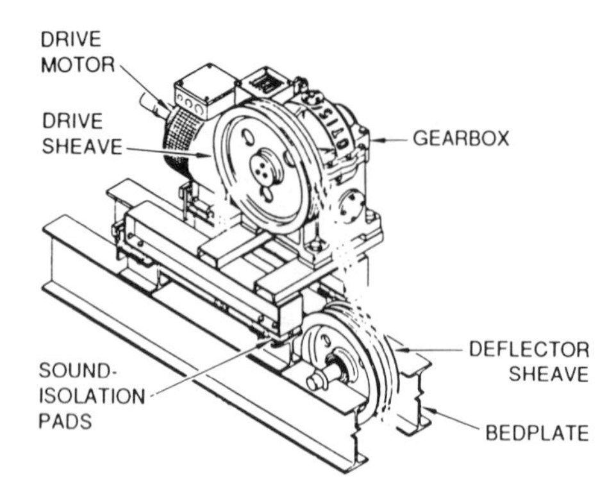

FIGURE 3-47. Geared Elevator Machine.

Hydraulic elevators (Fig. 3-49) use a radically different system to move the cab. (Consult manufacturer's literature for maximum distance.) Their speed ranges from 75 to 150 feet per minute. Hydraulic machines are comprised of a motor, pump, and valve assembly connected to the "jack" or plunger assembly that moves the elevator cab. Different-size pumps deliver output capable of providing the required speed for the designated load capacity. In cases where the plunger casing cannot be placed under the pit, a two-plunger arrangement inside the hoistway is available (Fig. 3-50). Roger M. Sigler, sales services coordinator at Dover Elevators, advises that "hydraulic elevators should be used in buildings with seven landings or less. Geared elevators for twenty-seven landings or less. Anything above this should be gearless."

3.12.3 Machine Rooms

Traction elevator machine rooms (Fig. 3-51) contain the following components:

1. The hoisting machine, which may be geared or gearless (gearless shown).
2. The car controller, which starts and stops the car, initiates door operation, and communicates with the passengers

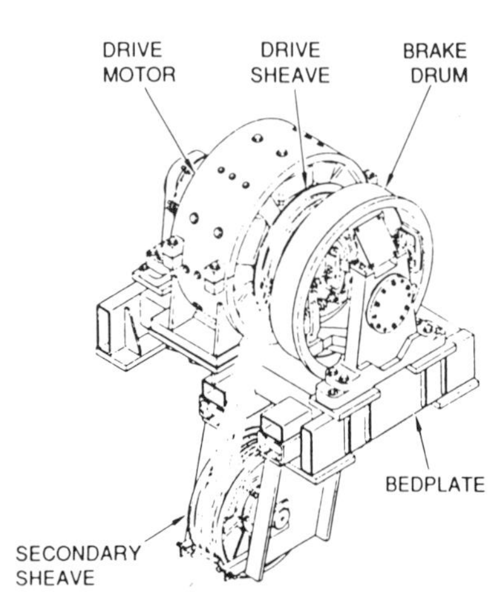

FIGURE 3-48. Gearless Elevator Machine.

Hydraulic Elevators

Hydraulic elevators have a **plunger**(s), to provide the lifting force, and **piping** to the hydraulic oil supply installed in the pit. In conventional systems, the plunger is lowered into a hole or casing in the center of the pit floor. A pair of spring-type **buffers** are installed on either side of the plunger.

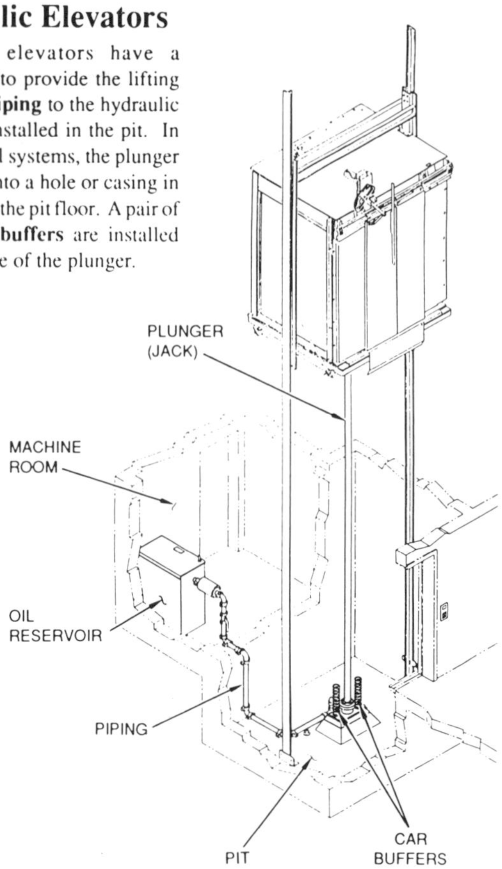

PLUNGER (JACK)
MACHINE ROOM
OIL RESERVOIR
PIPING
PIT
CAR BUFFERS

FIGURE 3-49. Hydraulic Elevator.

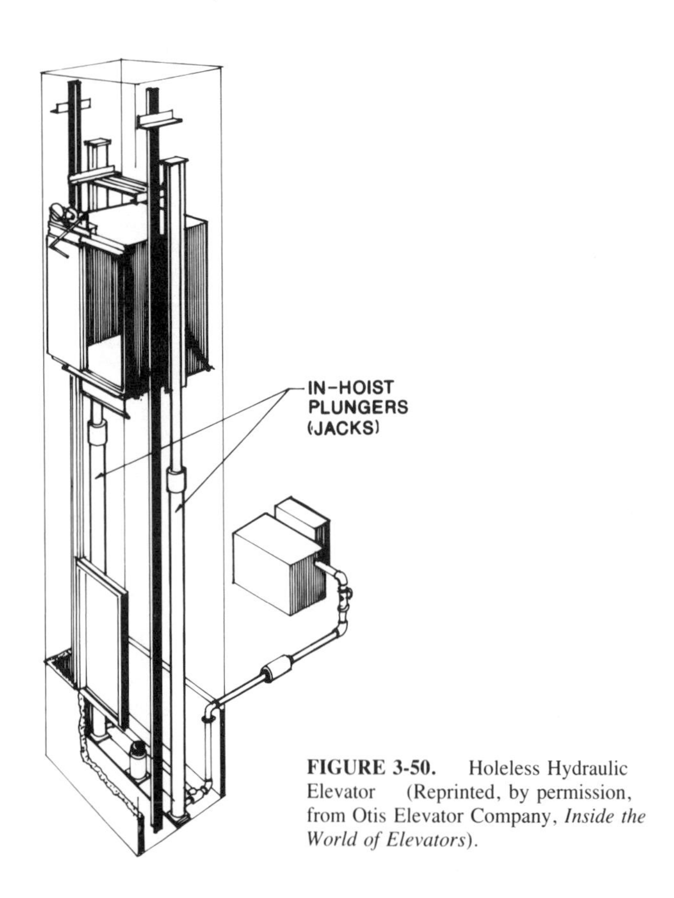

FIGURE 3-50. Holeless Hydraulic Elevator (Reprinted, by permission, from Otis Elevator Company, *Inside the World of Elevators*).

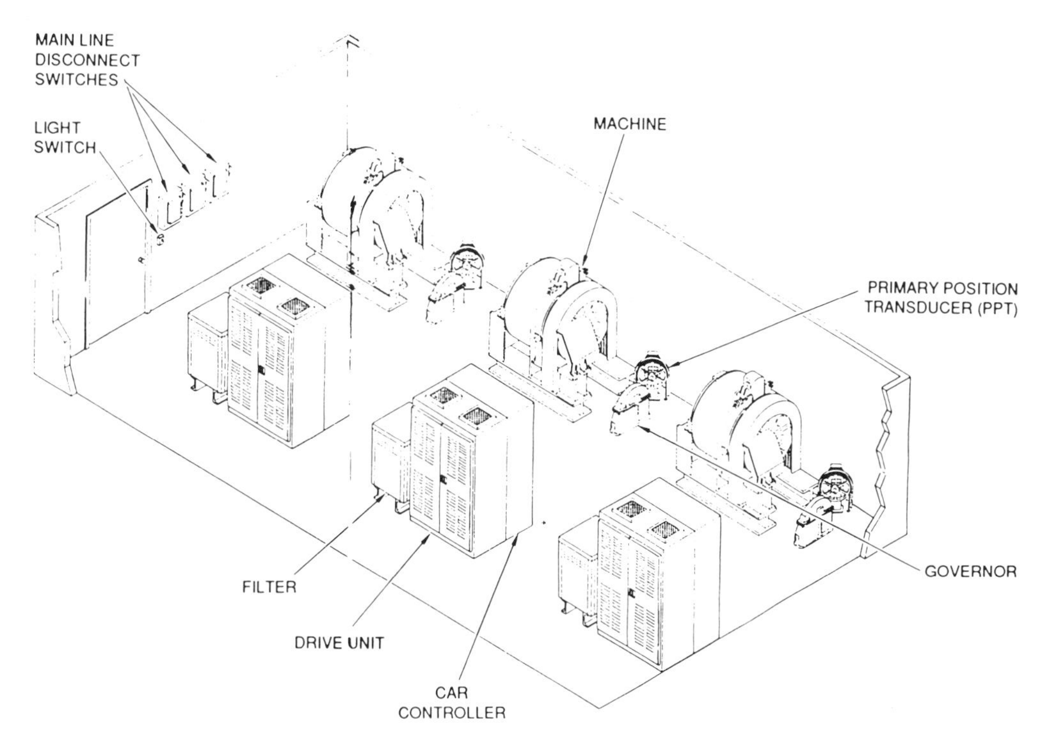

FIGURE 3-51. Traction Elevator Machine Room.

by providing car position information visually accompanied by an audible tone. It also determines whether operation of the car is safe.

3. The drive unit is required for high-rise, high-speed elevators only. It contains high-voltage, high-current equipment for each car.
4. The governor is a speed-sensing device. If the car overspeeds, the governor overspeed switch turns off power to the machine and activates the brakes attached to the machine. If the car speed continues to increase, the governor trips the safety rail arrestors to bring the car to a stop.
5. The primary position transducer (PPT) is a device that allows the car to stop smoothly and level with the landing. Transducers differ, depending on the kind of installation.
6. Main-line disconnect switches permit all power to be removed from the system in emergencies. These switches must be located within reach of the access door.

The size and layout of the elevator machine room are determined by each manufacturer. Be sure to consult the catalogs of several manufacturers before determining the dimensions of the room. Because the height requirement from the top floor landing to the top of the shaft is usually higher than the height of a typical floor, the machine room floor may be located several feet above the roof level. This information should be transmitted to the structural engineer together with information about the weights imposed on the machine room floor, the railings, and the hoist beam located above the machine. Power requirements should also be given to the electrical engineer.

The hydraulic elevator machine room (Fig. 3-49) is located adjacent to the pit. It includes a hydraulic fluid reservoir containing the pump that drives the fluid to the plunger or "jack" that sets the car in motion. The machine room walls are usually fire-rated. Traction elevator machine rooms may be placed next to the pit if dictated by special circumstances. Consult the elevator representative for cost differences, which are quite substantial, and space implications. This arrangement is sometimes referred to as an underslung elevator.

3.12.4 Elevator Pits

Elevator pits contain car and counterweight buffers that absorb the shock should a car fall to the bottom of the shaft. The drawing shown in Fig. 3-52 represents a high-rise installation. Pit depths for traction elevators vary from 4′-6″ to 13′-0″ or more. The higher the travel distance, the deeper the pit and the more potent and elaborate the buffer. The pit is usually 4′-0″ deep for hydraulic elevators.

In addition to the buffers, the pit contains a pit ladder for each elevator (Fig. 4-23), an emergency stop switch on one side of the pit for access and safety, and a sheave to keep the tension on the governor rope (traction elevators only). High-rise installations have other sheaves. To detail the pit, draw a plan showing the location of the ladder(s), pit floor slope, sump (if required), and location of the door(s).

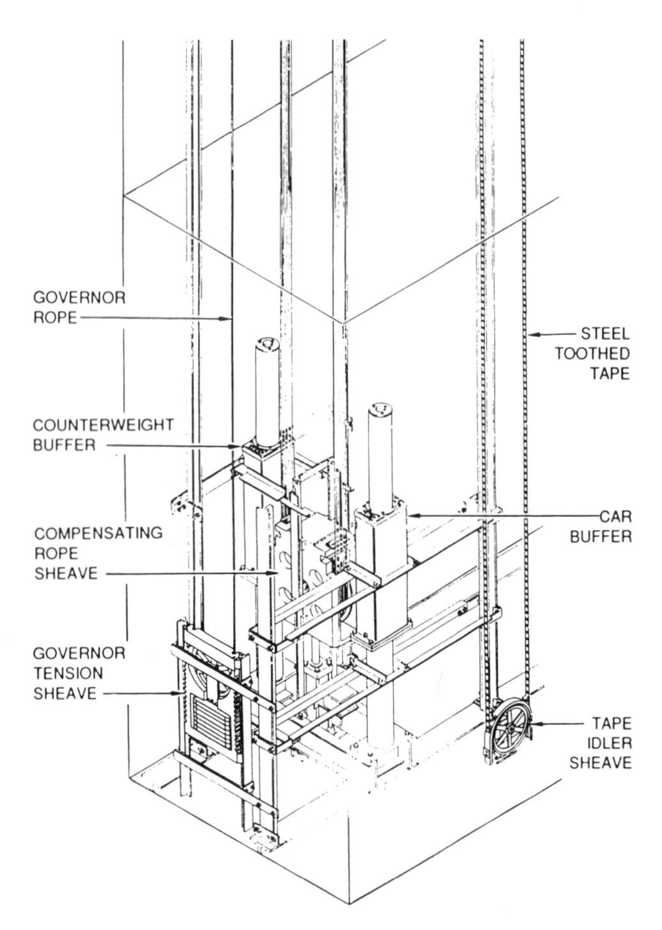

FIGURE 3-52. High-Rise Elevator Pit.

3.12.5 Hoistways

The hoistway or elevator shaft is surrounded by a fire-rated wall designed to resist the air pressure created by the movement of the car(s). Shaft Wall manufacturers provide tables relating wall thickness to elevator speeds (see Sec. 4.2).

3.12.6 Future Developments

One of the most exciting prospects for the future is a system developed by Nippon-Otis Elevator Company. The system places a linear induction motor in the counterweight assembly. The advantages are that the system does away with the costly machine room and uses less energy.

3.12.7 Acknowledgments

Description of components in this section are based on information taken from *Inside the World of Elevators*, Otis Elevator Company, 1990.

References

American Architectural Manufacturers Association. 1979. *Aluminum Curtain Wall Design Guide Manual*. Des Plaines, IL: American Architectural Manufacturers Association.

The American Institute of Architects. 1980. *Document D101*. Washington, D.C.: The American Institute of Architects.

American Society of Heating, Refrigeration and Air Conditioning Engineers, Inc. 1989. *ASRAE Handbook of Fundamentals*. Atlanta, GA: American Society of Heating, Refrigeration and Air Conditioning Engineers.

Backenstow, Don. "White Roofs vs Black: Which Saves More Energy?" *Exteriors* (Spring 1988): 44–47.

Brick Institute of America. 1989. *Principles of Brick Masonry*. Reston, VA: Brick Institute of America.

Brick Institute of America. 1987. *Technical Notes on Brick Construction*. 28B Revised II. Reston, VA: Brick Institute of America.

Building Owners and Managers Association. 1980. *Standard Methods for Measuring Floor Area in Office Buildings*. ANSI Z65.1. Reprinted May 1981. Washington, D.C.: BOMA.

Coursey, Richard. "Keeping Wind From Raising the Roof." *Architecture* (August 1988): 102–103.

Dalgliesh, W. A., and Garden, G. K. 1968. *Influence of Wind Pressures on Joint Performance*, CIB Symposium on Weathertight Joints for Walls, Oslo, Norway, September 1968. NBRI Report 51C. "Weathertight Joints for Walls," No. 26D. (NRC 9873).

Elmiger, A. 1976. *Architectural & Engineering Concrete Masonry Details for Building Construction*. Herndon, VA: The National Concrete Masonry Association.

Gordan, Douglas E., and Stubbs, M. Stephanie. "Longevity and Single-Ply Roofing." *Architecture* (January 1988): 95–99.

Griffin, C. W. "Flashing, The Designer's Toughest Task." *Roof Design* (June 1983): 50–60.

Griffin, C. W. "Reducing the Risk of Splitting in BUR." *Roof Design* (December 1983): 32–41.

Hardy, Steve. "Evaluating Roof Insulation Systems." *Architecture* (August 1989): 103–105.

Maslow, Philip. 1982. *Chemical Materials for Construction*. New York: McGraw-Hill, Inc.

McDonald, Timothy B. "Flashing For Built-Up Roofs." *Architecture* (October 1987): 105–106.

McQuiston, Faye C., and Parker, Jerald D. 1988. *Heating, Ventilating and Air-Conditioning, Analysis and Design*, 3rd ed. John Wiley & Sons, Inc.

Mobay Chemical Corporation. 1978. *Urethane Insulation Energy Saver Manual for Commercial, Institutional, Industrial and Residential Construction*. Pittsburgh, PA: Mobay Corporation.

Moreno, Elena Marcheso. "Roofing That Responds to Specific Climatic Conditions." *Architecture* (September 1987): 116–18.

National Roofing Contractors Association. "Built-Up Roofing System Guidelines." *Roof Design-Handbook* (1985): 26–36.

Otis Elevator Company. 1990. *Inside the World of Elevators*. Farmington, CT: Otis Elevator Company.

Precast/Prestressed Concrete Institute (PCI). 1989. *Architectural Precast Concrete*, 2nd ed. Edited by Sidney Freedman and Alan R. Keeney. Chicago: PCI.

Russo, Michael. "A Critical Look at Thermoplastic Roofing." *Exteriors* (Winter 1987): 24–27.

Shipp, Paul H., and Marchello, Maurice J. 1989. "What You Ought to Know About Air Barriers and Vapor Retarders." *Form & Function*. Construction Technology Laboratory, USG Research Center, United States Gypsum Corporation.

The Single-Ply Roofing Institute. "Guide to Specifying Single-Ply Roofing Systems." *Roof Design-Handbook* (1985): 20–24.

Solomon, Nancy B. "Roofing Primer, A Guide to Designing and Specifying Low-Slope Commercial Roof Membranes." *Architecture* (April 1991): 103–105.

Stephenson, Fred. "Design Considerations For Standing Seam Roofs." *Roof Design* (September 1982): 46–52.

Stephenson, Fred. "Designing With Metal Roof Systems." *Roof Design-Handbook* (1985): 38–45.

Stephenson, Fred. "Specifying Standing Seam for Conventional Buildings." *Roof Design* (March 1984): 44–52.

Szoke, Stephen, and McDonald, Hugh C. "Combining Masonry and Brick." *Architecture* (January 1989): 103–106.

Tobiasson, Wayne. 1989. *Vapor Retarders for Membrane Roofing Systems*. Misc. Paper 2489. Reprinted from the Proceedings of the 9th Conference on Roofing Technology with permission from the National Roofing Contractors Association by U.S. Army Corps of Engineers, Cold Regions Research and Engineering Laboratory (CRREL), Hanover, NH.

Unimast, Inc. 1991. *Steel Framing Systems: Technical Information*. Franklin Park, IL: Unimast.

Wagner, Michael. "Making It To The Top." *Architecture* (May 1990): 84–89.

Warsech, Karen. "Why Sealant Joints Fail." *Architecture* (December 1986): 100–103.

STANDARD DETAILS

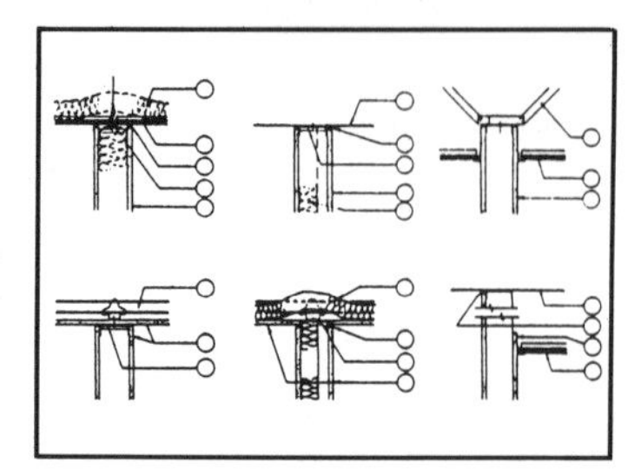

Good detailing practice requires that the architect take into consideration the effect natural forces have on buildings. Details must allow for expansion and contraction caused by temperature changes by providing expansion and control joints conforming to standards set by each manufacturer. They must provide tolerances that allow for imperfections resulting from misalignments, imperfect execution in the field, and settlements. Details must also conform to code requirements and be based on materials and assemblies that resist water, air penetration, vapor diffusion, and the effects of the freeze-thaw cycles. The following factors affect the detailing process:

Cost

Staying within the project budget must always be in the back of the mind of the person developing the details. First-class materials require commensurate expenditure. Overdesigned assemblies affect the overall cost of the project.

Sometimes it is a good idea to use expensive materials in areas with high visibility and balance them with less costly construction in unimportant parts of the project.

Durability and Weatherability

Products and manufacturers with a good track record backed by good warranties are preferred over new, untried products. Keeping abreast of technical developments is essential for a better understanding of what to include in the details to postpone the deterioration caused by the passage of time and the forces of nature.

Maintenance

Accessibility is necessary for maintenance of items that require periodic servicing to prevent corrosion. Items that cannot be accessed, such as masonry ties, should be manufactured from durable materials.

Compatibility With Adjacent Materials

Different kinds of metals when placed against each other in the presence of moisture cause corrosion of the weaker metal through a process called galvanic action or electrolysis. Special coatings or tape must separate these metals. Sealants must be compatible with the materials they come in contact with. Consult with the manufacturers before including these items in the specifications.

Redundancy

This is sometimes referred to as the "belt and suspenders" approach. It is a kind of insurance against failure. Providing gutters under a multiple skylight roof mullions is an example. If the skylight gaskets develop a leak, the gutters prevent the water from dripping into the space below.

Energy Conservation

Using double-glazing in thermally broken metal frames and adequate insulation in the walls is an example of energy conservation, especially in cold regions.

The choice of environmentally safe materials is the duty of every architect.

Conformance to Industry Standards

There is an association, institute, or other organization for just about every material used in the construction industry. These organizations issue informative publications that are very useful to the individual entrusted with developing the details. Some of these are listed under Sources of More Information in Chapter 3.

Aesthetics

It is easy to get carried away by the many factors that shape the details and end up with something that looks entirely different from the design intent. One must always strive to satisfy the aesthetics set by the designer without losing sight of the cost and technical considerations.

Sequencing

It is good to know which component or assembly is normally set in place first to avoid ending up with a non-buildable detail. Try to avoid developing details that require a trade to come back repeatedly to the job site to do work that could have been done without interruption. For example, in a building with a steel frame skeleton, designing a raised elevator machine room floor slab to be supported on CMU walls requires that the main floor slab be poured first. This would be followed by the masons (who may have already left the job) to build the supporting walls, then the concrete subcontractor is summoned back to pour the raised floor. It is better to design the steel frame to support the raised slab and pour the concrete without interruption. Field observation is helpful in that regard.

Current Methods of Producing Details

Large, well-organized offices have a distinct advantage over most smaller offices when it comes to saving time on details. Over time, they have accumulated a large number of details, organized them into an easy-to-scan-and-retrieve library that can be used on a project with a few modifications or none at all.

This chapter attempts to help smaller offices compete on a more even footing by providing them with the nucleus of a library of standard details that are almost identical in each project. Examples of these details are stair and elevator details, partition types, toilet room, and door details.

Most offices use one of the following methods to produce details:

Method 1

Develop details from scratch for each project. This is the old "reinventing the wheel" method of detailing. It is time-consuming, tedious, and repetitive.

Method 2

Copy details from past projects. While this method is an improvement over Method 1, it still requires time to modify the detail and change dimensions. In the end, the net time-saving may be negligible because it takes time to look for the detail in addition to the required alterations. Add to this the hazard that some vestige of the old project that does not apply to the project at hand may be overlooked, resulting in embarrassment and costly mistakes.

Method 3

Establish a library of generic details with an easy-to-use well-indexed retrieval system. This method avoids the shortcomings of the preceding ones. It is well suited to both CADD and manual methods of drafting. It is a true time-saver because the details, in most cases, do not need alterations and the indexed retrieval system makes it easy and fast to find the required details. This is the method used in this chapter.

Indexing

Indexing can be done using one of the following methods:

Method 1

The most advanced method of indexing is to store all the standard details in the computer, numbered according to the CSI (Construction Specifications Institute) numbering system. To retrieve the detail, the team member punches the number on the computer keyboard and includes it directly on the project sheet after modifying it, if necessary.

Method 2

All the standard details are reduced to one quarter size and placed on 8 ½″ × 11″ sheets according to their type—stair details, door details, etc. Each detail is given a number and each 8 ½″ × 11″ sheet is given an alphanumeric designation relating to the type of detail. For example, door detail sheets would be numbered D1, D2, etc., and individual details on the first door detail sheet would be numbered D1-1, D1-2, D1-3, etc. A brief description of the detail is then listed opposite each number on the index sheets. Tabbed dividers are placed between the categories for easy retrieval. All the details in this chapter are indexed according to this system in figures 4-89 through 4-95.

A list of suggested headings for a comprehensive standard detail library could be indexed as follows:

C	Ceiling	M	Miscellaneous
CO	Column	P	Partitions
D	Door	R	Roof
E	Exterior	S	Site
EL	Elevator	ST	Stair
G	Graphics	T	Toilet
I	Interior	W	Window

For offices that have not switched to CADD, it is recommended that a master copy of the details be stored separately as insurance against damage to or loss of the day-to-day use copies. A blank index sheet with tick marks for easy subdivision is included in the Appendix (A-12).

Mounting the Detail on the Project Sheet

After the detail is retrieved, a copy is made on the office copying machine using appliqué film, commonly referred to as sticky-back, to be adhered to the project sheet. It is highly desirable that a good-quality office copier be used. Some copiers have a tendency to distort the copy. The next step is to adhere the film to the detail sheet, making sure that it is located on the square designated by the mock-up set (see

Sec. 1.3.2). If changes are required after the detail is added to the detail sheet, use an X-Acto knife to cut the part to be deleted. Be sure not to cut through the project sheet. Erase the residual adhesive and then make the correction on the front of the project sheet.

Another method is to tape the top edge of each standard detail (which is drawn on film) to the project sheet and draw the rest of the details as they evolve. This enables the team to run diazo prints during the evolution of the detail sheet for use as check prints. When the sheet is ready to be issued, a wash-off Mylar is made of the completed sheet to include in the final set.

Personally, I prefer the first method because it is easier, faster, and less costly. Besides, the result is not much different from the wash-off Mylar method.

The details included in this section are drawn to scale. The notations are chosen carefully. I have omitted most of the notations or dimensions that may differ from project to project. The user must check each detail carefully and make any necessary modifications or additions as well as add the missing information before using the detail. While the author has made every effort to make these details as dependable as possible, he and the publisher disavow any responsibility for any mistakes that may occur as a result of using them. The project architect must check them thoroughly before including them in the project.

Each detail section includes a perspective or axonometric drawing that gives an overview of the system, an introduction with some very useful information, in addition to guidelines, accompanying each detail. Some of the sections have structural information and a list of sources of information. The user must check the detail for conformance to the applicable code(s) and modify as needed.

To modify the detail, use a correction and cover-up tape such as Post-It brand tape by 3M (Fig. 4-1) to cover up the part that needs to be deleted before making the copy to be used on the project. Be sure to use the appropriate width tape.

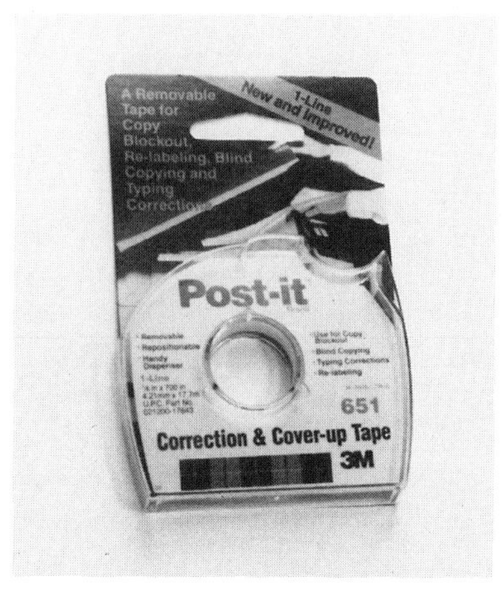

FIGURE 4-1. Correction and Cover-Up Tape.

At the end of this chapter, I have added all the details reduced to quarter size. These can be reproduced and used to paste on the mock-up sheet (see Sec. 1.3.2). An alternative method, to be used only on rush jobs, is to write the detail number on the designated square of the mock-up sheet. I prefer the first method because it provides a graphic way to check the set at a glance for completeness and it gives the team a clear picture of the project.

For offices that use CADD, it is anticipated that a disk containing all the details will be produced in the future if the demand warrants it.

4.1 STAIR, LADDER, AND ELEVATOR DETAILS

4.1.1 General

Stair and ladder details occur in most projects. Including them in the office standard details library is a must. Exit stairs or "exitways," as they are referred to in most codes, form a part of an evacuation system that provides the building occupants with a relatively safe route in case of fire or an emergency. Other components of the system are the "exitway access" (usually corridors leading to the exitway) and "exit discharge," which can be a door (when the stair opens directly to a street or open court) or a corridor (smoke stairs in a high-rise core must have a protected access to the street) or a lobby protected by a sprinkler system leading to the exterior. Read the code carefully for specific requirements.

The National Association of Architectural Metal Manufacturers (NAAMM) in its *Metal Stairs Manual* (1992) categorizes stairs into four classes as follows (reproduced with permission):

Industrial Class

Stairs of this class are purely functional in character and consequently they are generally the most economical. They are designed for either interior or exterior use, in industrial buildings such as factories and warehouses, or as fire escapes or emergency exitways. They do not include stairs which are integral parts of industrial equipment.

Industrial class stairs are similar in nature to any light steel construction. Hex head bolts are used for most connections, and welds where used, are not ground. Stringers may be either flat plate or open channels; treads and platforms are usually made of grating or formed of floor plate, and risers are usually open, though in some cases filled pan-type treads and steel risers may be used. Railings are usually of either pipe, tubing, or steel bar construction.

When used for exterior fire escapes the details of construction are similar, except that treads and platforms are of open design, usually grating or perforated floor plate. Also, the dimensions, methods of support and other details are usually dictated by governing code regulations.

Service Class

This class of stairs serves chiefly functional purposes, but is not unattractive in appearance. Service stairs are usually located in enclosed stairwells and provide a secondary or emergency means of travel between floors. In multi-storied buildings they are commonly used as egress stairs. They may serve employees, tenants, or the public, and are generally used where economy is a consideration.

Stringers of service stairs are generally the same types as those used on stairs of the industrial class. Treads may be one of several standard types, either filled or formed of floor or tread plate, and risers are either exposed steel or open construction. Railings are typically of pipe construction or a simple bar type with tabular newels, and soffits are usually left exposed. Connections on the underside of the stairs are made with hex head bolts, and only those welds in the travel area are smooth.

Commercial Class

Stairs of this class are usually for public use and are of more attractive design than those of the service class. They may be placed in open locations or may be located in closed stairwells, in public, institutional or commercial buildings.

Stringers for this class of stairs are usually exposed open channel or plate sections. Treads may be any of a number of standard types, and risers are usually exposed steel. Railings vary from ornamental bar or tube construction with metal handrails to simple pipe construction, and soffits may or may not be covered. Exposed bolted connections in areas where appearance is critical are made with countersunk flat or oval head bolts; otherwise hex head bolts are used. Welds in conspicuous locations are smooth, and all joints are closely fitted.

Architectural Class

This classification applies to any of the more elaborate, and usually more expensive, stairs; those which are designed to be architectural features in a building. They may be wholly custom designed or may represent a combination of standard parts with specially designed elements such as stringers, railings, treads or platforms. Usually this class of stair has a comparatively low pitch, with relatively low risers and correspondingly wider treads. Architectural metal stairs may be located either in the open or in enclosed stairwells in public, institutional, commercial, or monumental buildings.

The materials, fabrication details, and finishes used in architectural class stairs vary widely, as dictated by the architect's design and specifications. As a general rule, construction joints are made as inconspicuous as possible, exposed welds are smooth and soffits are covered with some surfacing material. Stringers may be special sections exposed, or may be structural members enclosed in other materials. Railings are of an ornamental type and, like the treads and risers, may be of any construction desired.

NAAMM in its *Pipe Railing Manual* (1992) provides the following insights about pipe rails (reproduced with permission):

Provisions also must often be made to accommodate the physically handicapped. ANSI A117.1, *Specifications for Making Buildings and Facilities Accessible To and Usable by Physically Handicapped Persons* describes the requirements for handrails for areas to be used by physically handicapped persons. It specifies that handrails shall have a diameter of 1¼″ to 1½″ and that the clearance between the handrail and the wall shall be 1½″. Both the 1¼″ pipe with its 1.66″ outside diameter and the 1½″ with its outside diameter of 1.90″ are considered acceptable.

Local codes should always be checked for their specific requirements, but government regulatory bodies, such as OSHA, should also be checked. In certain instances the requirements of the latter may be more rigorous than those of the local jurisdiction.

The axonometric drawing shown in figure 4-2 identifies exit stair components in a steel frame building. This drawing should be reproduced, the missing dimensions filled out according to the applicable code, and the updated drawing distributed to the staff. It will serve as a graphic code reference that supplies information at a glance. This saves the time spent on looking for the code and searching for the applicable section.

4.1.2 Required Drawings

The following drawings are required to convey the necessary information to the stair fabricator:

1. A plan drawing shown individually (Fig. 4-3) or included in a core plan enlarged to a ¼″ = 1′-0″ scale and keyed to the floor plan.
2. A stair section drawn to the same scale (Fig. 4-4). For manually drawn stairs, the thing to remember is that these drawings are targeted to the stair fabricator, who will base the shop drawings on the architectural drawings. It is not necessary to meticulously draw every step and handrail run as shown in Fig. 4-5.

   The method of subdividing the distance between landings is shown in Fig. 4-6. This method is also used to subdivide the flights shown on the plan.
3. Stair details as shown in Fig. 4-7 through Fig. 4-16. Theses details lend themselves to use as standard drawings after all the dimensions are filled out and rail information designed to resist the forces mandated by the code are filled in. These details are suitable for flights and landings up to 5′-0″ wide.

The architect does not have to draw the stairs to the last detail. NAAMM recommends the following (reproduced with permission):

Generally speaking, the architect should be concerned, in his drawings, only with the conceptual design and the provision of sufficient details to clearly explain the materials to be used and the esthetic effect desired. If he provides complete details of all structural parts and their connections, such details must meet load requirements of governing codes or special conditions. Detailing is often left to the fabricator, and will be shown on the shop drawings which he submits for the architect's approval. Although metal stairs of all types are essentially custom designed, each stair manufacturer has his own preferred and

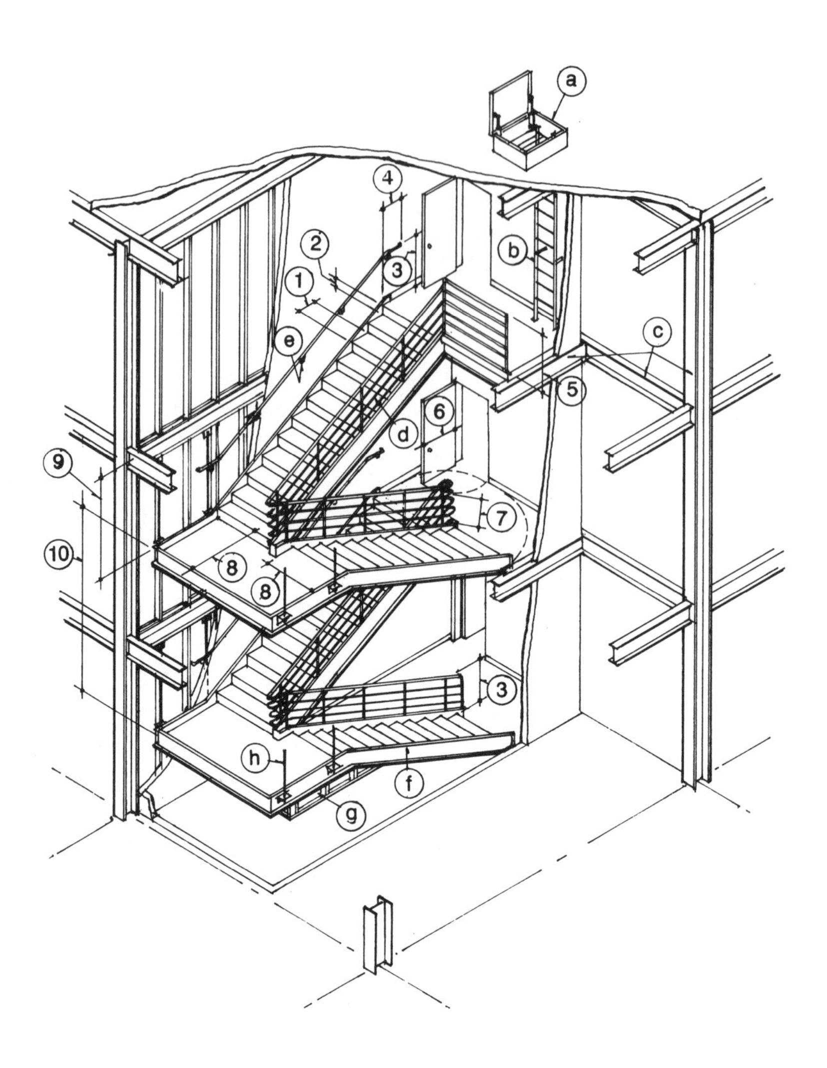

STAIR COMPONENTS

a. Roof hatch, sometimes referred to as roof scuttle.
b. Ladder to roof (not always required).
c. Stair supporting structural frame. Most codes require that the supporting structure have at least the same firerating as the stair enclosure.
d. Handrail. Most codes require that it be designed so that no article 6" or larger in diameter can pass through it.
e. Wall bracket. Specify that it be spaced to sustain code-mandated loads.
f. Stringer, sometimes referred to as "string." It is usually a channel or flat plate placed at both sides of the treads and designed to carry the loads.
g. Furring under stair. Not always used, although it is a good idea to prevent storage which is prohibited by code under the stair and prevent dirt and dust accumulation.
h. Hanger rods. Placed inside the wall (Fig. 4-11 & Fig. 4-12) to carry intermediate landings.

CRITICAL DIMENSIONS (See Section 3.1)

1. Minimum tread width
2. Maximum riser height
3. Handrail height Min. Max.
4. Handrail extension beyond top tread = bottom tread =
5. Guard rail height
6. Door width Min. Max.
7. Minimum clearance between door swing and handrail
8. Minimum flight and landing width (based on units of exit-width)
9. Maximum height of flight
10. Minimum headroom

FIGURE 4-2. Stair Components and Critical Dimensions.

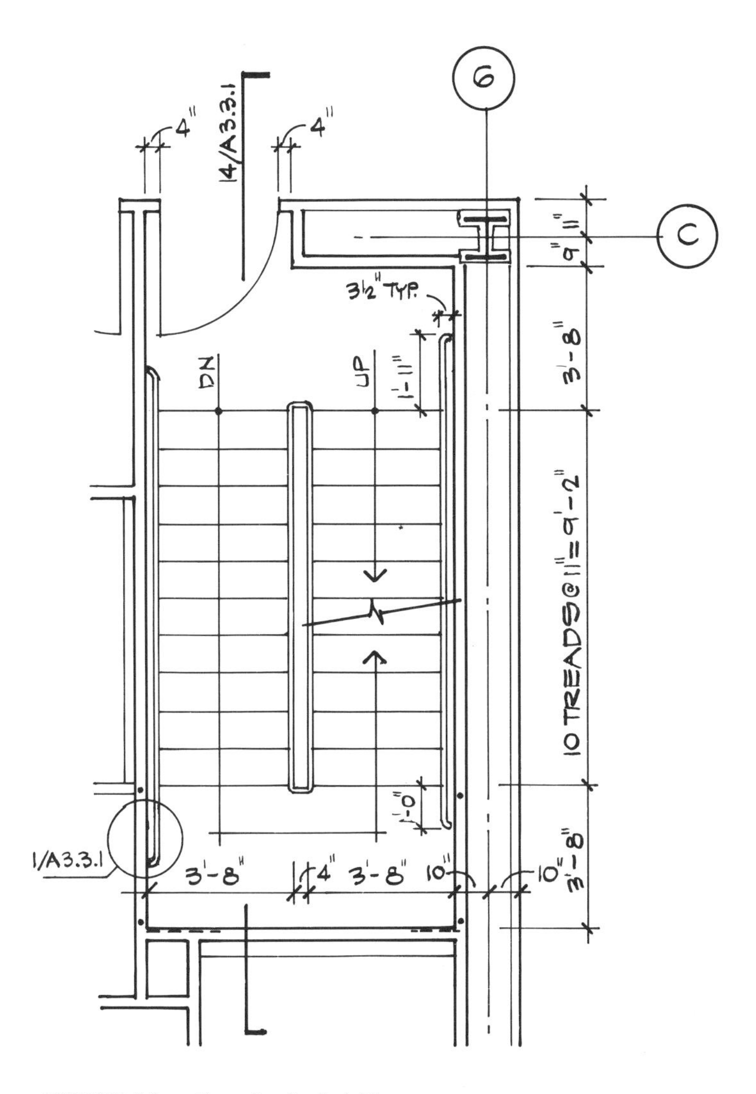

FIGURE 4-3. Example of a Stair Plan.

proven methods of fabricating typical repetitive parts, especially on the more common types of stair. What may be the best detail or connection method in the opinion of one manufacturer is not necessarily consistent with the practices of another. And when the architect is contemplating the use of special design features, he should contact one or more fabricators early in the design stage to avail himself of any suggestions which may result in better or more economical design.

4.1.3 Structural Considerations

Cast-in-place concrete stairs are designed by the structural engineer. Apart from the ¼" plans, section, and handrail details, the only details required are the wall bracket and tread nosing details (Fig. 4-13 and 4-14). A rule of thumb for drafting only is that the slab thickness in inches (the distance between the bottom of the slab and a line connecting the bottom of the risers) is half the flight span in feet. For instance, if the flight is 8'-0" long, draw the slab thickness at 4". Do not put that dimension on the drawing. The engineer will show all dimensions and reinforcement.

To determine the size of stair components, refer to Figure 4-17. Here again, these sizes are to be used for drafting only. They are not to be called out on the drawings. The stair fabricator prepares shop drawings detailing all the components. These must be checked for conformance to design dimensions by the architect and for structural integrity by the engineer.

4.1.4 Pre-Engineered Stairs

There are national stair manufacturers that fabricate and assemble stairs from components designed specifically for stairs. They usually bid for projects over four stories in height. Because of their mass-production techniques and CADD capabilities, they can underbid local fabricators for most high-rise projects.

Pre-engineered stairs may either be carried by the structure at each floor or, in some low-rise structures, freestanding. Treads may be fabricated from steel checker plate, concrete-filled pans, or precast concrete treads. Handrails are manufactured in several styles and finishes.

Show the least number of steps (manual drafting)

Do not show repetitious identical floors on multi-story projects, even if there is space to include them.

THIS DRAWING SHOWS ALL THE PERTINENT INFORMATION AT A FRACTION OF THE TIME IT TAKES TO DRAW THE ONE SHOWN IN FIG. 4-5.

Show intermediate railings at the top and bottom of stair and any atypical conditions only (manual drafting).

14 SECTION THRU STAIR#2 — 1/4 = 1-0 — 2/A2.3.1 9/A2.3.2 4/A2.3.4

FIGURE 4-4. Example of a Stair Section.

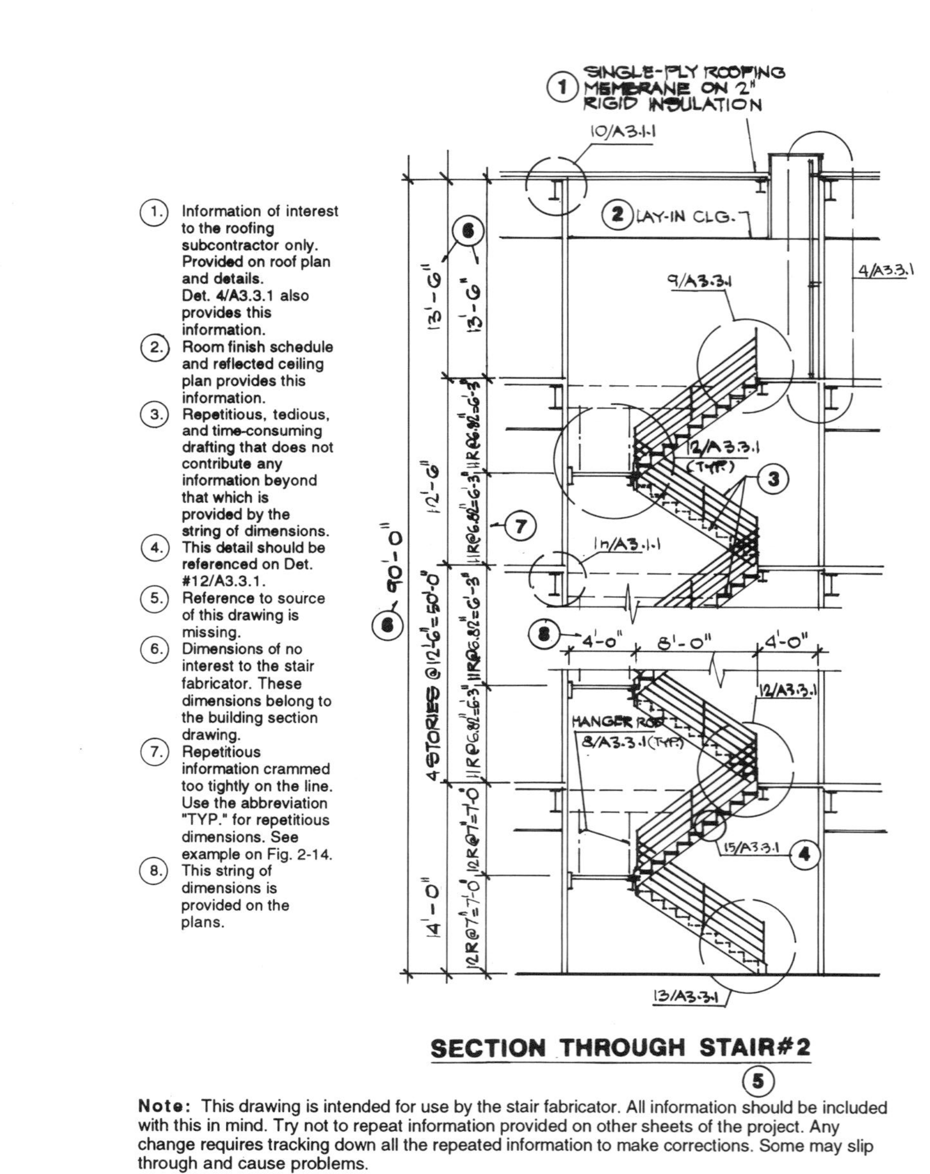

1. Information of interest to the roofing subcontractor only. Provided on roof plan and details. Det. 4/A3.3.1 also provides this information.
2. Room finish schedule and reflected ceiling plan provides this information.
3. Repetitious, tedious, and time-consuming drafting that does not contribute any information beyond that which is provided by the string of dimensions.
4. This detail should be referenced on Det. #12/A3.3.1.
5. Reference to source of this drawing is missing.
6. Dimensions of no interest to the stair fabricator. These dimensions belong to the building section drawing.
7. Repetitious information crammed too tightly on the line. Use the abbreviation "TYP." for repetitious dimensions. See example on Fig. 2-14.
8. This string of dimensions is provided on the plans.

SECTION THROUGH STAIR#2

5

Note: This drawing is intended for use by the stair fabricator. All information should be included with this in mind. Try not to repeat information provided on other sheets of the project. Any change requires tracking down all the repeated information to make corrections. Some may slip through and cause problems.

FIGURE 4-5. Example of Overdrafting (see Fig. 4-4 for proper drafting).

PLACE THE SCALE ON THE FIRST RISER (OR LANDING) AS SHOWN THEN TILT IT UNTIL THE REQUIRED NUMBER OF TREADS (OR RISERS) COINCIDES WITH THE RUN (OR RISE) LIMIT.

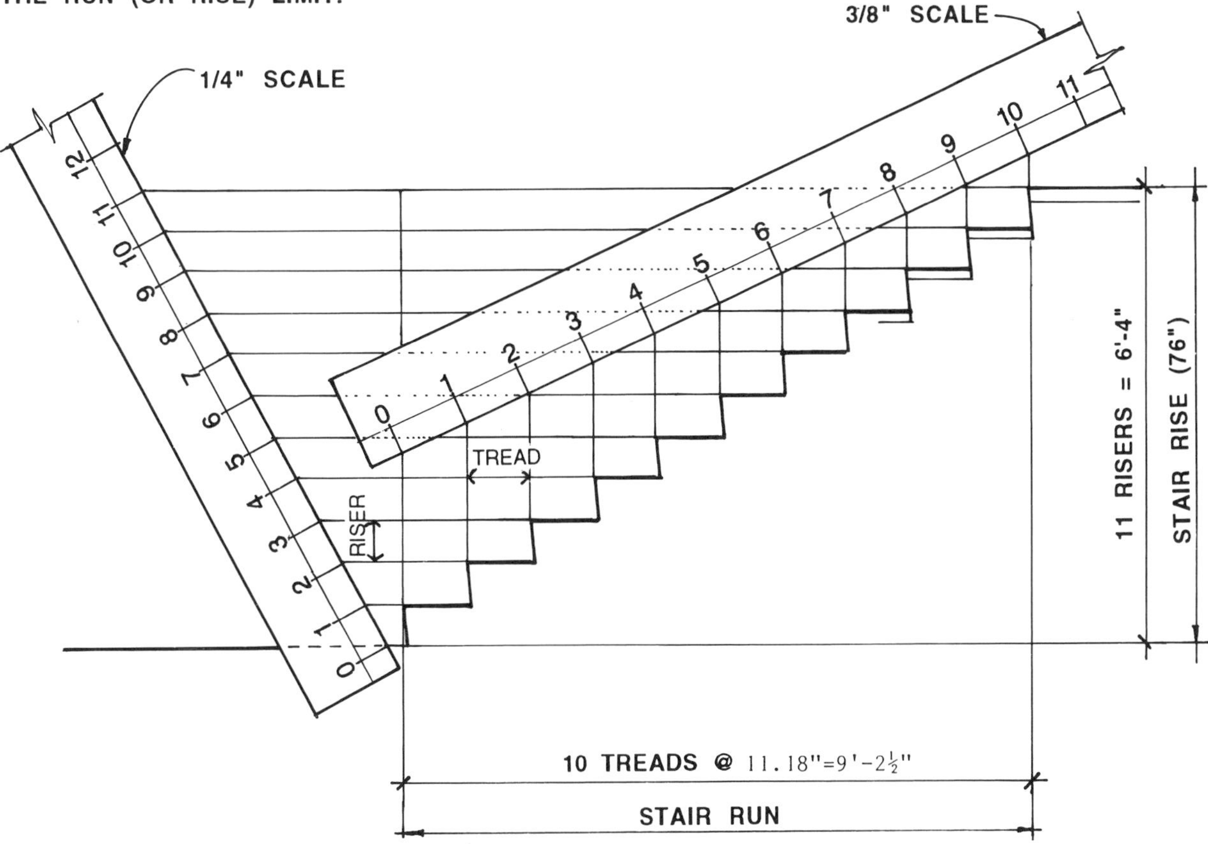

Guidelines:

1. Determine height of flight (6'-4" in the example above) and subdivide into risers. Make sure risers do not exceed the maximum height mandated by the code.
2. Calculate the width of a tread. Most codes rule that 2 risers + 1 tread = 24" to 25". In this case, riser height $= \frac{76}{11} = 6.91"$.

   2 risers + 1 tread = 25" max. $2 \times 6.91" + X = 25$, $X = 25 - (2 \times 6.91") = 11.18"$

FIGURE 4-6. Method of Drawing Stair Sections.

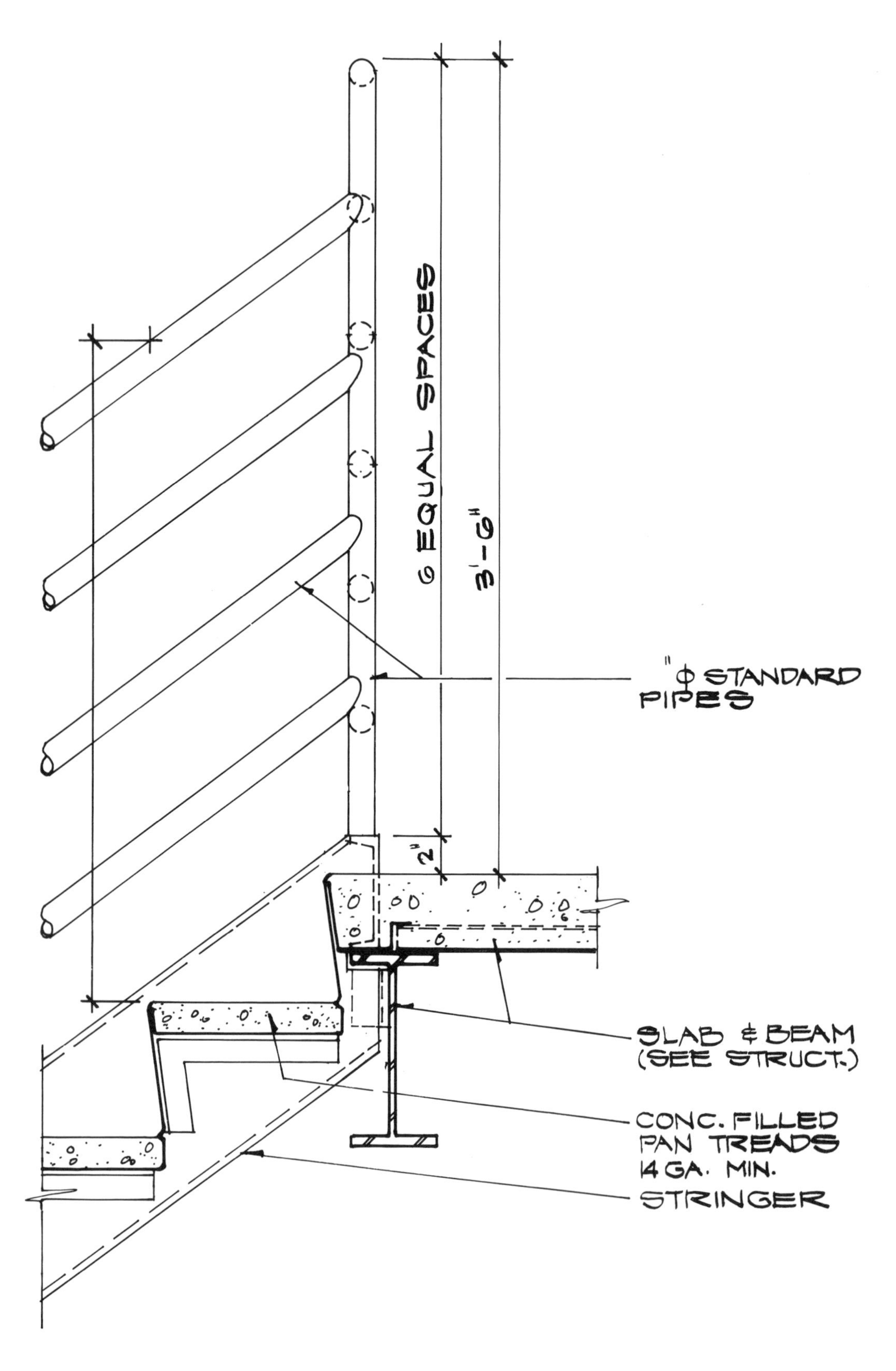

FIGURE 4-7. Detail at Top of Stair.

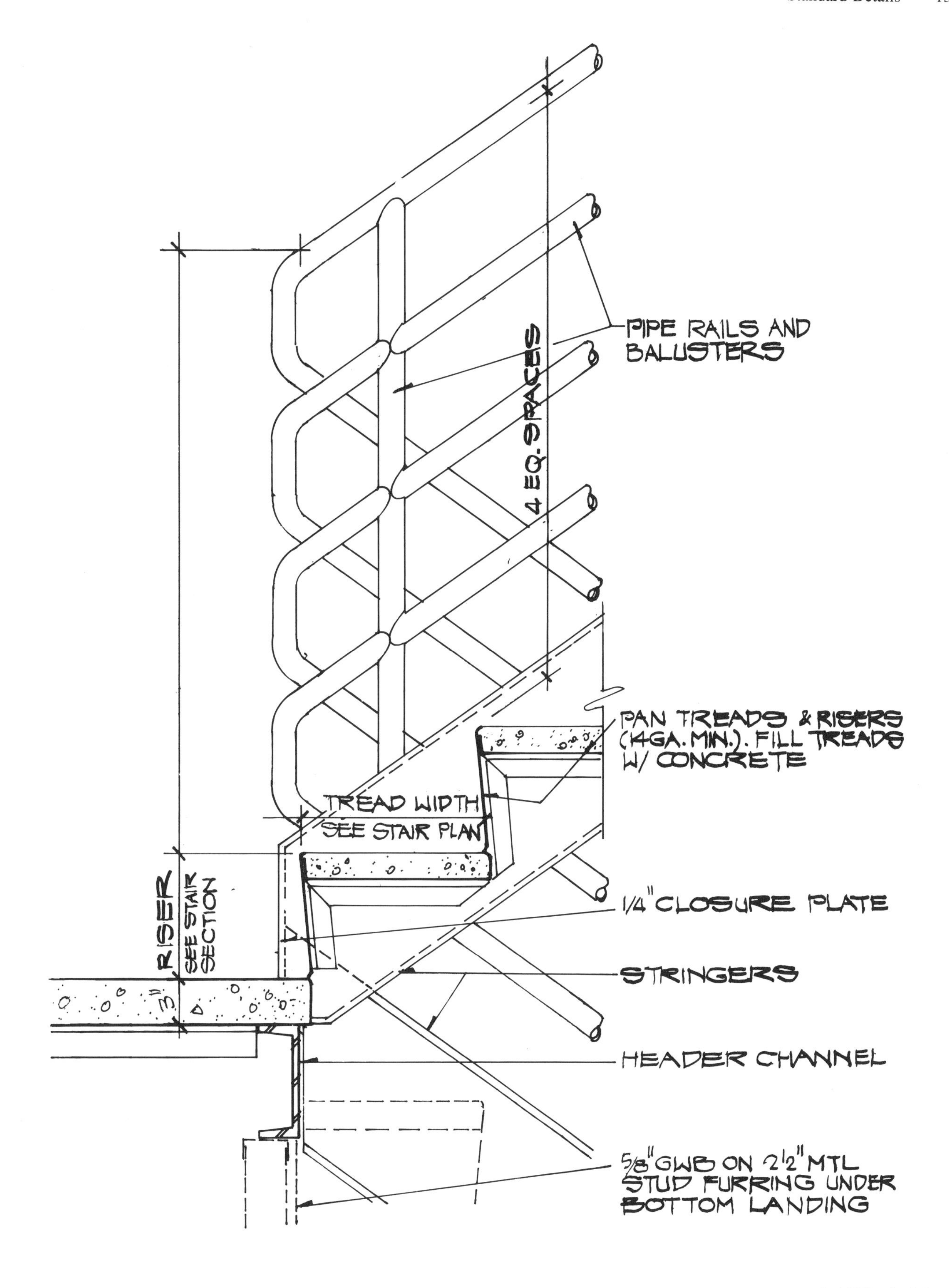

FIGURE 4-8. Detail at Intermediate Landing.

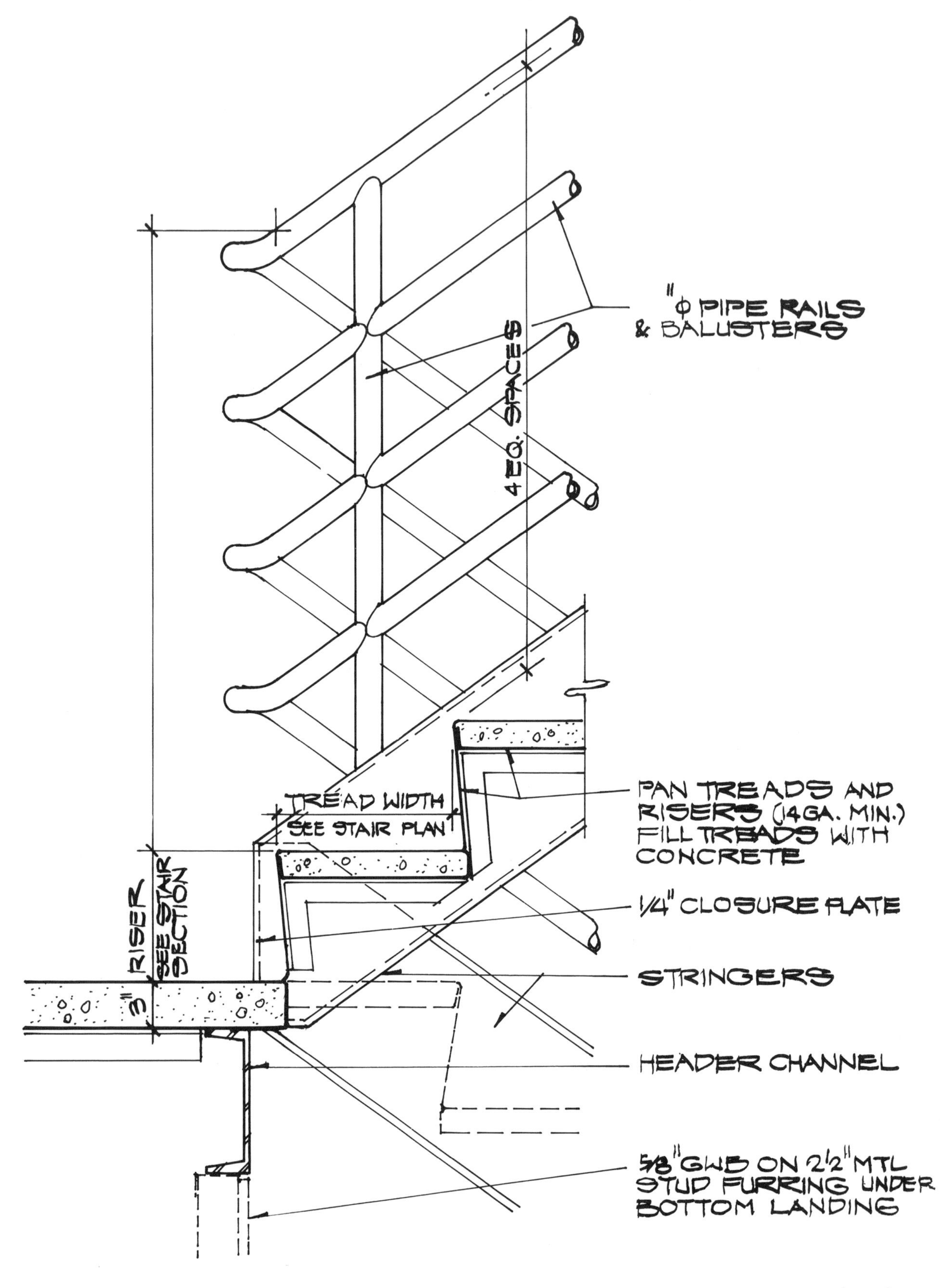

FIGURE 4-9. Detail at Intermediate Landing (Alternate).

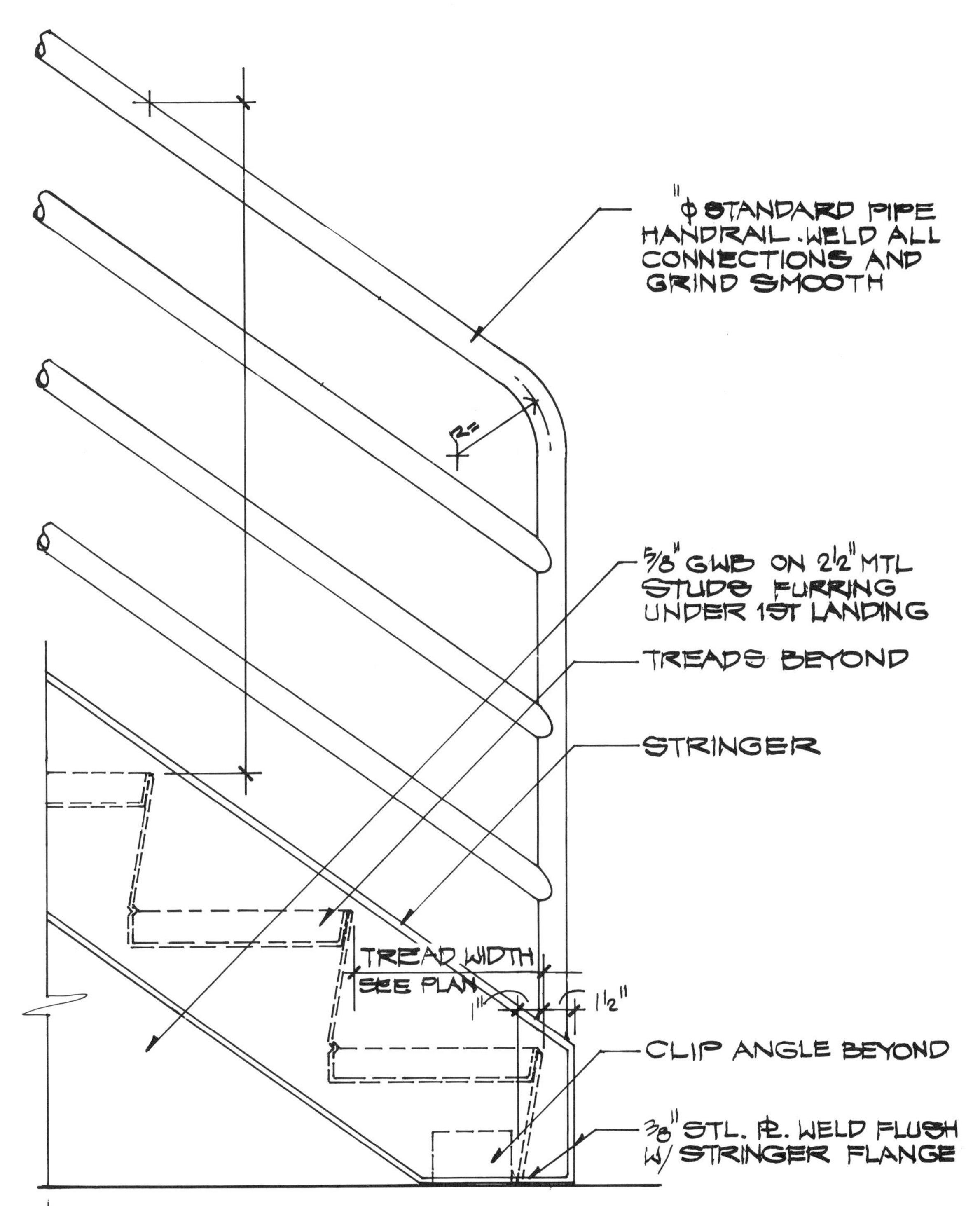

Guidelines:

For Fig. 4-7 , Fig. 4-8 , Fig. 4-9 , and Fig. 4-10.

1. Slab and beam design varies. The beam may be a channel, a tube or concrete.
2. The 2" barrier at the edge of the landing in Fig. 4-17 is required to prevent items from rolling and falling off the top landing onto somebody using the stair below.
3. Stairs wider than 5'-0" should be designed by the engineer.
4. Figure 4-9 shows a staggered step arrangement that results in a more comfortable rail transition between flights. It may require a longer stair enclosure.
5. The railing must be designed to withstand code-mandated loads. All that is needed is for a structural engineer to design it once to be used on all projects where most of the work done by the office is located. The design should include maximum baluster spacing, pipe diameter, and weight - STD, XS, XXS. The top rail should also have a diameter conforming to that required by the code for the physically handicapped.

1 1/2"=1'-0"

FIGURE 4-10. Detail at Bottom of Stair.

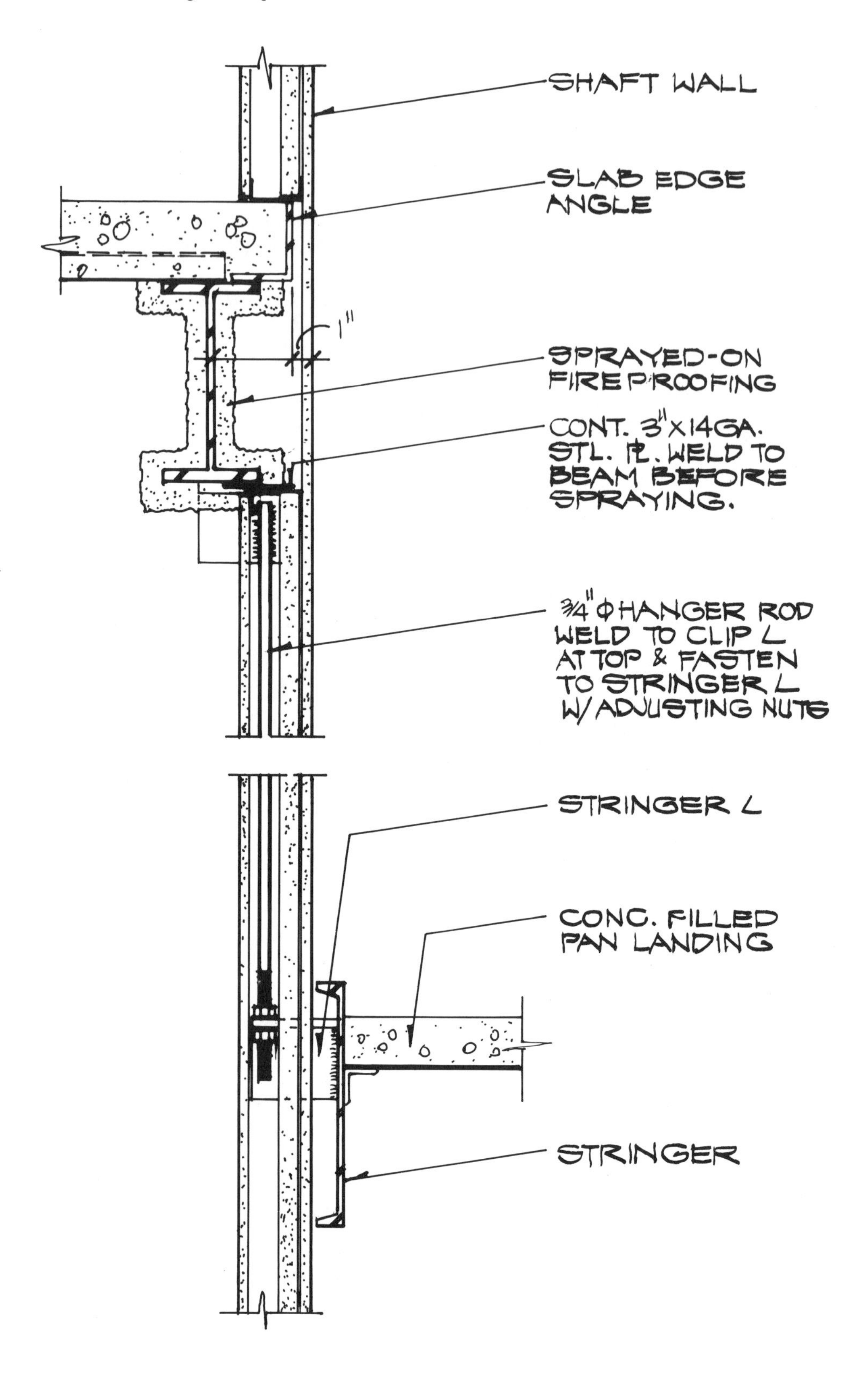

1 1/2"=1'-0"

FIGURE 4-11. Detail of a Hanger Rod in a Shaft Wall.

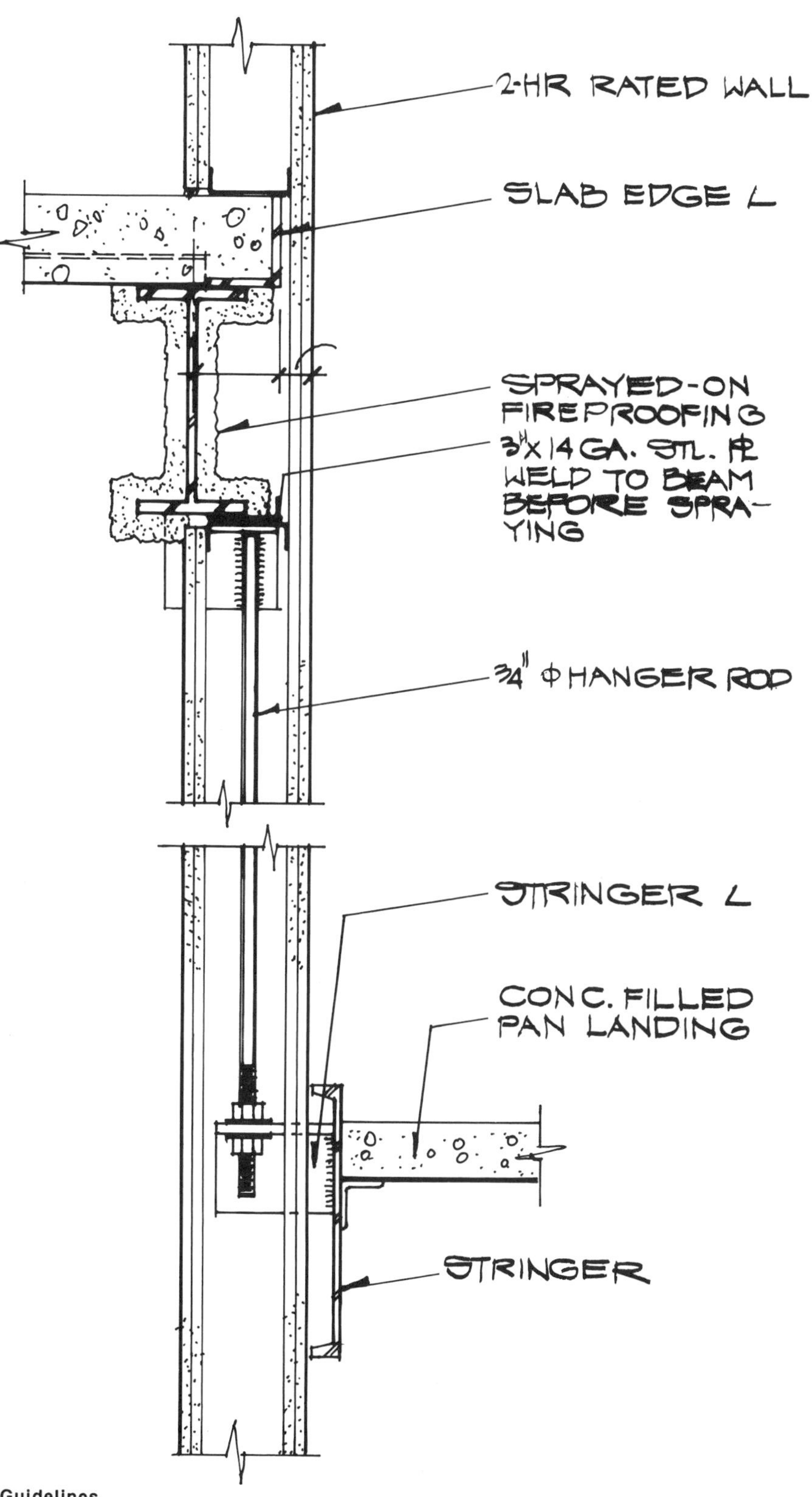

Guidelines
for Fig. 4-11 and Fig. 4-12

1. These two figures show hanger rod details in 2-hour shaft wall and a 2 hour standard partition respectively.
2. Floor structure varies. Modify the drawing as needed.
3. If the beam depth is such that the distance between the stud runner below the beam and the one above the slab is more than 24", provide appropriate support for the GWB layer spanning between the two.

1 1/2"=1'-0"

FIGURE 4-12. Detail of a Hanger Rod in a Stud Wall.

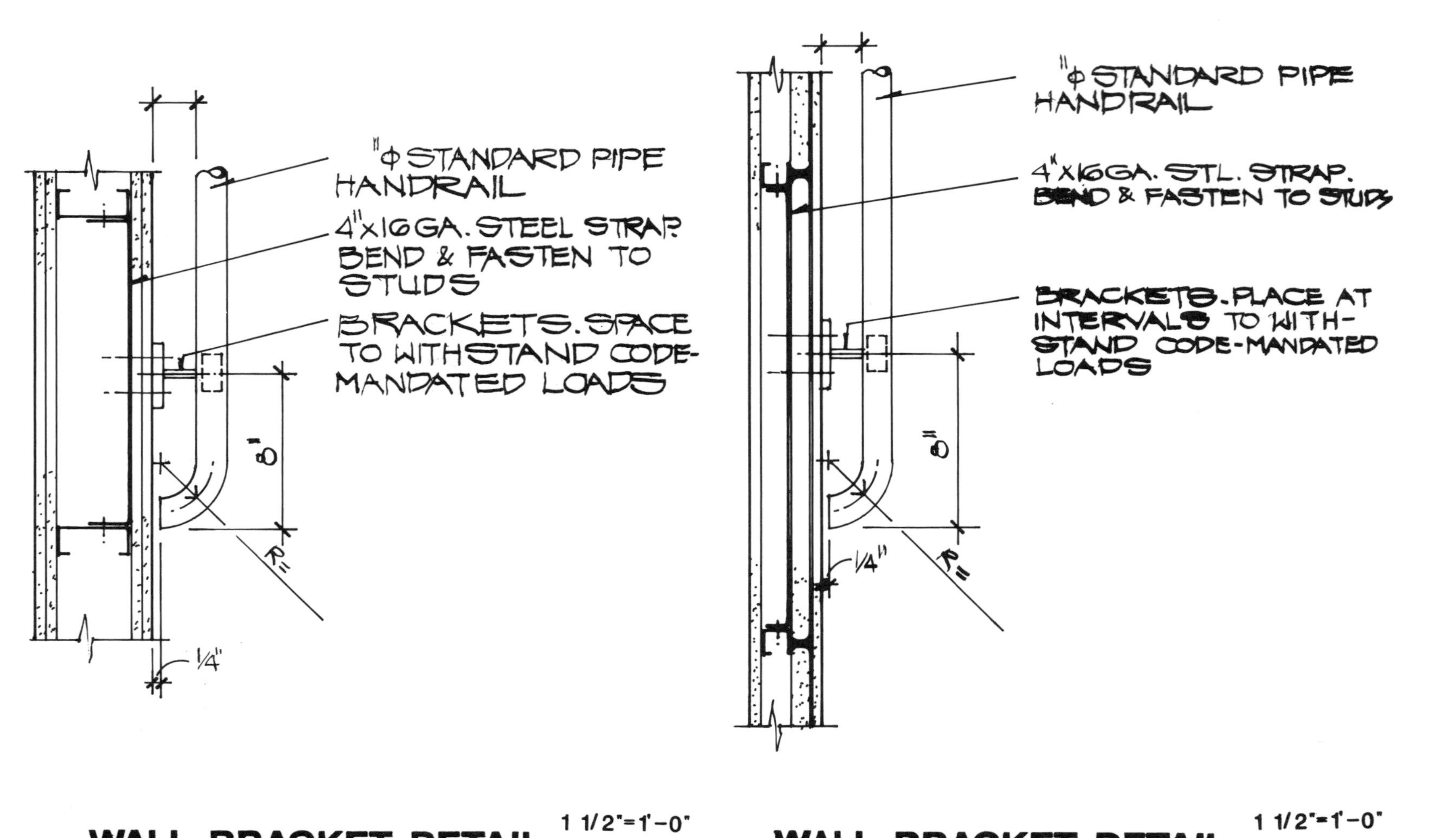

WALL BRACKET DETAIL 1 1/2"=1'-0"

WALL BRACKET DETAIL 1 1/2"=1'-0"

Guidelines:

1. Detail on the left shows a 2-hour stud partition. The detail on the right is a 2-hour shaft wall.
2. Check the building code for handrail clearance.
3. Make sure that specifications mandate conformance to all applicable codes.
4. Wall thickness is shown on the partition types. Do not repeat on these details.

FIGURE 4-13. Wall Bracket Details.

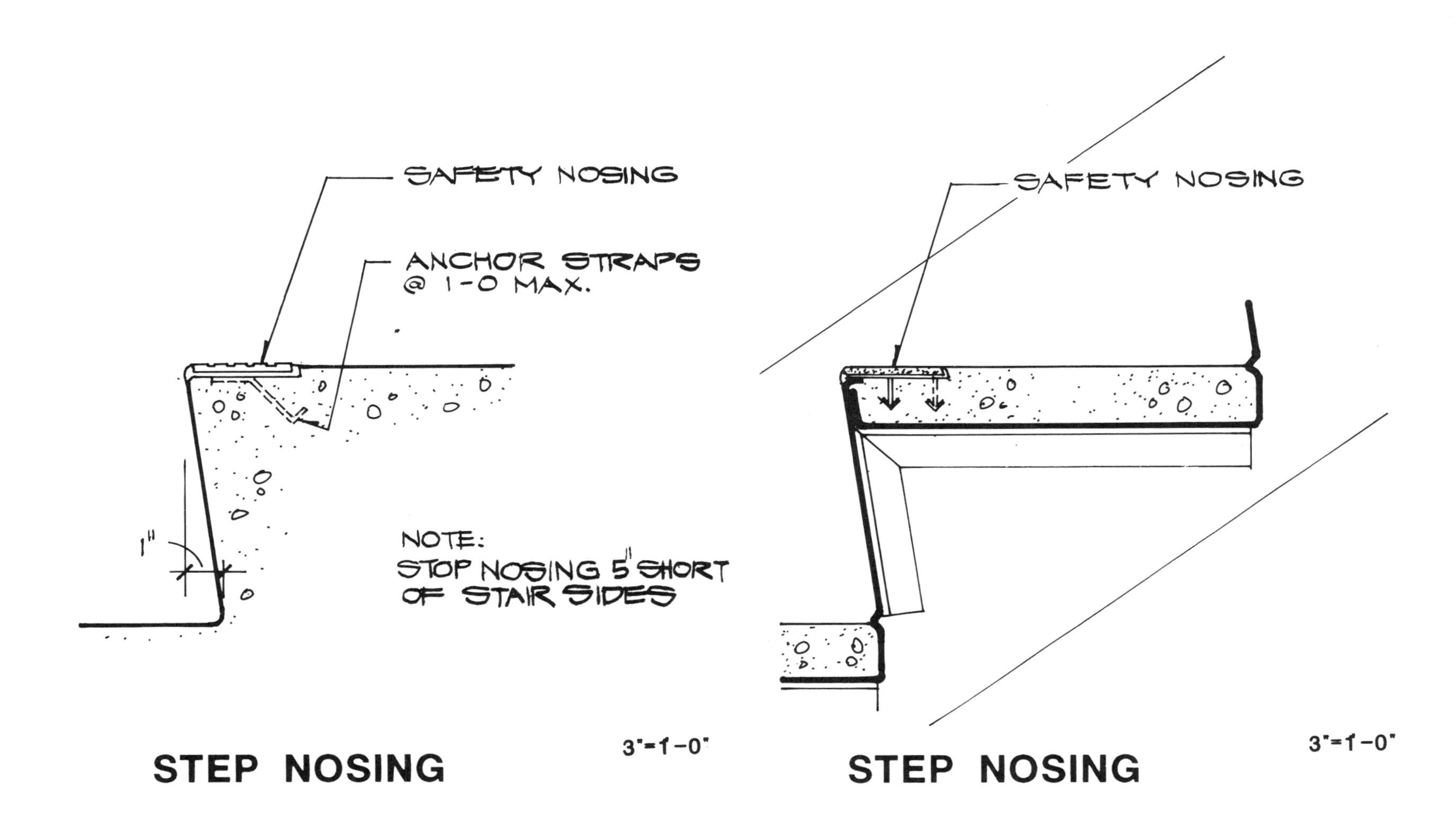

Guidelines:

1. For exterior stairs, slope the tread 1/4" toward the nosing.
2. Make sure that the specs agree with the detail.

FIGURE 4-14. Step Nosing Detail.

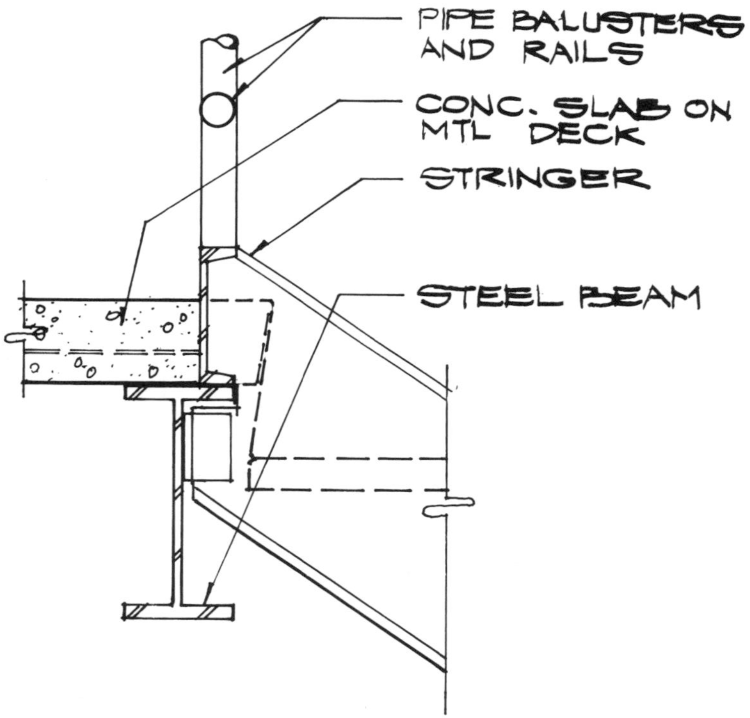

PIPE BALUSTERS AND RAILS
CONC. SLAB ON MTL DECK
STRINGER
STEEL BEAM

SECTION b–b

1 1/2"=1'–0"

Guidelines:
1. Detail at bottom of top floor guardrail shown. See Fig. 4-16.
2. Check struct. for slab and beam sizes.

FIGURE 4-15. Guardrail Detail.

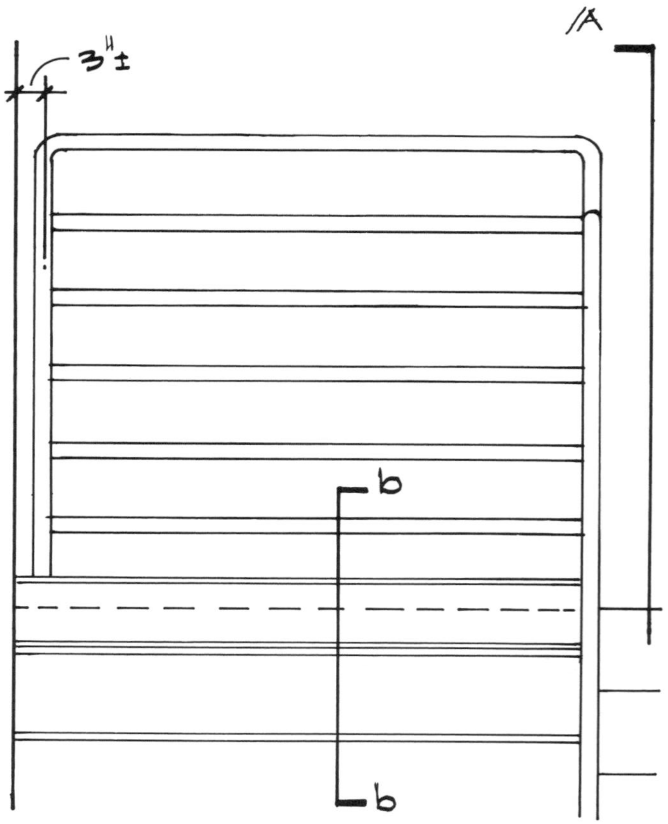

3"±
/A
b
b

GUARDRAIL DETAIL

3/4"=1'–0"

Guidelines:
See Fig. 4-15 for guardrail detail (detail at right) and Fig. 4-7 for other detail.

FIGURE 4-16. Guardrail Detail.

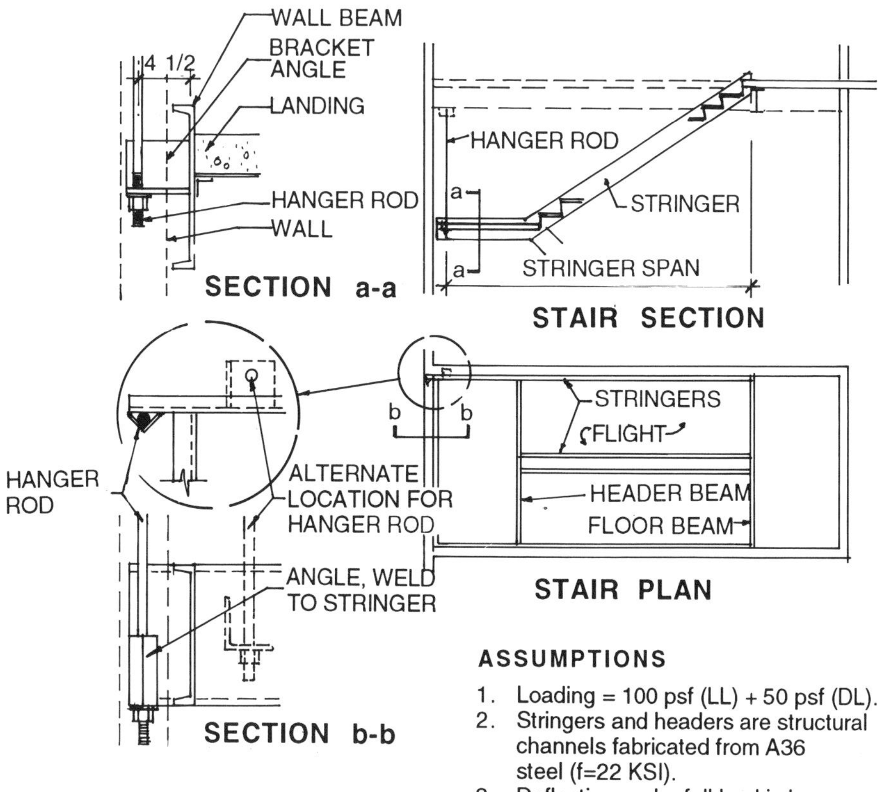

ASSUMPTIONS

1. Loading = 100 psf (LL) + 50 psf (DL).
2. Stringers and headers are structural channels fabricated from A36 steel (f=22 KSI).
3. Deflection under full load is less than 1/360 of span.

FLIGHT WIDTH	STRINGER SPAN (UP TO)	STRINGER SIZE	HEADER SPAN	HEADER SIZE	HANGER ROD SIZE	BRACKET ANGLE
UP TO 3'-8"	12-0	MC 10 X 8.4	8-0	C6 X 8.2	5/8 ø	4 x 4 x 3/8
	14-0	MC 10 X 8.4	8-0	C6 X 8.2	3/4 ø	4 x 4 x 3/8
	16-0	MC 12 X 10.6	8-0	C7 X 9.8	3/4 ø	4 x 4 x 3/8
UP TO 4'-0"	12-0	MC 10 X 8.4	8-6	C6 X 8.2	3/4 ø	4 x 4 x 3/8
	14-0	MC 12 X 10.6	8-6	C6 X 8.2	3/4 ø	4 x 4 x 3/8
	16-0	MC 12 X 10.6	8-6	C7 X 9.8	3/4 ø	4 x 4 x 3/8
UP TO 5'-0"	12-0	MC 12 X 10.6	10-6	C8 X 11.5	3/4 ø	4 x 4 x 3/8
	14-0	C9 X 13.4	10-6	C8 X 11.5	3/4 ø	4 x 4 x 3/8
	16-0	C10 X 15.3	10-6	C9 X 13.4	3/4 ø	4 x 4 x 1/2

Guidelines:

1. The information shown in this table is for drafting guidance only. Do not include channel size on the drawings. The fabricator indicates all sizes and details on the shop drawings. The structural engineer checks these drawings for structural integrity.
2. Check applicable codes for stair flight dimensions,clearances, and rail heights.

FIGURE 4-17. Stair Component Sizes (Inspired by a detail by CTJ & D Architects, Houston, TX).

4.1.5 Ladders and Ship's Ladders

Ladders may be prefabricated or custom built. They may be assembled from steel or aluminum components. Figure 4-23 shows an example of a steel elevator pit ladder.

The Occupational Safety and Health Administration (OSHA) requires that ladders 20′-0″ to 30′-0″ in height must have a cage around them for safety. Beyond 30 feet, the ladder must be offset at a landing. A safety device incorporating a life belt, friction brakes, and sliding attachments may be substituted for the cage and platform requirements in some applications.

Ship's ladders are canted at 58° to 68°. They are usually used to service mechanical equipment and allow the user to carry tools and parts more easily than ladders. One type, the LAPEYRE (Fig. 4-18), features alternating treads that are deeper than a regular ship's ladder. The manufacturer claims this enables the user to climb and descend more comfortably. There is one catch, however; this is a single-source product that may stand in the way of competitive bidding. The cost implications must be researched before specifying this type.

FIGURE 4-18. The Lapeyre Ship's Ladder.

FIGURE 4-19. Standpipe, Fire Hose Valve, and Emergency Telephone at Stair Landing.

4.1.6 Guidelines

The following are points to be taken into consideration in stair design:

1. If the handrail is bent to a radius at the junction between flights, it is recommended that the gap between flights be no less than 4″ to fashion the complex curves required at these locations.
2. Standpipes, where required, may be placed at the floor-level landing or at the corner of the intermediate landing (Fig. 4-19). This latter alternative allows firemen to use the two-way valve to serve two floors. An emergency telephone is required in stairs serving high-rise buildings. Most codes require one every fifth floor. Telephones may also be required for use by the physically handicapped. Check the ADA (Americans With Disabilities Act) requirement.
3. Wall-mounted handrails are recommended to be continuous around the intermediate landing at locations used by the elderly.
4. If space allows, offset the steps as shown in Fig. 4-9. This allows for a more comfortable handrail configuration. This is also especially desirable in main lobby and ceremonial stairs.
5. If the building frame is steel sprayed with fireproofing, it may be preferable to construct the floor landing as part of the stair rather than part of the floor. This latter option would leave the sprayed beam exposed to view (if it is part of a beam extending beyond the confines of the stair enclosure). It also requires that the enclosing wall fill the

FIGURE 4-20. View of the Underside of a Floor Landing (Note the unsightly fire safing insulation in the deck flutes and the exposed sprayed-on fire-proofing.)

gaps between the steel deck flutes (Fig. 4-20), which is difficult to achieve. An alternative solution is to enclose the beam and deck with GWB furring.

6. Many factors affect stair costs. NAAMM lists the following: a. The welding of treads and risers directly to stringer (unit construction) will eliminate the need for carrier angles or bars and will sometimes reduce costs. b. The use of floor plate or tread plate for treads and risers usually results in maximum economy. c. The most economical type of rail for a stair is a steel pipe rail connected at the ends by standard terminal castings to a square or rectangular tube newel.

4.1.7 Elevator Details

Details include dimensioned plans (pit, typical hoistway plan, and machine room plan), hoistway section (see sheet A3.4.1 in the mock-up example in Chapter 1) showing pit depth, floor-to-floor height, travel (distance between first and top landing), and overhead (distance between top landing and top of hoistway). In addition, details include head and jamb details (Fig. 4-21), sill detail (Fig. 4-22), pit-ladder detail (Fig. 4-23), venting louver details (where required by code) and special cab, hall lantern details if custom designed, as well as main lobby and typical entrance elevations.

4.1.8 Sources of More Information:

Callender, John Hancock, ed. *Time-Saver Standards*, latest edition. New York: McGraw-Hill, Inc.

National Association of Architectural Metal Manufacturers. 1992. *Metal Stairs Manual*. 5th ed. 600 S. Federal St., Chicago, IL 60605. 312-922-6222.

National Association of Architectural Metal Manufacturers. 1985. *Pipe Railing Manual*, 2nd ed. 600 S. Federal St., Chicago, IL 60605. 312-922-6222.

Ramsey, Charles G., and Sleeper, Harold R. Latest edition, *Architectural Graphic Standards*. New York: John Wiley & Sons, Inc.

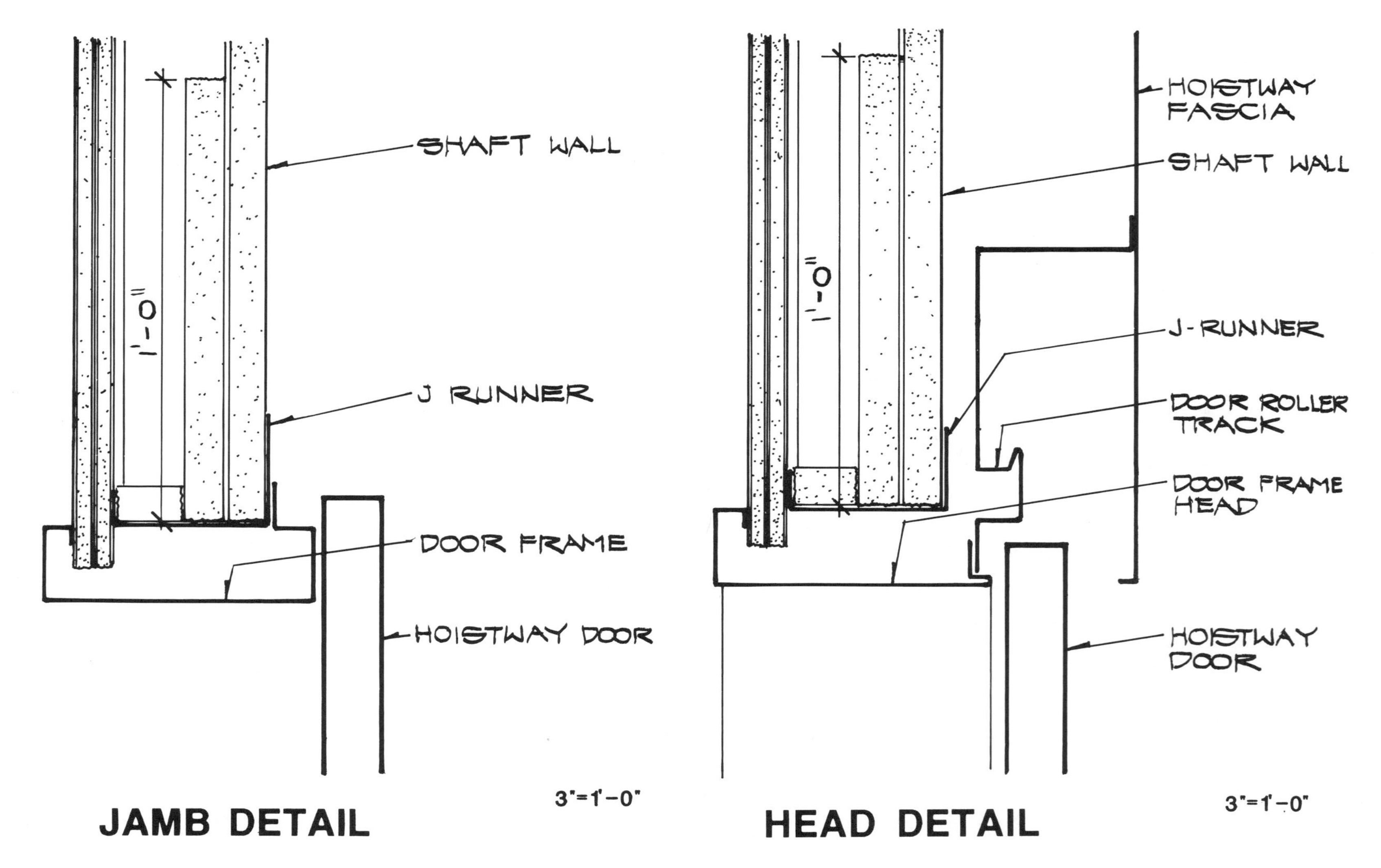

Guidelines:

1. Thickness of wall is determined by partition height, elevator speed and bracing (if any). See Section 3.12.
2. Door frame profile may be customized at extra cost. Contact representative for more information.

FIGURE 4-21. Hoistway Door Details.

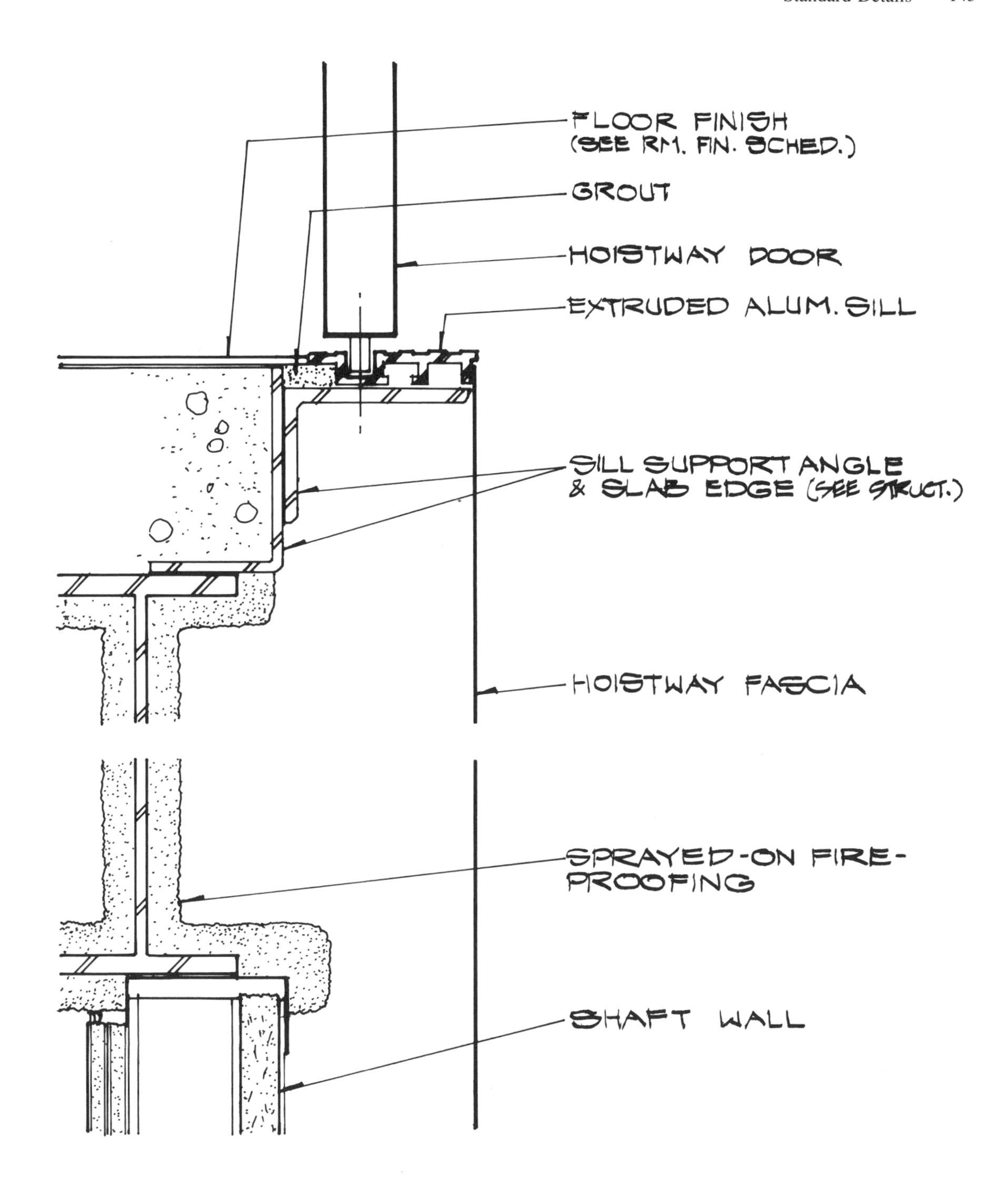

SILL DETAIL

Guidelines:

1. Some elevator consultants recommend that the sill support angle be included in the elevator spec.
2. Runner is attached to beam before spraying. Allow for beam deflection.

FIGURE 4-22. Hoistway Doorsill Detail.

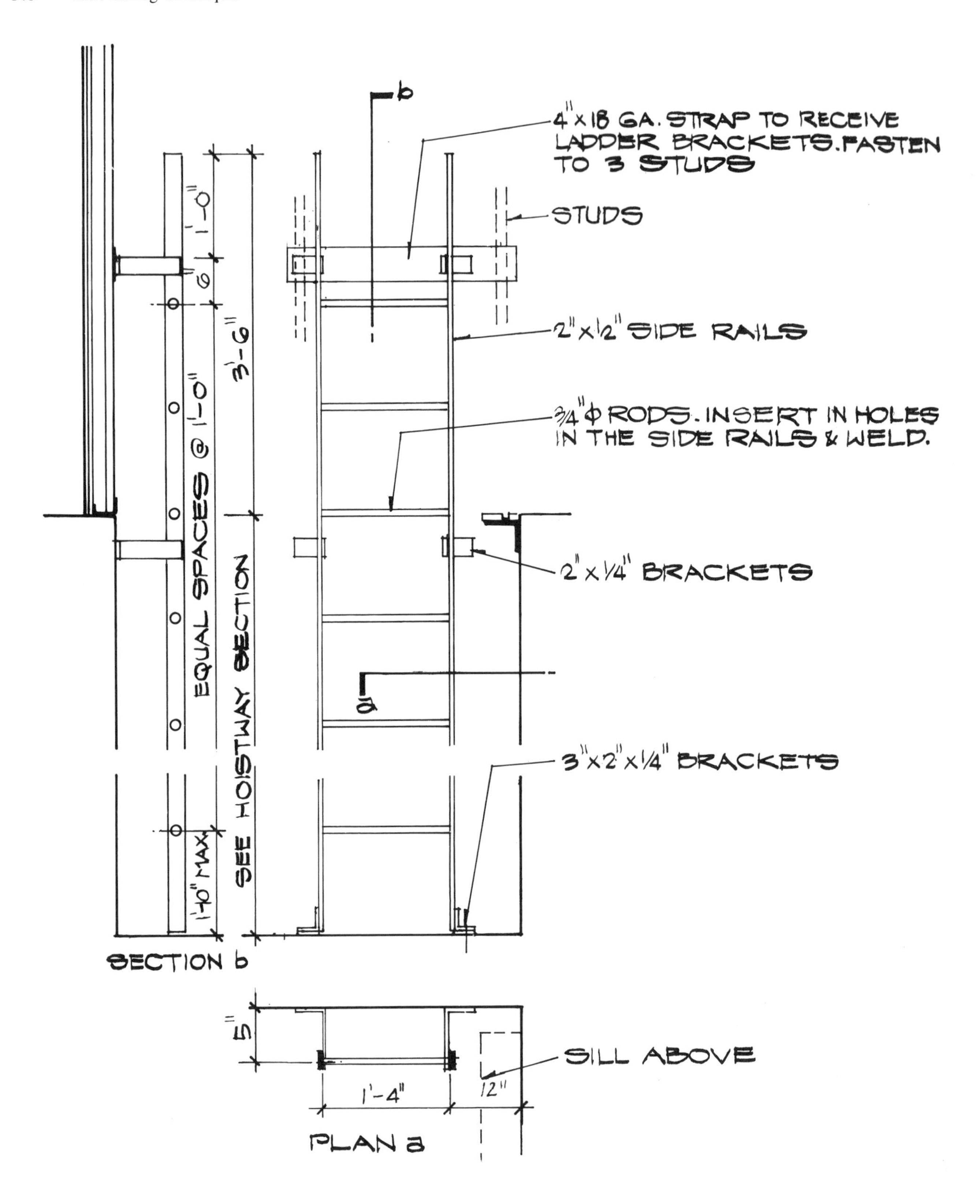

3/4"=1'-0"

PIT LADDER DETAIL

FIGURE 4-23. Elevator Pit Ladder Detail.

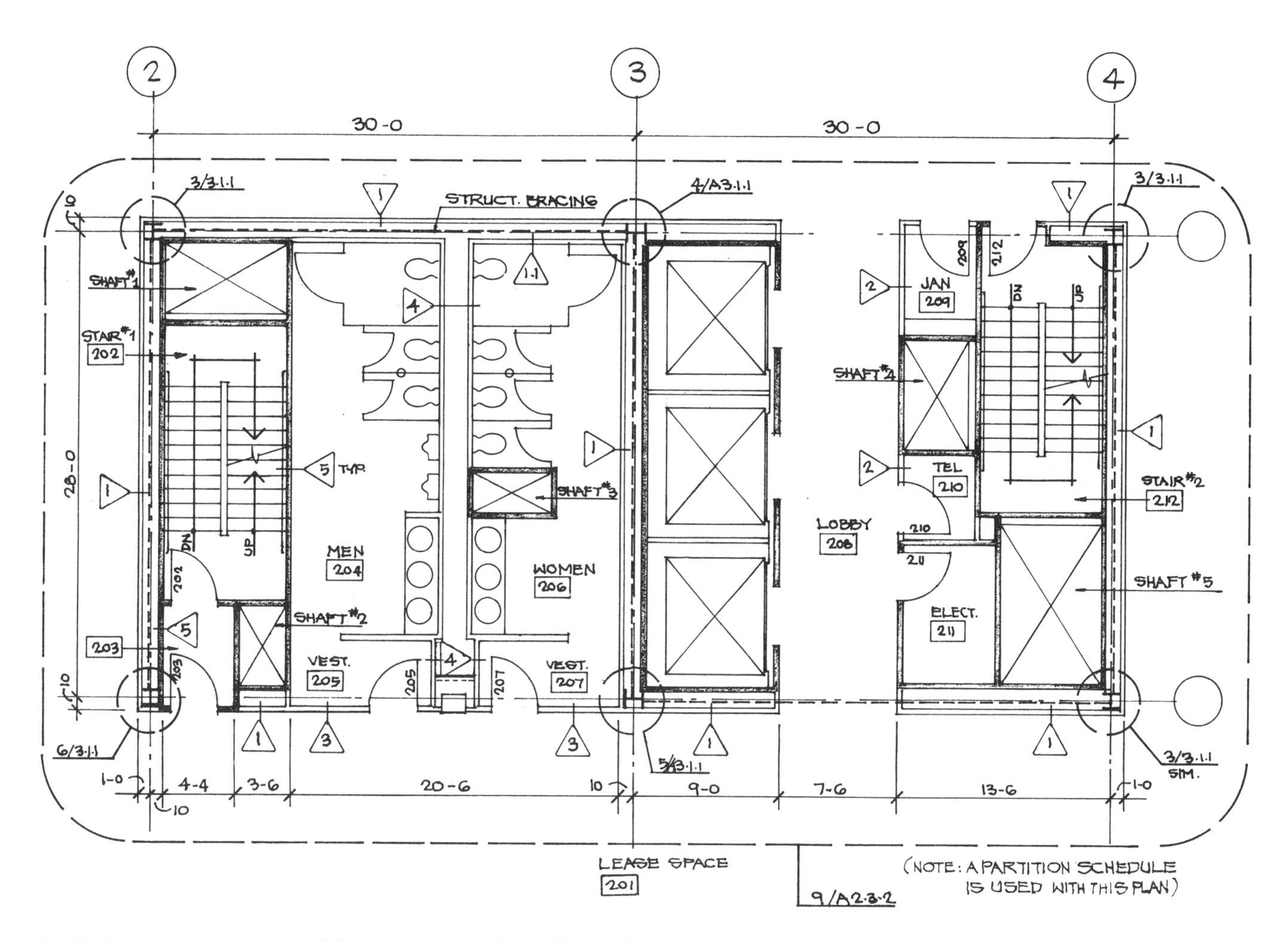

FIGURE 4-24. Example of an Office Building Core Plan (Note: A Partition Type Schedule is used with this core.)

4.2 Partitions

4.2.1 General

Interior walls are usually referred to as partitions. In addition to subdividing the space, partitions serve many functions. They protect the occupants from noise, fire, smoke, and, with lead shielding, protect against radiation. They also provide security and privacy.

Gypsum dry wall or gypsum wallboard partitions (GWB) and concrete masonry unit (CMU) partitions are the types most commonly used to subdivide interior spaces in buildings. This section does not attempt to include details of all available partition types. To do so is beyond the scope of this book. I have narrowed the selections to the more common types, stating the fire and sound-transmission class (STC) ratings of each. In addition, I have added useful hints under each detail to help the team members understand its limitation.

Each partition-type detail is identified with numbers keyed to a legend (Fig. 4-48). Offices that use the ConDoc method of identifying materials by CSI numbers or conventionally with notations can cut the numbers out and add the CSI numbers or notations as the case may be.

Refer to the code analysis for the project to identify the partitions that are required to be rated and to Architectural Graphic Standards for the recommended acoustical ratings between spaces and choose the type that best satisfies these requirements. Be sure to check the guidelines under the detail for more information, including upgrades for added acoustical and fire protection. Then fill out the fire and STC ratings accordingly. Figure 4-24 shows an example of partition-type identification on an office building core plan.

A novel time-saving way to represent partitions in schedule form is shown in Chapter 2. Partitions are a seemingly simple but complex task. To do this task properly, each partition must conform to the code to provide fire or smoke protection, provide acoustical isolation where required, be strong enough to span between supports, be sealed properly around the perimeter, and be readily identifiable on the drawings. This section strives to provide as much of this information as possible. Other sources of information are also listed to aid the reader in adding to the types included here.

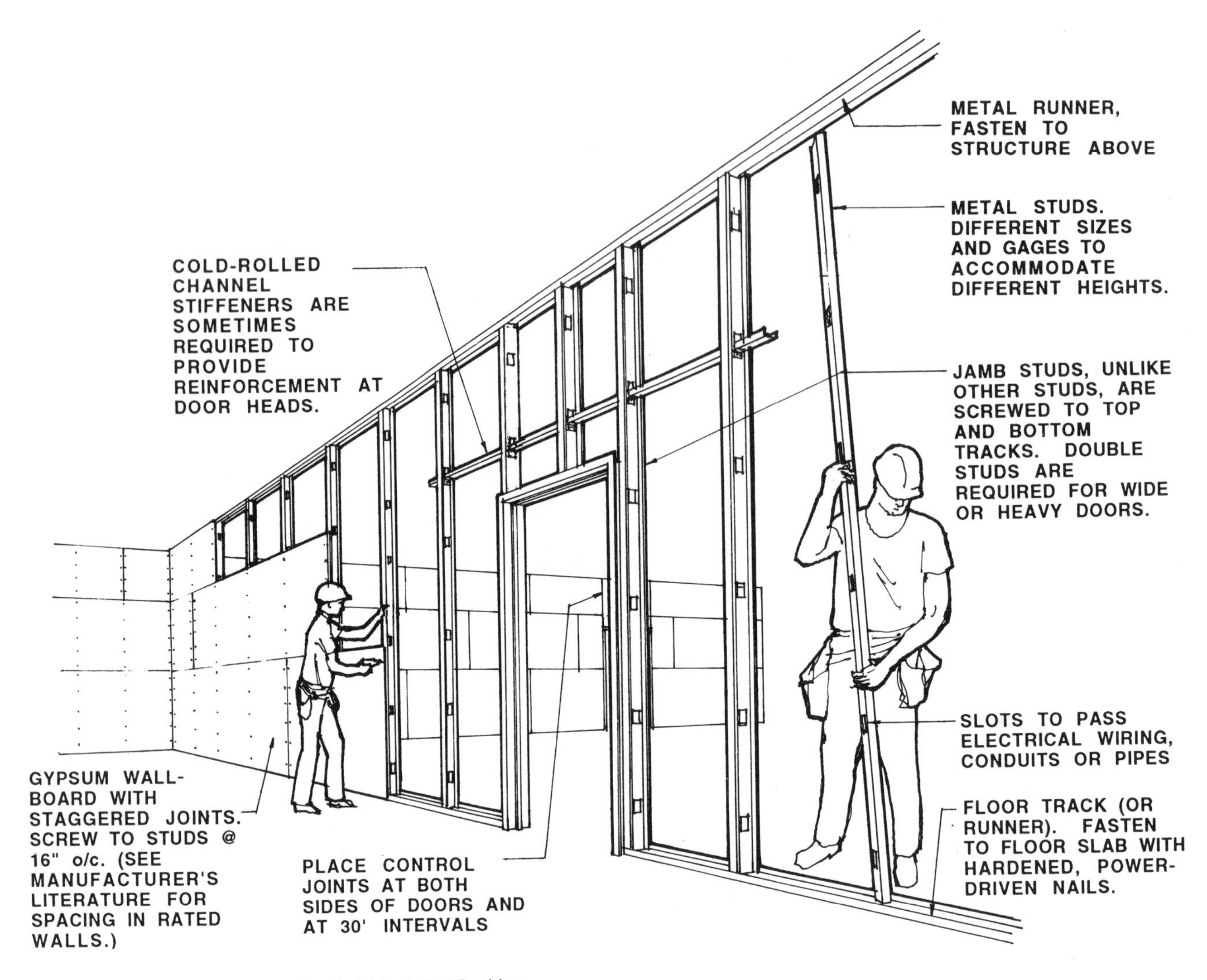

FIGURE 4-25. Perspective of a Standard Metal Stud Partition.

4.2.2 Standard Gypsum Wallboard (GWB) Partitions

These partitions (Fig. 4-25) are the most commonly used types. Partitions used in office buildings and retail spaces to divide between tenants are usually referred to as demising walls. They fall into this category. Type "X" gypsum wallboard is used in fire-rated construction. Depending on the number of GWB layers and the thickness of the boards, partitions can provide 1, 2, 3, or 4-hour fire protection. The following set of details represents some of the types most commonly used in commercial and public buildings (Fig. 4-26 to 4-34).

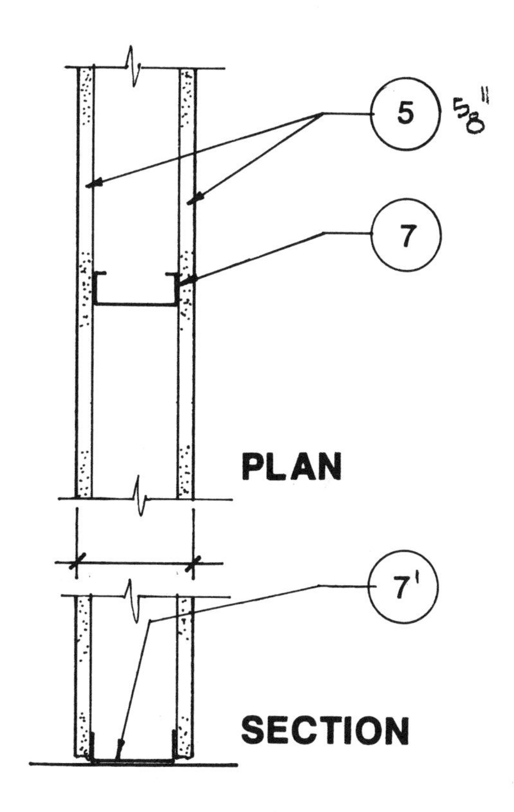

FIRE RATING:
FIRE TEST:
STC RATING:

TYPE:

Guidelines:

1. Nonrated wall shown. One-hour rating can be achieved with Type "X" GWB and perimeter caulking. Fire tests: UL Des. #465 and other tests.
2. Height limitation:
   13'-6" for 25 GA. 3 5/8" studs @ 24" o/c
   16'-0" for 25 GA. 3 5/8" studs @ 16" o/c
3. STC rating:
   40 as shown
   49 with 3" sound attenuation blankets

FIGURE 4-26. GWB Partition Detail.

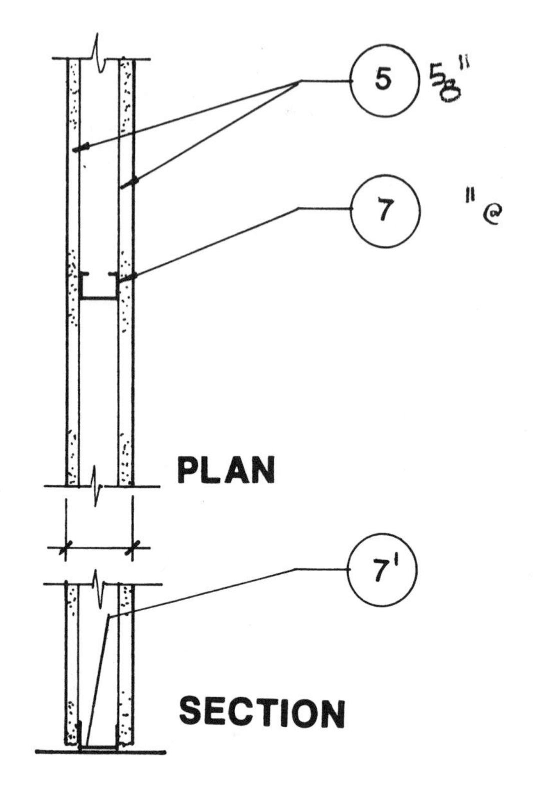

FIRE RATING:
FIRE TEST:
STC RATING: 35-39

TYPE :

Guidelines:

1. Not fire-rated as shown. Use Type "X" GWB and perimeter caulking for 1-hr. rating. Fire test: UL Des. #U448 with 2 1/2" studs. Other tests for 1 5/8" studs.
2. Height limitation:
   10'-9" with 25 GA. studs @ 24" o/c
   12'-6" with 25 GA. studs at 16" o/c

FIGURE 4-27. GWB Partition Detail.

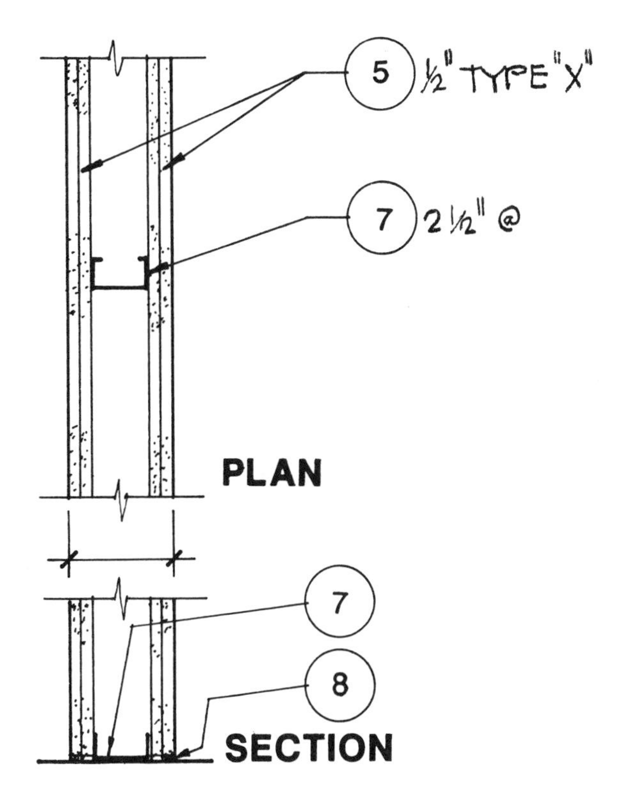

FIRE RATING: 2 HRS.
FIRE TEST: UL Des. #U412
STC RATING:

TYPE :

Guidelines:

1. Face layer is attached with screws in the design shown. Fire Test OSU T3218, 9-17-65 is similar but with an adhered face layer.
2. Height limitation:
   11'-3" with 2 1/2" studs @ 24" o/c
   13'-6" with 2 1/2" studs @ 16" o/c
3. STC 54 with 1 1/2" mineral fiber insulation.

FIGURE 4-28. GWB Partition Detail.

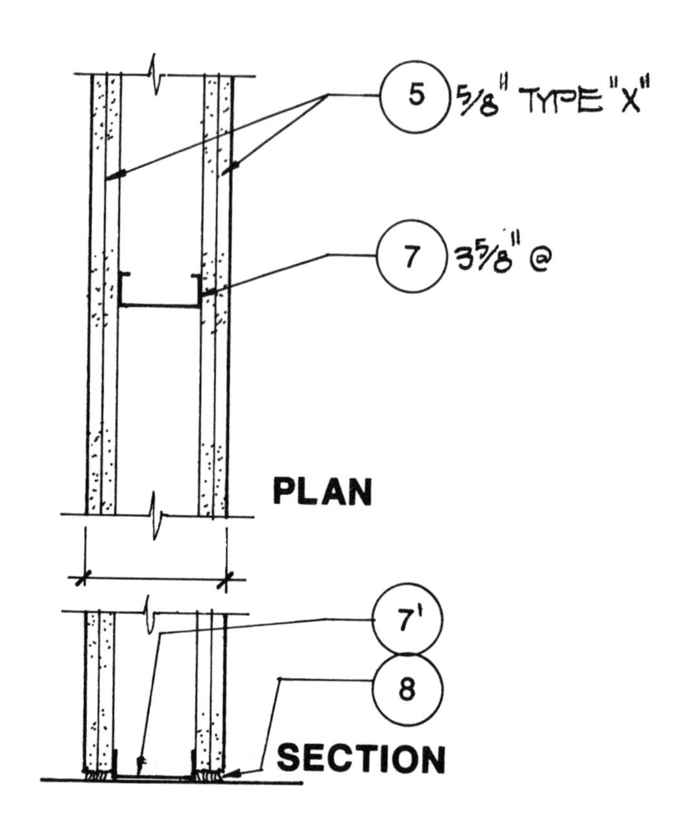

FIRE RATING: 2 HRS.
FIRE TEST: U.L. DES. #U411
STC RATING:

TYPE :

Guidelines:

1. STC rating:
   48 as shown
   56 with 2" sound attenuating fire blanket
2. Height limitation:
   13'-6" with 25 GA. studs @ 24" o/c
   16'-9" with 25 GA. studs @ 16" o/c
   Use heavier studs to increase height.

FIGURE 4-29. GWB Partition Detail.

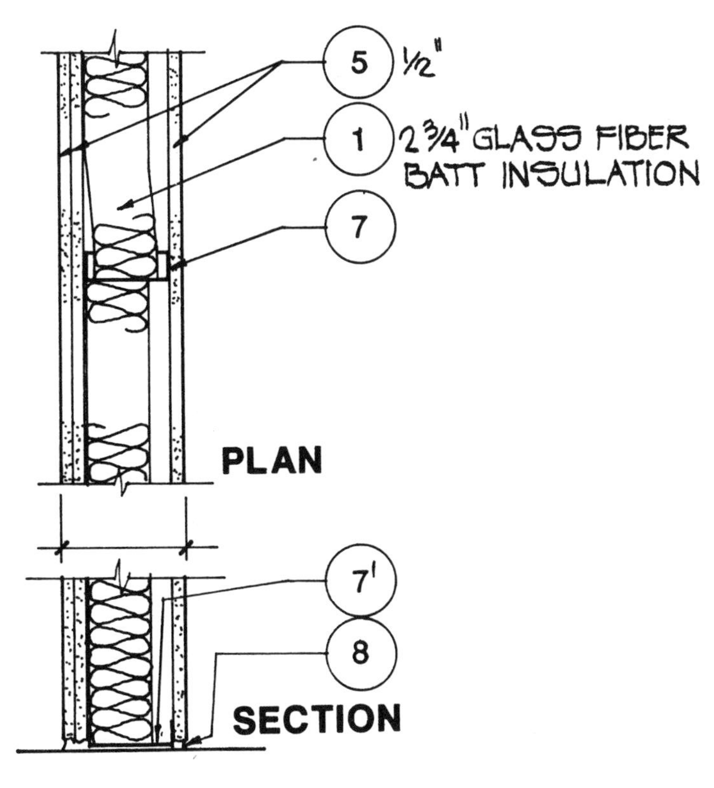

FIRE RATING:
FIRE TEST:
STC RATING:

TYPE :

Guidelines:

1. For 1-hr. fire rating, add the designation TYPE "X" beside #5 above. Fire test: FM WP733, 12-3-84.
2. Height limitation:
   13'-5" with 2 1/2" studs @ 24" o/c
3. Fill out wall thickness based on stud size.
4. STC rating varies; consult with manufacturers and write insulation information beside designation #1.

FIGURE 4-30. GWB Partition Detail.

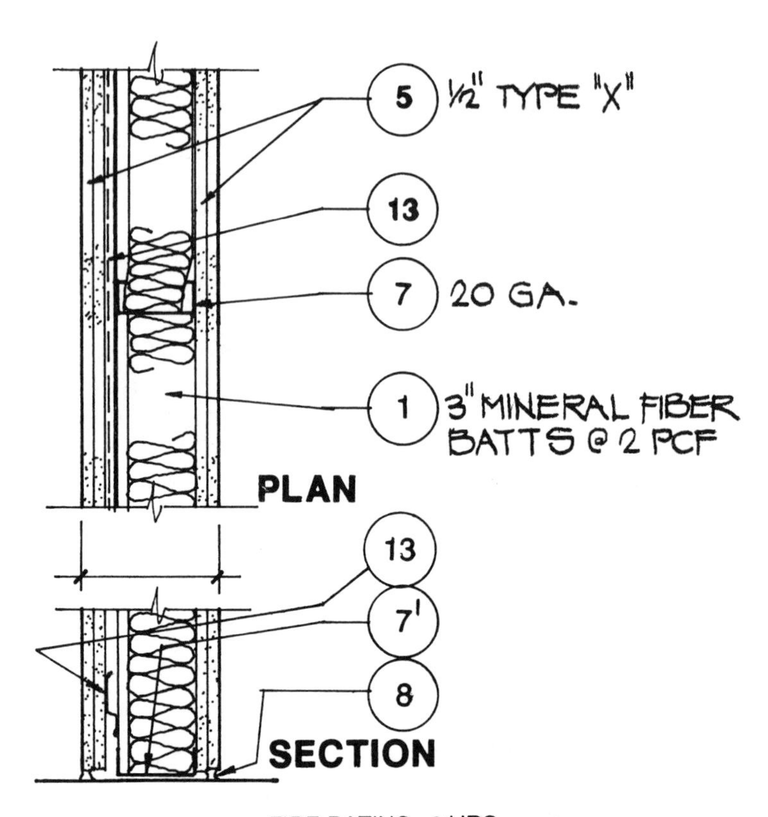

FIRE RATING: 2 HRS.
FIRE TEST: U.L. DES. #U454
STC RATING: 60
: 61 W/ 5/8" GWB

TYPE :

Guidelines:

1. This is a USG proprietary design. Other manufacturers have similar designs with a lower STC rating.
2. Height limitation:
   17'-03" with 3 1/2" studs spaced @ 24" o/c.
   Height may be increased. Consult with manufacturer.

FIGURE 4-31. GWB Partition Detail.

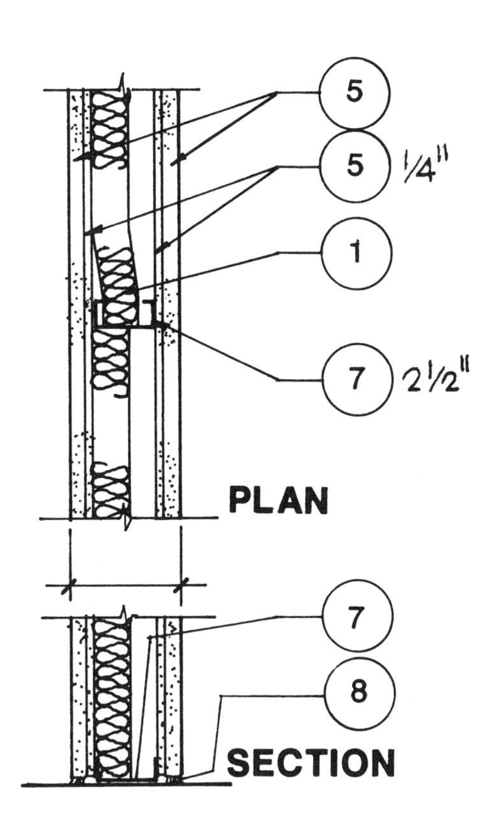

FIRE RATING:
FIRE TEST:
STC RATING: 55-59

TYPE :

Guidelines:

1. For 1 hr. fire-rating, add the designation - TYPE "X" - beside #5 above. Fire test : T-1174-0SU
2. For STC 50-54, use 1/2" GWB and 2" glass fiber insulation. For STC 55, use 1 1/2" mineral fiber insulation.
3. Height limitations:
   11'-3" with 25 GA. studs @ 24" o/c
   13'-6" with 25 GA. studs @ 16" o/c
4. Fill in the missing dimension.

FIGURE 4-32. GWB Partition Detail.

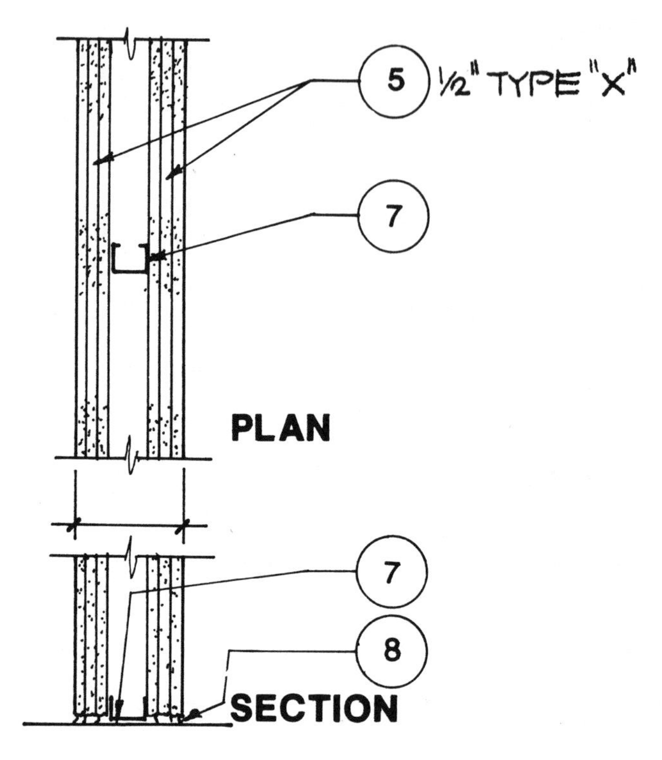

FIRE RATING: 3 HRS.
FIRE TEST: U.L. DES. #U435
STC RATING:

TYPE:

Guidelines:
1. This type is usually used as a fire separation wall.
2. Height limitation:
   12'-4" with 1 5/8" studs @ 24" o/c.
3. STC rating is:
   59 with 1 1/2" mineral fiber insulation
   50-54 with 1 1/2" glass fiber insulation

FIGURE 4-33. GWB Partition Detail.

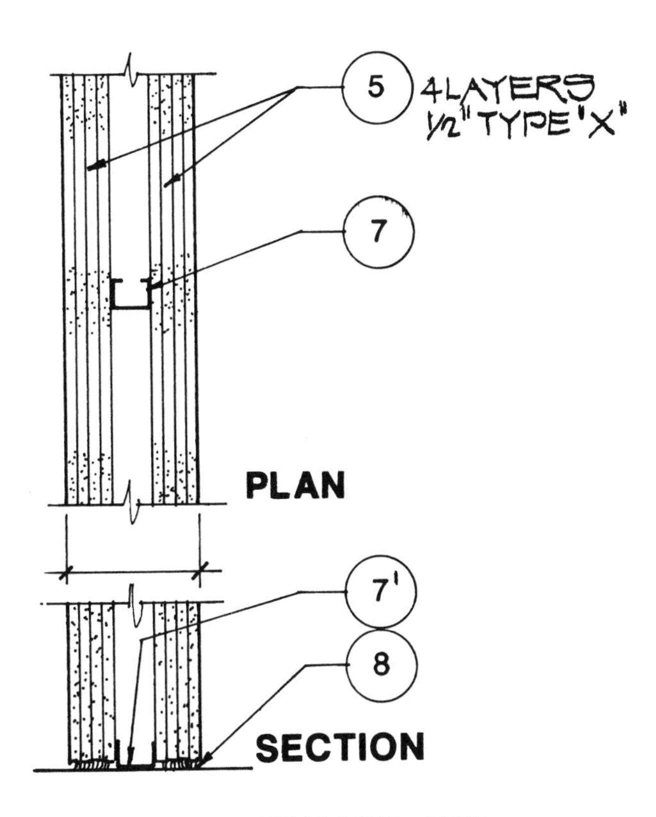

FIRE RATING: 4 HRS.
FIRE TEST: U.L. DES. #U435
STC RATING:

TYPE :

Guidelines:

1. This type is usually used as a fire separation wall.
2. Height limitations, same as Fig. 4-33.
3. STC rating:
   62 with 1 1/2" sound attenuation blanket
   55 with 1 1/2" mineral fiber insulation
4. Thickness depends on stud size.

FIGURE 4-34. GWB Partition Detail.

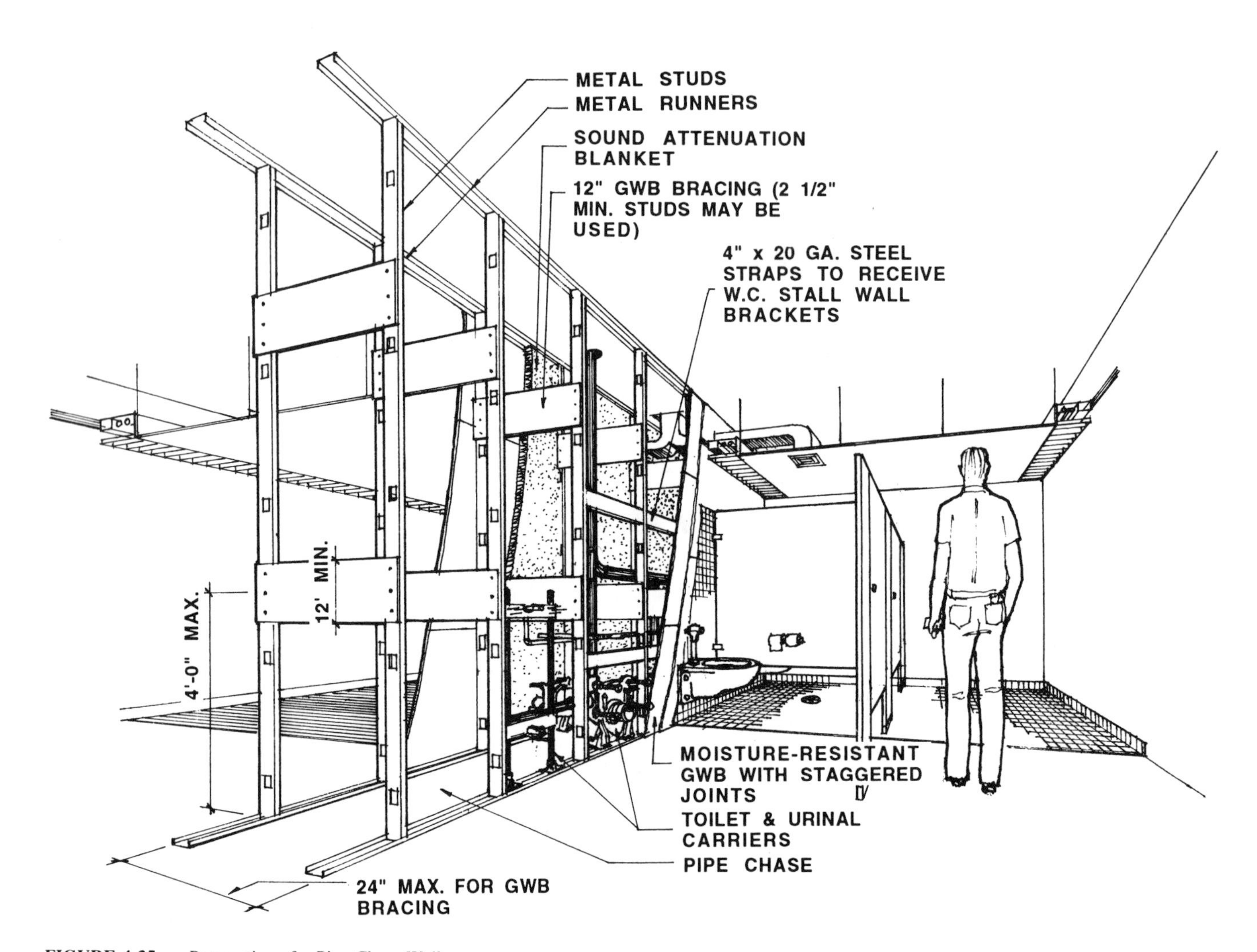

FIGURE 4-35. Perspective of a Pipe Chase Wall.

4.2.3 Pipe Chase Walls

As the name implies, these walls are designed to enclose pipes (Fig. 4-35). They must be constructed like a standard partition (instead of a chase wall) if the pipes are relatively narrow and the number of fixtures is limited. Because carriers for wall-mounted toilets and urinals cannot be accommodated in that type of partition, a double wall arrangement is required. Refer to Architectural Graphic Standards for recommended chase widths and check with the plumbing engineer to make sure that the pipe runs, and vertical risers will fit inside the space between the studs.

To provide sound isolation, chase walls should continue to the structure. In addition, sound attenuation blankets must be placed in at least one side of the chase as well as around the perimeter of the rest room. Toilet accessories such as flush-mounted feminine napkin disposals mounted at the back of the stall should be sealed properly to prevent the creation of acoustic flanking paths. A better solution is either to mount them on stall partitions or, if placed at the back of the stall, use the surface-mounted type. Figures 4-36 and 4-37 are examples of chase walls.

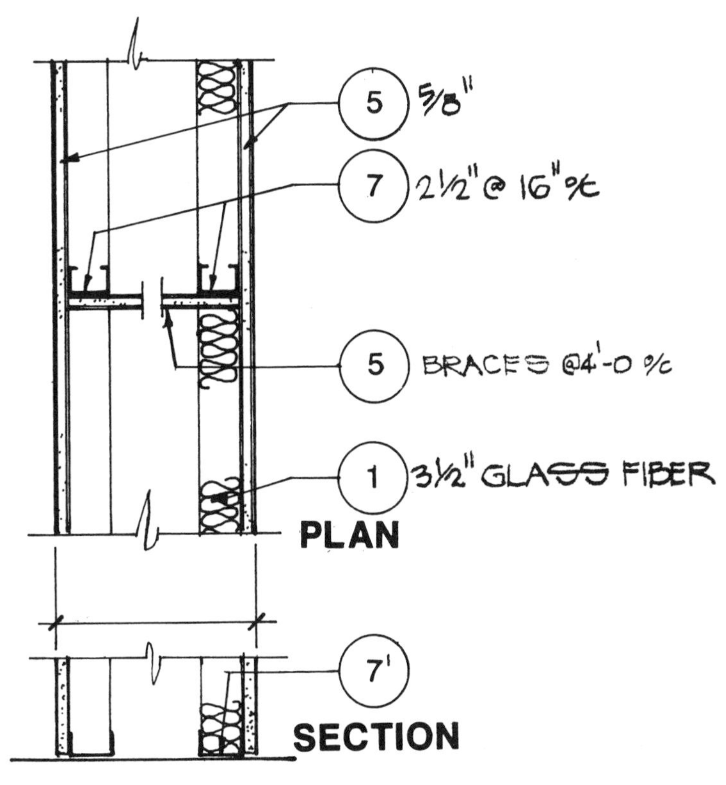

FIRE RATING:
FIRE TEST:
STC RATING: N/A

TYPE :

Guidelines:

1. For 1-hr. fire rating, add the designation Type "X" beside #5. Fire test: U.L. Design #U420.
2. Height limitation:
   11'-6"
   If more height is required, brace at 11'-6" max. or consult with manufacturer.
3. Use studs as bracing if chase width is over 24". Fill in width dimension or write "width varies, see plan" if more than one chase width is used on the project.
4. This is a pipe chase partition, see Fig. 4-35.

FIGURE 4-36. Chase Wall Detail.

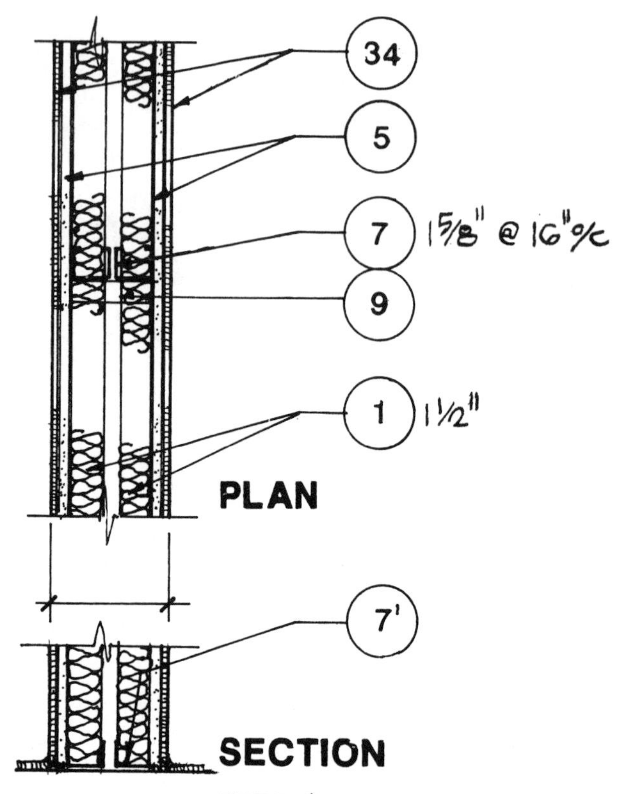

FIRE RATING: 1 HR.
FIRE TEST: U.L.DES. #U445
STC RATING: 60

TYPE :

Guidelines:

1. This is a USG proprietary design for a chase wall. Use only if sound isolation is very important.
2. Height limitation:
   13'-3"
3. Ceramic tile may be installed on one side only provided the opposite side is 3/8" Type "X" GWB.
4. Width is 5" as shown, 4 5/8" if guideline #3 is used.

FIGURE 4-37. Chase Wall Detail.

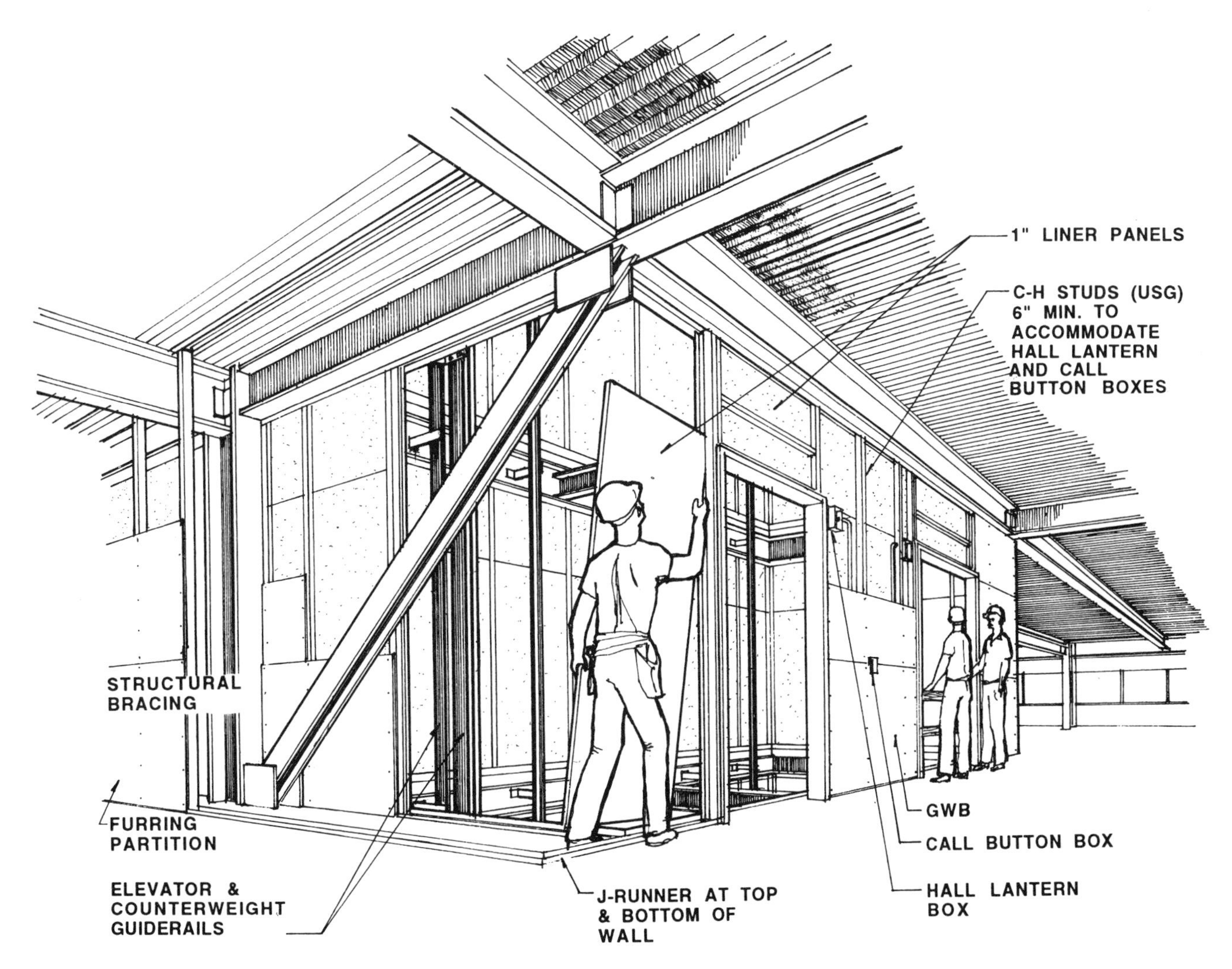

FIGURE 4-38. Perspective of a Shaft Wall.

4.2.4 Shaft Walls

Shaft walls (Fig. 4-38) were invented to facilitate the construction of enclosures around floor openings such as elevator and mechanical shafts. They are designed to be built from one side only, thus obviating the need for scaffolding to support the person attaching the wallboard on the side of the partition facing the floor opening.

Shaft walls are usually more expensive than standard GWB partitions, but this is offset by the savings in the time and labor required to erect the scaffolding. For that reason, their use should be restricted to the aforementioned locations and similar enclosures such as egress stairs.

To size the studs around elevator shafts, observe the following procedure:

1. Find the elevator speed and the number of elevators in the shaft. This is usually determined by the elevator representative in small projects or the elevator consultant in high-rise buildings.
2. Refer to the manufacturer's shaft wall tables and find the pressure on the wall resulting from the movement of the elevator (Fig. 4-39).

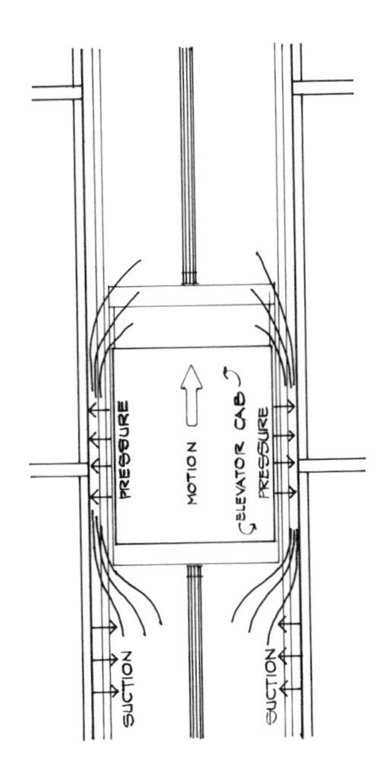

FIGURE 4-39. Pressures on a Shaft Wall.

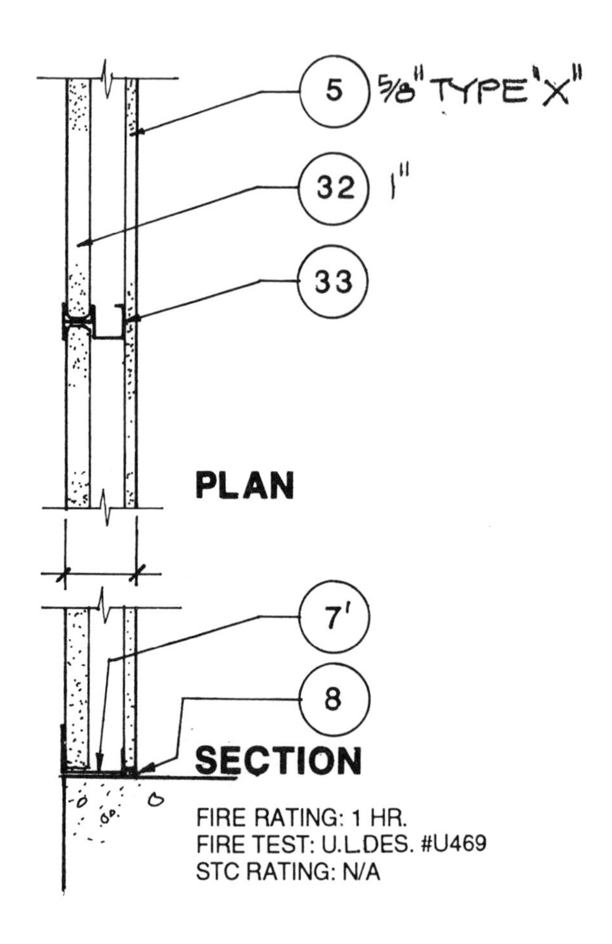

TYPE :

Guidelines:

1. In stair enclosures, attach an extra 1/2" layer of GWB to the 1" liner and push the slab edge back 1/2".
2. Height limitation:
   14'-0" with 25 GA. studs @ 24" o/c
   18'-0" with 22 GA. studs @ 24" o/c
   19-9" with 20 GA. studs@ 24" o/c
3. See Section 4.2.4 for information about elevator shaft walls including limiting heights.
4. STC rating:
   35 as shown
   39 with 1" mineral fiber insulation

FIGURE 4-40. Shaft Wall Detail.

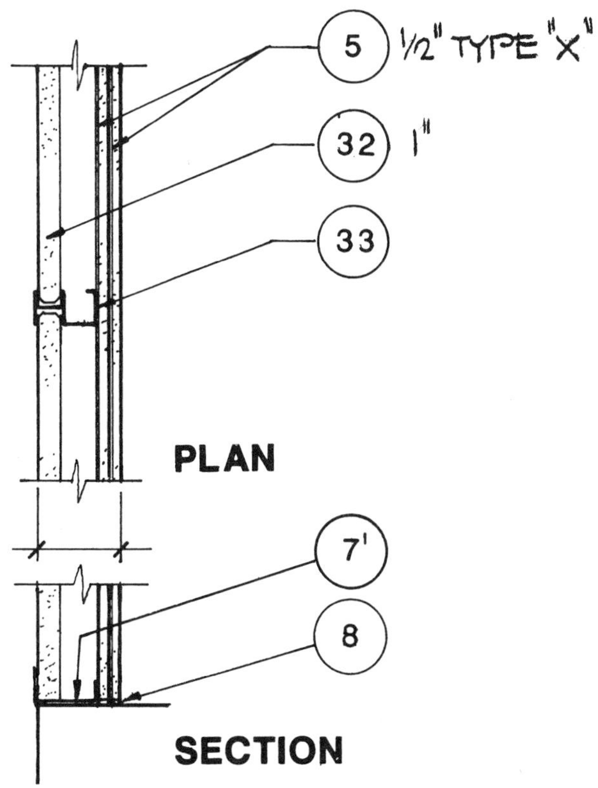

FIRE RATING: 2 HRS.
FIRE TEST: U.L.DES. #U438
STC RATING:

TYPE:

Guidelines:

1. See manufacturer's tables for height limitations.
2. Refer to manufacturer's tables for height limitations at elevator shafts.
3. STC rating:
   39 as shown
   47 with 1" sound-attenuation fire blanket
4. Wall thickness varies depending on stud size. Minimum size is 2 1/2" for a wall thickness of 3 1/2". Height limit may be increased with 4" or 6" studs.

FIGURE 4-41. Shaft Wall Detail.

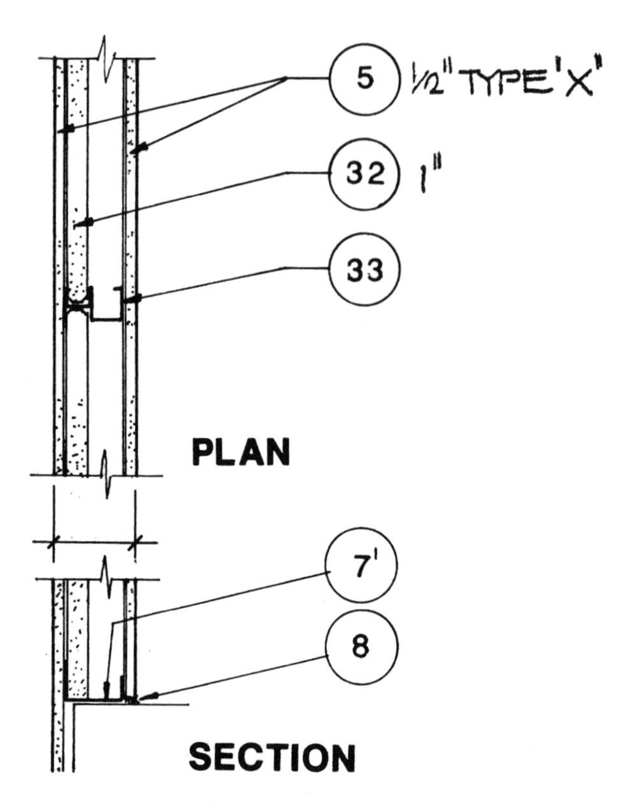

TYPE:

FIRE RATING: 2 HRS.
FIRE TEST: U.L. DES. #U467
STC RATING:

Guidelines:

1. See notes #1 & #2, Fig. 4-41.
2. STC ratings:
   39 without insulation
   50 with 1 1/2" sound-attenuation fire blankets.
3. This partition may be used as an area separation wall.

FIGURE 4-42. Shaft Wall Detail.

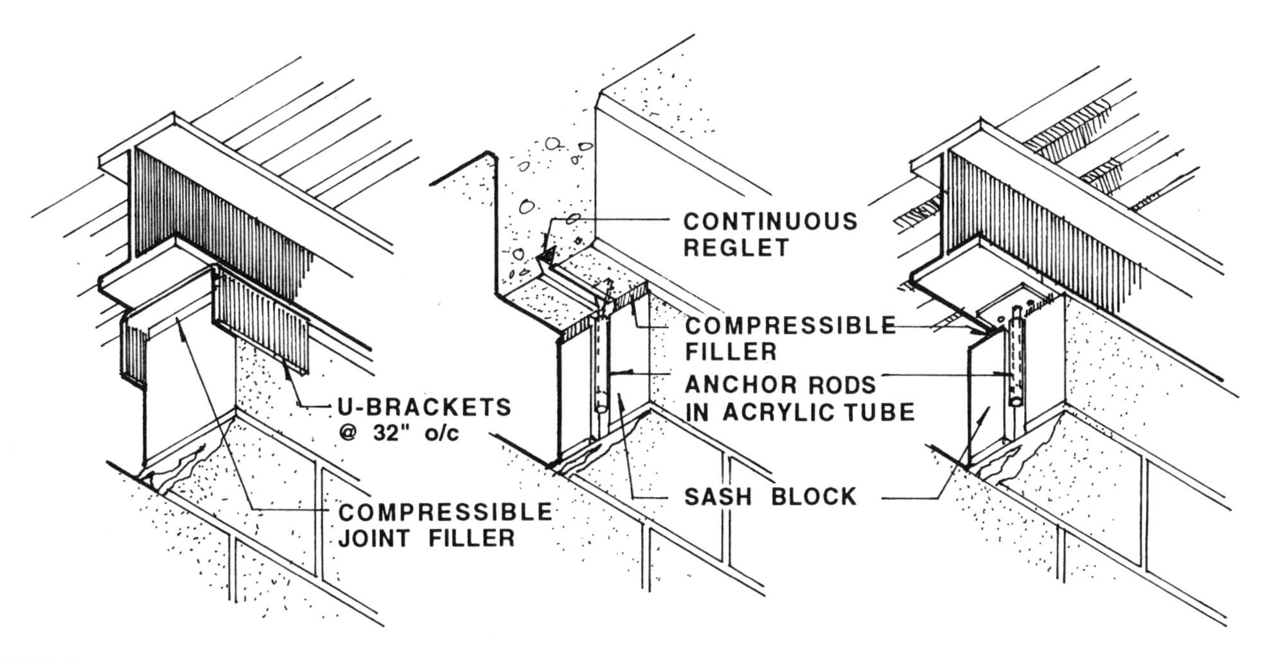

FIGURE 4-43. Some Methods for Anchoring CMU Walls.

3. From the table choose a stud size corresponding to the height of the partition. Use the L/120 column for plain shaft walls, the L/240 column for shaft walls with veneer (marble, granite, or similar materials). Use the L/360 column for brittle veneers that can tolerate very little movement without cracking, such as plaster. Figures 4-40, 4-41, and 4-42 show shaft wall details.

4.2.5 CMU Partitions

Concrete Masonry Unit (CMU) partitions are usually used in locations where standard GWB partitions may be subject to damage from bumps, collision, or dampness. Dock areas and electric vaults are examples of areas where CMU is the preferred material for partitions.

Building codes stipulate that CMU partitions must span either vertically or horizontally. To figure the wall thickness, codes specify a maximum ratio of $L \div T$ where L is the span and T is the nominal thickness. The National Concrete Masonry Association recommends a ratio of 36 for non-bearing walls. For example, if the span from the floor to the bottom of the structure is 12′ (144″) and $L \div T = 36$, then, $144 \div T = 36$ or $T = 144 \div 36 = 4''$. The structural engineer should be involved in the determination of the thickness because the wall may be subjected to wind pressure or impact forces. He or she may fill the cells with grout and add reinforcement to the courses subject to impact. Figure 4-43 shows some methods of attaching the top of the wall to the structure. There again the structural engineer should make the determination as to the method of attachment. Walls may also span horizontally between pilasters or intersecting walls. The shortest vertical or horizontal span is used in determining the wall thickness.

4.2.6 Partition Top Details

Because a partition type may extend to the structure in one location and end at the ceiling or just above it at other locations, the details showing top of partitions are drawn separately and given separate designations (Fig. 4-44 to 4-48) so that they may be coupled separately in the partition-type designation to describe these different conditions.

For example, the partition shown in Figure 4-27 may be coupled with any of the partitions shown in Figure 4-44 to represent any of these six conditions. Figure 4-49 shows an example of general notes for partitions.

4.2.7 Miscellaneous Partition Types

In addition to the partition types mentioned in the previous subsections, there are other types less commonly used. These include:

1. Demountable or relocatable partitions (Fig. 4-50). These are used mostly in office buildings that require frequent changes due to tenant moves or reorganization within large companies. The manufacturer claims that these partitions can be dismantled and erected in a new location without damage to the components. Many developers

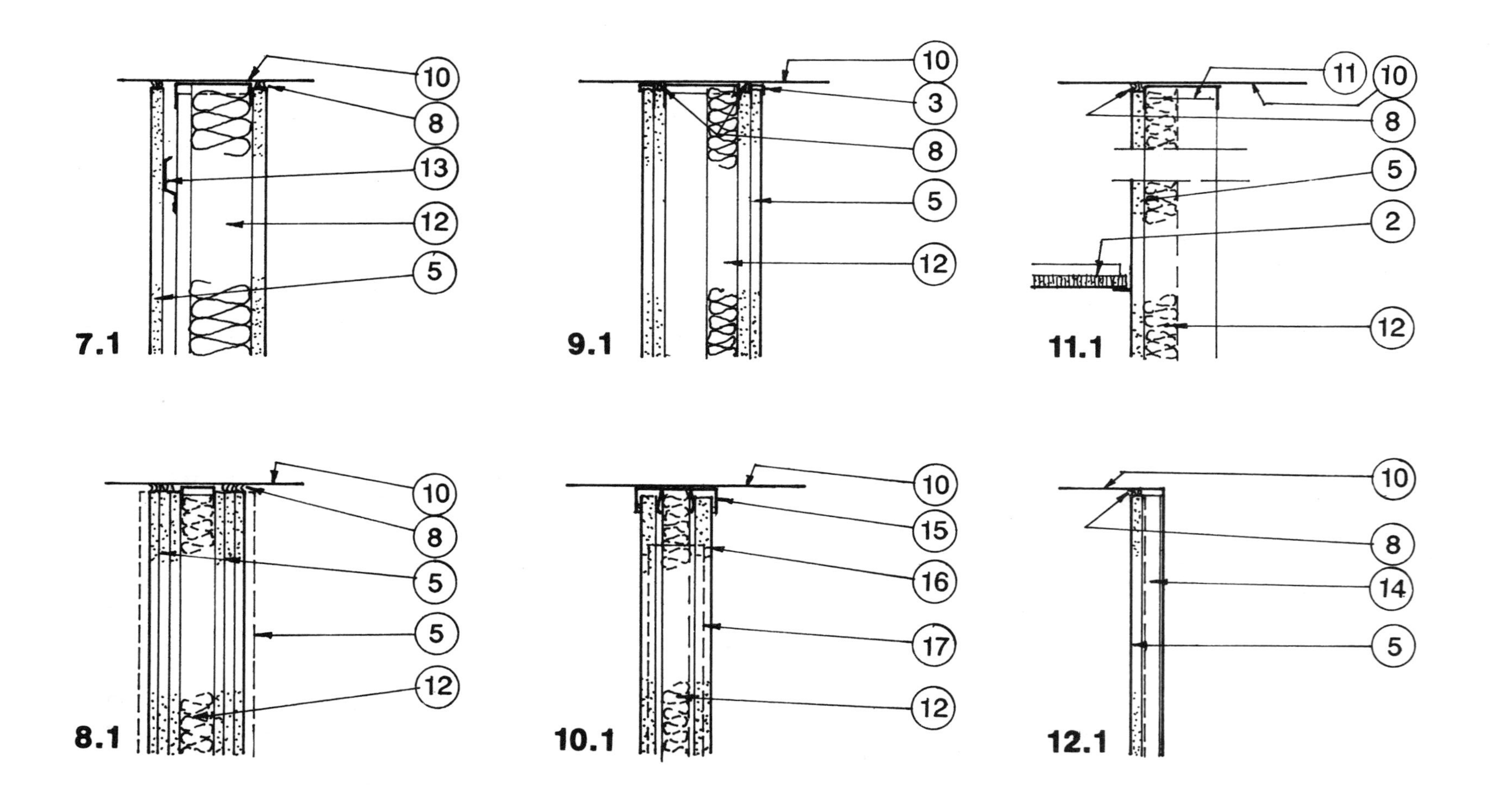

TOP OF WALL DETAILS (GWB)

NOTE #1:
IF TOP OF WALL DETAILS DIFFER FROM PARTITION TYPES, PARTITION TYPES SHALL PREVAIL.
NOTE #2:
SEE GUIDELINES IN FIG. 4-44.

FIGURE 4-45. Top of Wall Details (cont.)

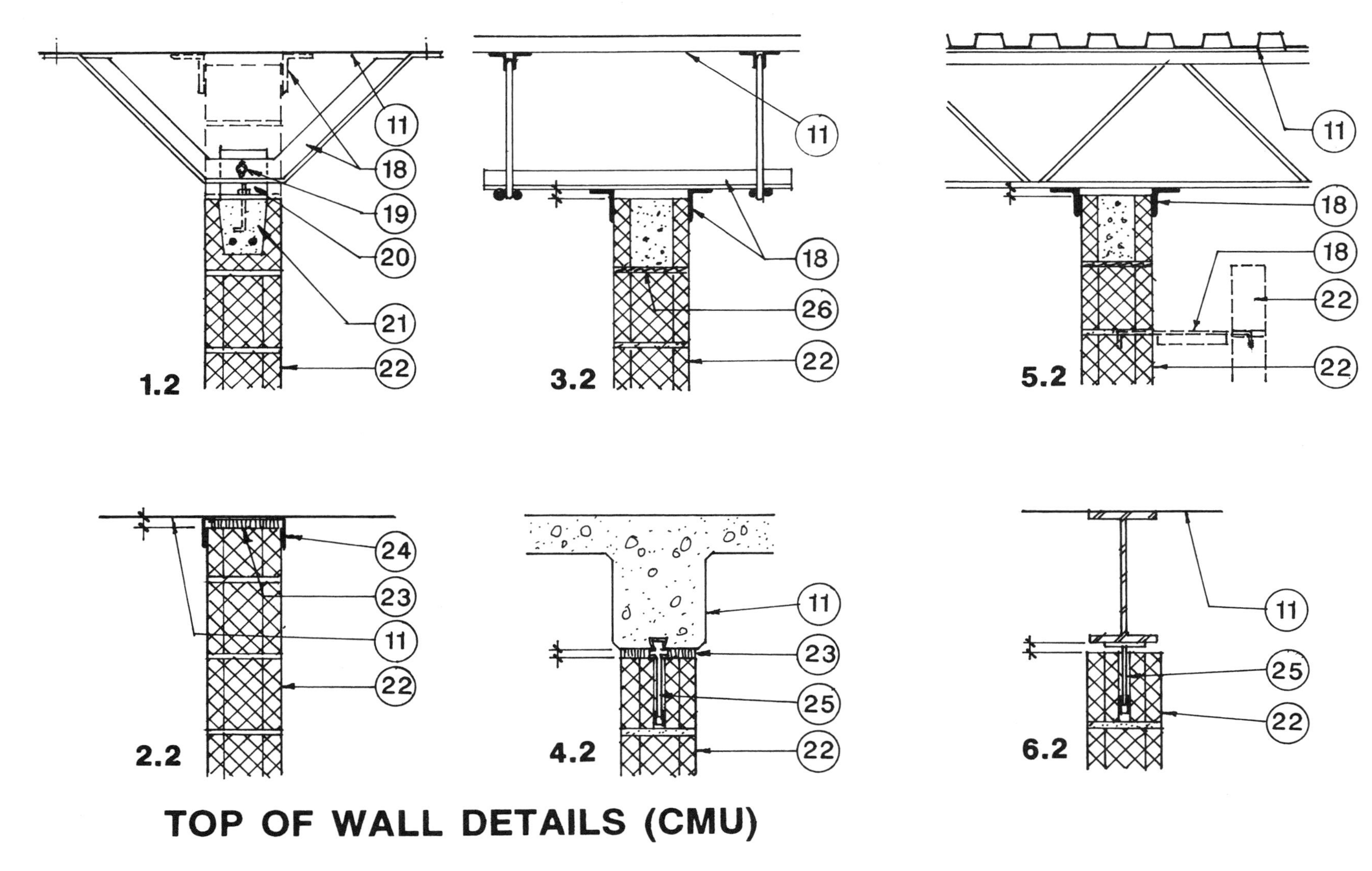

TOP OF WALL DETAILS (CMU)

Guidelines:

1. Fill in missing clearance dimensions in consultation with the structural engineer based on the amount of deflection allowed in the the design.
2. Hardware indicated by #24 and #25 is manufactured by Hohmann (Div. 4 in *Sweet's Catalog*).
3. See legend for notes and explanation of partition numbering system. Note that different numbers are given to the CMU details to allow both the GWB & CMU details to be added-to while remaining separate.

FIGURE 4-46. Top of Wall Details (CMU).

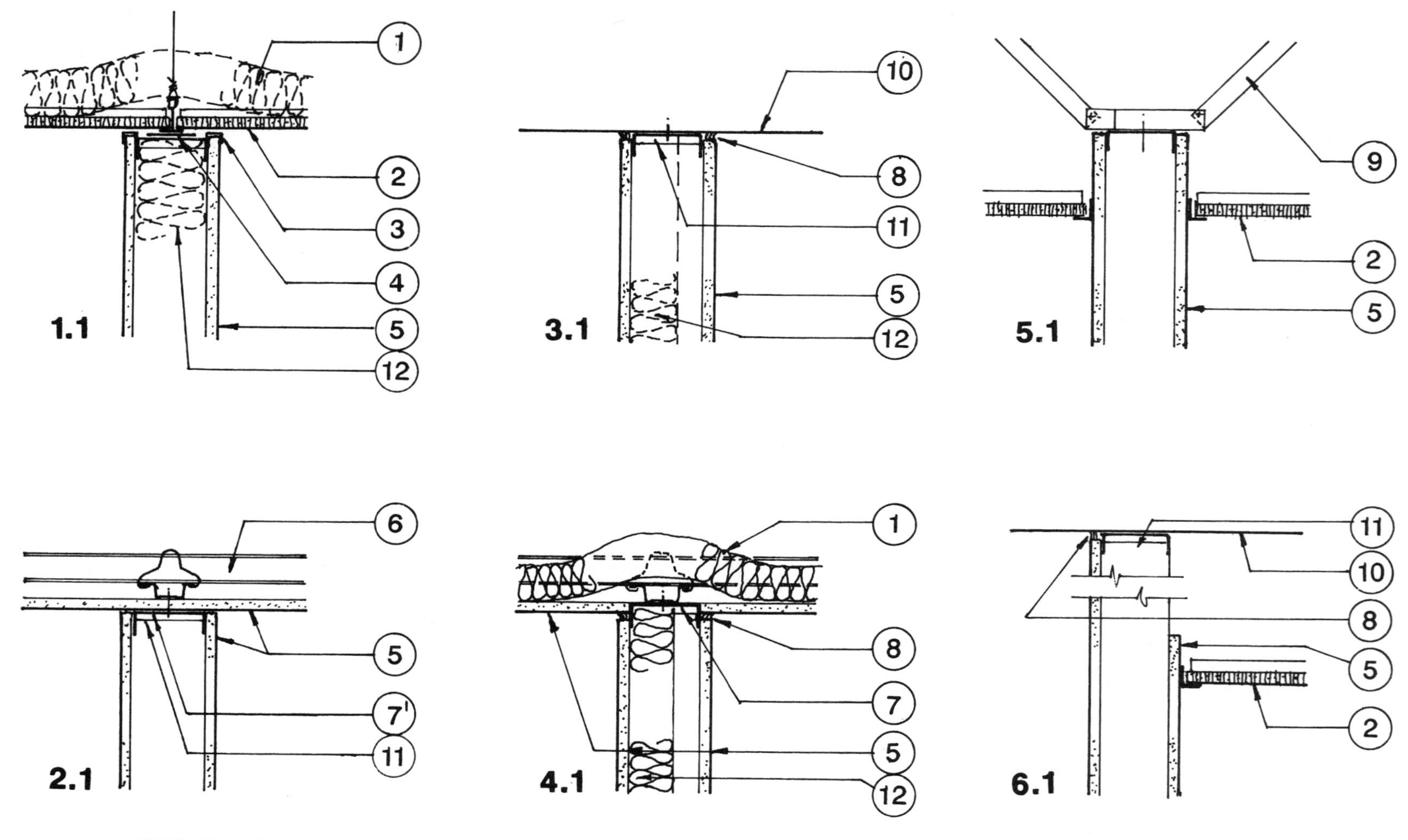

TOP OF WALL DETAILS (GWB)

Guidelines:

1. Details shown above are generic details to be used on all projects. Details not applicable to the project at hand are to be labeled "NOT USED." These details are to be added to Fig. 4-45, Fig. 4-46, Fig. 4-47, and Fig. 4-48. Add details as required.
2. See legend (Fig. 4-48) and Fig. 4-49 for general notes.
3. Add insulation as required. Modify details to agree with partition types as much as possible (see Note #1 in Fig. 4-45).

FIGURE 4-44. Top of Wall Details (GWB).

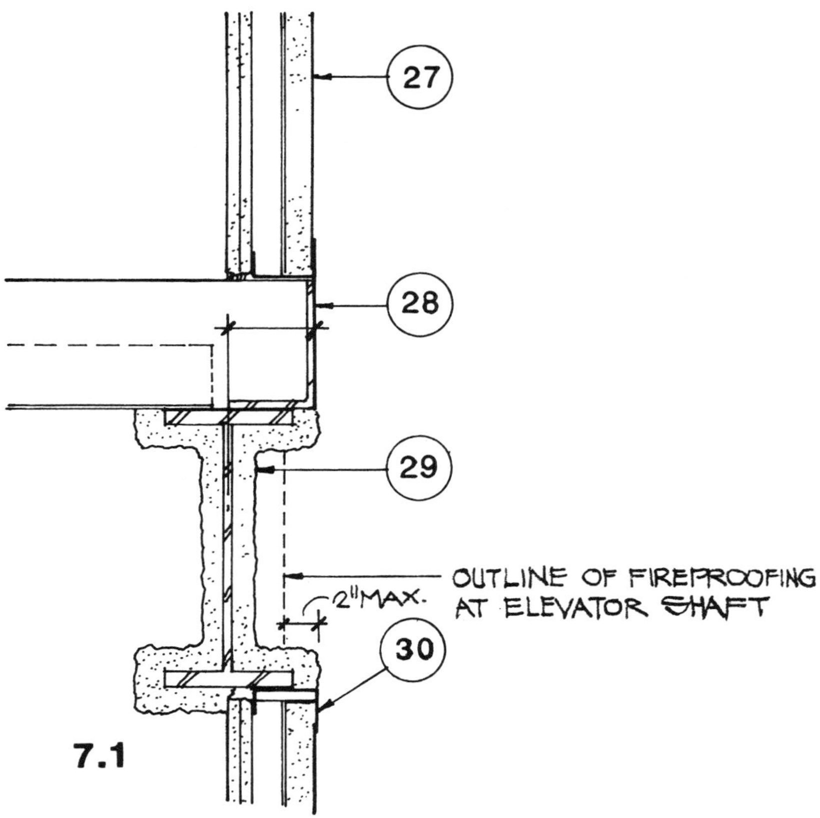

MECH.& ELEVATOR SHAFTS 1 1/2"=1'-0"

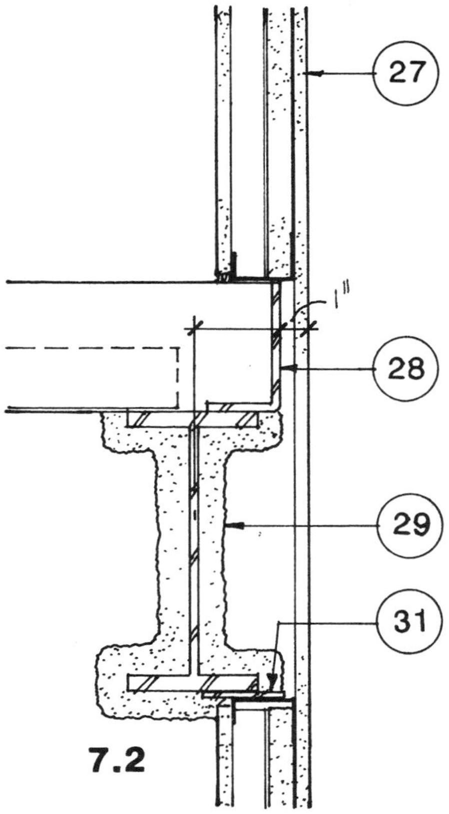

DET.@ STAIR WELL 1 1/2"=1'-0"

Guidelines:

1. Two-hour shaft wall shown. Modify if wall rating is different.
2. Fill in missing dimensions and hour rating for sprayed-on fire proofing.
3. Elevator codes usually require placement of a 75° cant over any ledge that exceeds 2". By bringing the spray outline to within 2" of the face of the wall, no cant is needed.
4. If distance between top and bottom runners in stair detail exceed 2'-0", provide support for GWB.

FIGURE 4-47. Top of Wall Details (Shafts and Stairs).

PARTITION MATERIALS KEYING LEGEND

1. ACOUSTIC BLANKET, EXTEND 4'-0" BEYOND EACH SIDE OF PARTITION
2. LAY-IN CEILING
3. VINYL TRIM, FINNED
4. COMPRESSIBLE GASKET
5. GYPSUM WALLBOARD (GWB) OR AS SHOWN
6. 1 1/2 " COLD-ROLLED CHANNEL
7. STUD

7^1 STUD RUNNER

8. SEALANT
9. METAL RUNNER BRACING @ 4'-0" o/c MAX. FASTEN TO STRUCTURE ABOVE
10. STRUCTURE
11. 1/2" GAP BETWEEN TOP OF STUD AND RUNNER
12. INSULATION BATTS
13. RESILIENT CHANNELS
14. FURRING CHANNELS
15. SPECIAL RUNNER
16. SPECIAL STUDS
17. VINYL-CLAD WALLBOARD
18. STEEL ANGLES (SEE STRUCT.)
19. SLOTTED CONNECTION
20. STEEL TEE WITH ANCHOR
21. LINTEL BLOCK (SEE SCHED. FOR REINF.)
22. CMU WALL
23. COMPRESSIBLE JOINT FILLER
24. U-BRACKETS @ 32 IN. o/c
25. ANCHOR. PLACE IN GROOVE OR SASH BLOCK
26. EXPANDED METAL
27. SHAFT WALL
28. SLAB EDGE ANGLE AND DECK (SEE STRUCT.)
29. SPRAYED-ON FIREPROOFING
30. FASTEN RUNNER TO BEAM BEFORE SPRAYING
31. CONT. 14 GA. STEEL PLATE. WELD TO BEAM
32. LINER PANELS
33. METAL SHAFT WALL STUDS
34. CERAMIC TILE
35. PLYWOOD

Guidelines:

If outer layer of partition is a material other than GWB or CMU, identify on the detail beside #5. Other materials may be cement board, moisture-resistant wallboard, plywood, etc.

FIGURE 4-48. Partition Materials Keying Legend.

GENERAL NOTES:

1. ALL PARTITIONS ADJACENT TO OR LOCATED AT THE BACK OF PLUMBING FIXTURES SHALL BE CONSTRUCTED WITH MOISTURE-RESISTANT GWB.
2. APPLIED FINISHES SUCH AS VINYL WALL COVERING, CERAMIC TILES, ETC. ARE NOT SHOWN ON PARTITION-TYPE DRAWINGS UNLESS THE FINISH IS PART OF A FIRE-RATED ASSEMBLY. REFER TO THE ROOM FINISH SCHEDULE FOR THIS INFORMATION.
3. JOINT TAPING MAY BE OMITTED IF UNEXPOSED TO VIEW, PROVIDED HORIZONTAL JOINTS ARE STAGGERED 24" ON OPPOSITE SIDES OF THE PARTITION OR IF THE PARTITION IS MULTILAYERED. TAPING MUST BE APPLIED IF THE PARTITION IS PART OF A SMOKE- OR SOUND-RATED ASSEMBLY.
4. TOP OF PARTITION DETAILS ARE INTENDED TO SHOW HOW THE PARTITION RELATES TO THE STRUCTURE ABOVE AND HOW IT IS SEALED, ETC. REFER TO PARTITION TYPES FOR THE ACTUAL CONSTRUCTION OF EACH PARTITION.

NOTE: Place these notes in the area reserved for notes in the title block. Add notes as needed.

FIGURE 4-49. General Notes for Partitions.

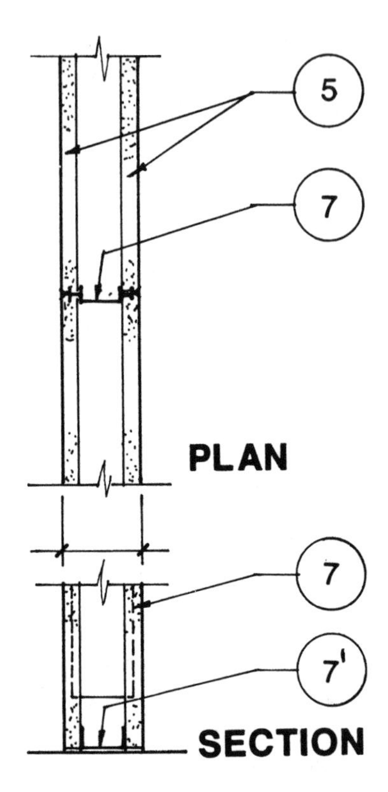

FIRE RATING:
FIRE TEST:
STC RATING:

TYPE :

Guidelines:

1. This is a movable wall (Ultrawall by USG). Similar designs are available from other manufacturers.
2. Height limitation:
   12'-8" with H-studs @ 24" o/c
   11'-10" with H-studs @ 30" o/c
3. STC rating:
   42 as shown
   47 with 1" thermafiber blankets
4. Consult with manufacturer for other STC ratings. Fire ratings are available in 0, 20 min., 1 and 2 hrs. depending on construction.

FIGURE 4-50. Demountable or Relocatable Partition Detail.

FIGURE 4-51. Open-Office Partitions (Westinghouse ASD Group Open-Office System shown).

and building owners, however, believe that standard drywall partitions are more convenient to tear down and re-erect in the new location. The higher initial cost of the demountable partition, the possibility of damage by unskilled workers, and the vinyl wall covering finish, which may not be approved by the lessee, may be factors in that thinking. They are, however, suitable for large companies that reorganize frequently.

Another demountable system is the one used in open office partitions (Fig. 4-51). These systems are part of interior furnishings that, in many cases, are not included in the construction contract.

2. Area separation walls (Fig. 4-52 and 4-53), also called fire walls or party walls, provide fire protection between separate residential units in multifamily developments. They provide a continuous barrier from the foundation to the underside of the roof deck where the roof is of noncombustible construction and is properly fire-stopped at the walls. They extend above the roof and must be enclosed in weatherproof or metal sheathing if the roof is combustible. The Building Code must be consulted for details. These walls are designed to prevent the spread of fire from one unit to the next.
3. Movable partitions belong in the gray area between doors and partitions. They are used to subdivide large spaces, such as conference or ballrooms, into smaller lecture or seminar rooms. The most common types are the familiar folding and accordion partitions. The sound isolation they provide should be researched carefully before a choice is made. Only the overhead track and the jamb condition need to be detailed. Inform the structural engineer about the weight of the partition and the electrical engineer about the power requirements if motor-operated.
4. Furring (Fig. 4-54) is not a partition, but is usually included in the partition types. It is a finish attached to masonry or concrete walls and in some cases, is free-standing (Fig. 4-55).

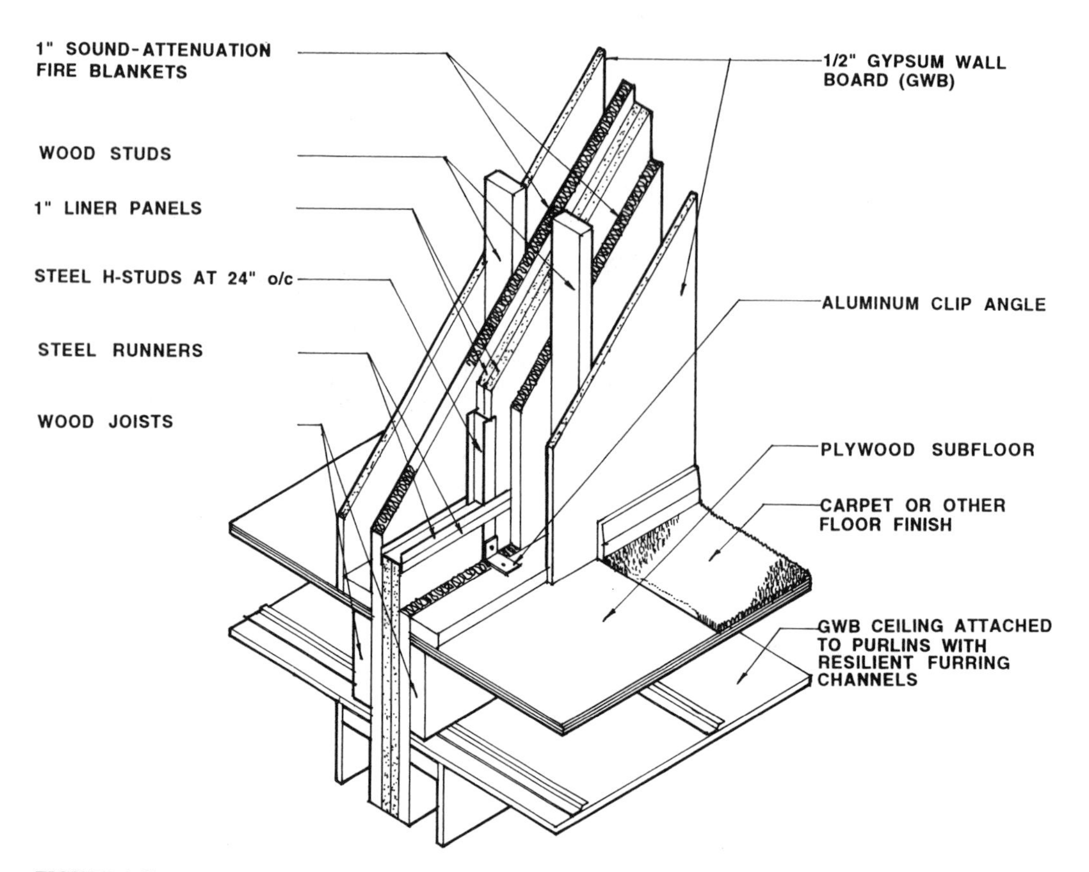

FIGURE 4-52. Axonometric Drawing of an Area Separation Wall.

4.2.8 Partition Schedule

The details shown in this section represent the method that is commonly used to detail partitions. Partition type numbers are usually keyed to the plan using a triangular symbol (Fig. 4-24). While this way of representing partitions has served the profession well for a long time, a faster and less labor-intensive way would be to use the same method used to represent doors and room finish. Using a schedule (Fig. 2-20) saves the time spent on drawing the details (for offices that draw partition types from scratch for each project) or trimming and positioning the sticky-backs (for offices that use standard details).

A schedule expedites the task of providing detailed information. The types most used by the office may be drawn only once on a schedule and made into a master sheet to be reproduced as a wash-off Mylar or sepia for each project. Each type used on the project is then marked in the "used/not used" column and new types added as needed. Such a schedule is ideal for CADD application.

4.2.9 Guidelines

The following guidelines should be observed for all GWB partitions:

1. Vertical control joints must be located no more than 30 feet apart. Doors extending to the ceilings act as control joints. Shorter doors must have a control joint aligned with each jamb.
2. Limiting heights must be checked in all locations of the project to make sure that the actual height is within the limit prescribed by the manufacturer. Height can be increased by using wider studs, heavier gauge studs, decreasing stud spacing, or by adding more layers of wallboard. Consult manufacturer's representatives, making sure that none of their solutions are peculiar to their product. Also consult the Fire Resistance Design Manual issued by the Gypsum Association in that regard.
3. Choose the least costly assembly (i.e., the one with the least number of wallboard layers and most common type of stud) to achieve the STC or fire rating if either is required.
4. Partitions subject to impact by moving equipment must be protected. One of two methods is usually used: Provide an impact-absorbing bumper attached to the wall or build the wall strong enough to take the impact while keeping its integrity. This latter approach can be attained by building the wall out of CMU and, if the expected impact is severe, reinforcing the wall and filling the cells with grout or building a concrete wall.
5. Because ducts penetrating rated walls may melt during a fire and allow the flame and smoke to penetrate through the wall, codes stipulate, in most cases, that fire dampers be installed at the point of penetration. Dampers are shutters that have different ratings to match the wall rating and are activated by a fusible link that releases the

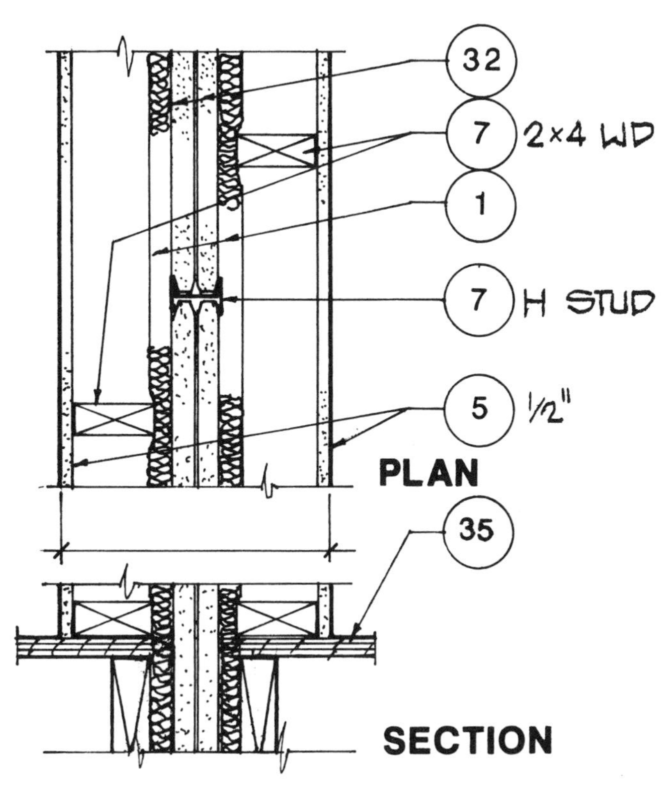

FIRE RATING: 2 HRS.
FIRE TEST: WHI 495-PSV-0245
STC RATING: 53

TYPE:

Guidelines:

1. This is a USG proprietary design for a party wall (see Section 4.2.6). Other manufacturers have similar designs.
2. Contact manufacturer's representatives for height limitations.

FIGURE 4-53. Area Separation Wall Detail.

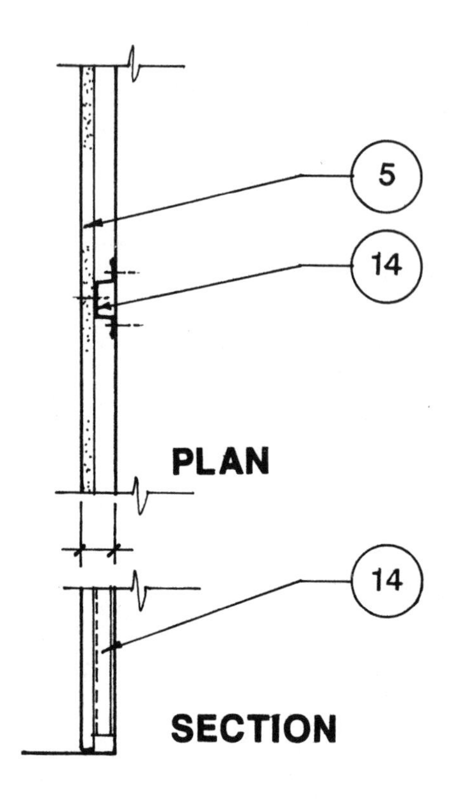

TYPE :

Guidelines:

This type is used as furring over concrete or masonry. If wall requires thermal or acoustical insulation, use "Z" furring strips sized to accommodate thickness of 1", 1 1/2", 2", and 3".

FIGURE 4-54. Furring Detail.

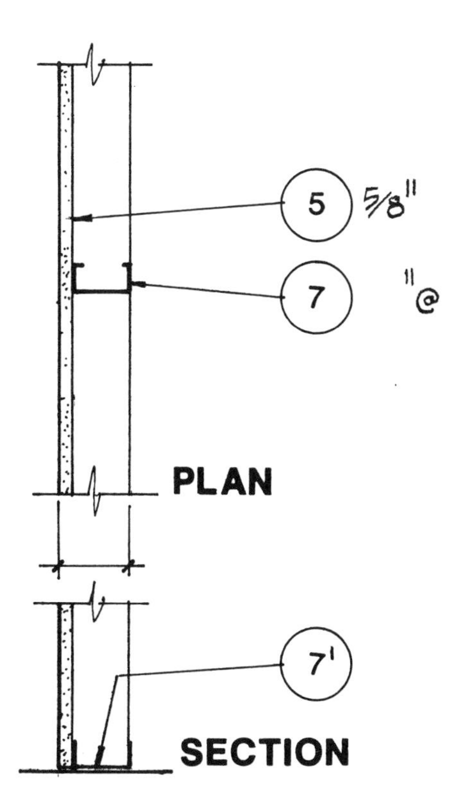

TYPE:

Guidelines:

1. This is a nonrated free-standing partition used as furring or as a demising partition for unfinished commercial spaces.
2. Height limitation:
   9'-9" with 25GA. 2 1/2" studs @ 24" o/c. Brace at midspan, if height is 14'-0 or less,or use wider studs or narrower spacing.

FIGURE 4-55. Furring Detail.

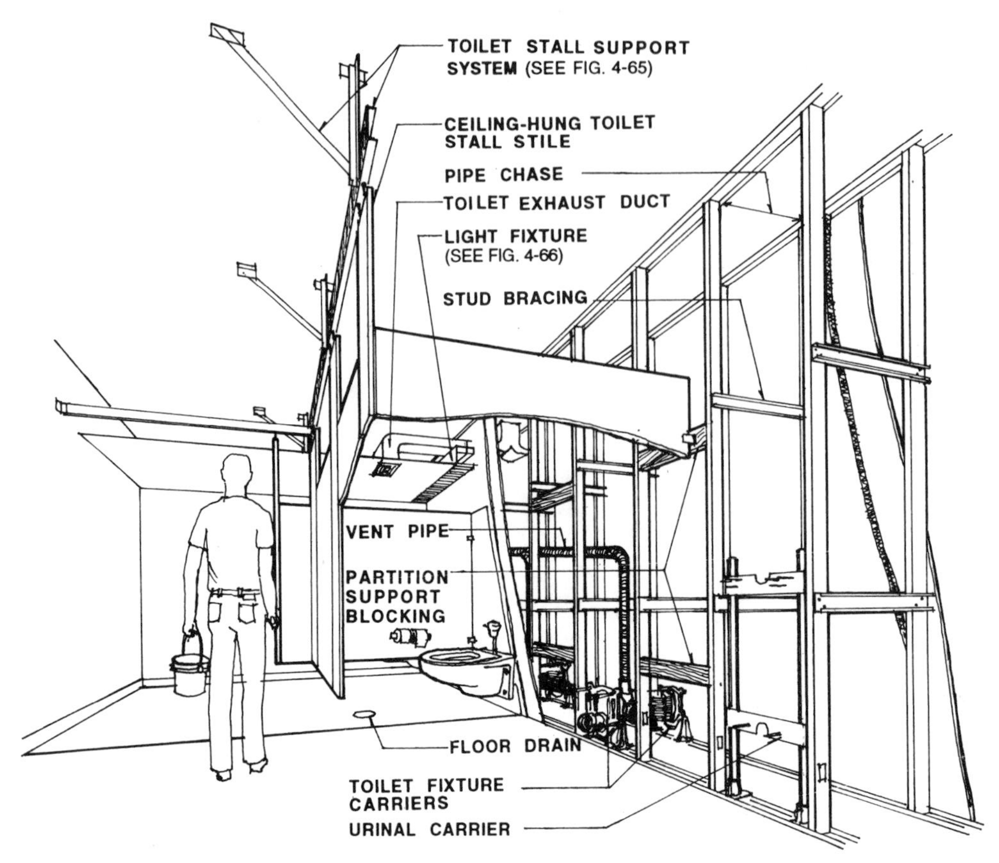

FIGURE 4-56. Cutaway Perspective of a Toilet Room.

shutter during a fire. Check the code and discuss the location with the HVAC engineer. An alternative method is to encase the ducts within a fire-rated enclosure.

6. It is usually a good idea to poshe fire-rated partitions to differentiate them from other walls. This simplifies checking for conformance to code and alerts the mechanical engineer to the required locations for fire dampers (Fig. 4-24).

4.2.10 Site Visits

Before visiting the site, the office representative must read the specifications sections that apply to partitions, sealants, and related items. He must also examine the drawings carefully and note any unusual conditions and details. The following are a few guidelines:

1. Studs must be plumb and sized according to the drawings. Wallboard must be the right thickness and type such as type "X," MR, cement board, or regular GWB.
2. Door openings must be either reinforced with cold-rolled channels or doubled studs at the jambs and fastened to top and bottom runners for heavy doors.
3. For acoustically rated partitions, conditions that allow passage of flanking sound must be disallowed. Make sure that the proper thickness and type of sound-attenuation blankets are used and placed properly.
4. Make sure that the wallboard is stored in a dry, well-ventilated enclosed space and that the joint cement application is done at temperatures above the minimum temperature specified.
5. Horizontal application of the wallboard is preferred because it provides a stronger installation. Fasteners should be started at the center of the sheet and continued toward the edges.

4.2.11 Sources of More Information:

Elmiger, A. 1976. *Architectural & Engineering Concrete Masonry Details for Building Construction*. Herndon, VA: The National Concrete Masonry Association.

Gypsum Association. 1988. *Fire-Resistance, Sound Control Design Manual*. 12 ed. 1603 Orrington Avenue, Evanston, Illinois 60201

Gypsum Manufacturers Catalogs

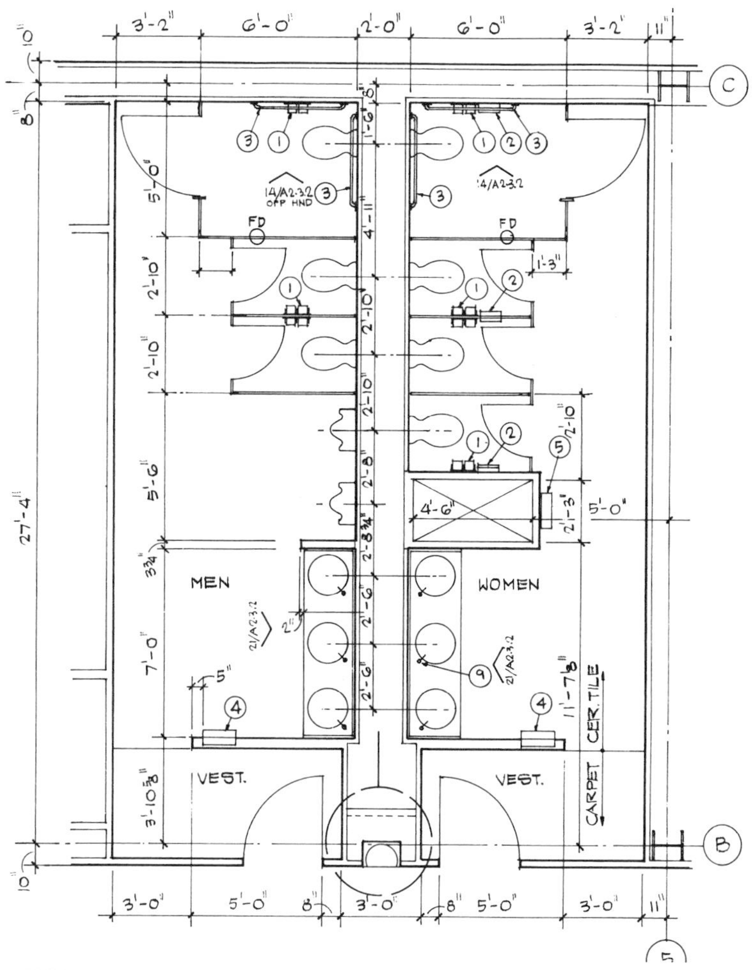

FIGURE 4-57. Example of a Toilet Room Plan.

4.3 TOILET ROOM DETAILS

4.3.1 General

Toilet rooms (Fig. 4-56) are included in all major projects. The set of drawings required to describe a toilet room include a plan (Fig. 4-57), a detail showing mounting heights for toilet fixtures and accessories (Fig. 4-58 and 4-59). These drawings along with the legend in Figure 4-60 can be used in most projects with or without minor modifications. Many offices also include interior elevations. In most cases, these elevations contain information already provided in the room finish schedule, the specifications, and the standard details. The room finish schedule indicates the wall, floor, and ceiling materials and finishes. Where ceramic tile is installed, the height can be shown in the "remarks" column of the schedule. The specifications describe stall partition material: metal, plastic laminate, marble, etc., and the type: ceiling hung, wall hung, or floor mounted. The plan and mounting heights drawings indicate the location and heights of the fixtures and accessories. The rest of the information can be provided by a detail of the lavatory counter (Fig. 4-61 and 4.62) as well as any special details such as the structural support required for ceiling-hung partitions, ceramic tile pattern (if of a special design), and special lighting (if custom designed).

4.3.2 Dimensioning

Toilet room plans are usually drawn at ¼-inch scale (Fig. 4-57). One string of dimensions indicates the center lines of all plumbing fixtures tied to a fixed reference point such as a column grid line. The plumber uses this information to locate the toilet and urinal carriers shown in the cutaway perspective (Fig. 4-56). These dimensions are also used to position the lavatories.

Toilet stall partitions are manufactured in standard and custom sizes. The person entrusted with dimensioning the partitions should strive to gear his dimensions to the standard stall depth of 4′-8″ and the standard stall widths, which are 32, 34, or 36 inches. Provision for the physically handicapped–accessible facility must be checked for conformance to the clearances required for operating a wheelchair. Refer to the applicable codes, both local and federal.

4.3.3 Number of Fixtures

The plumbing code sets the minimum number of fixtures required for each sex based on the number of occupants.

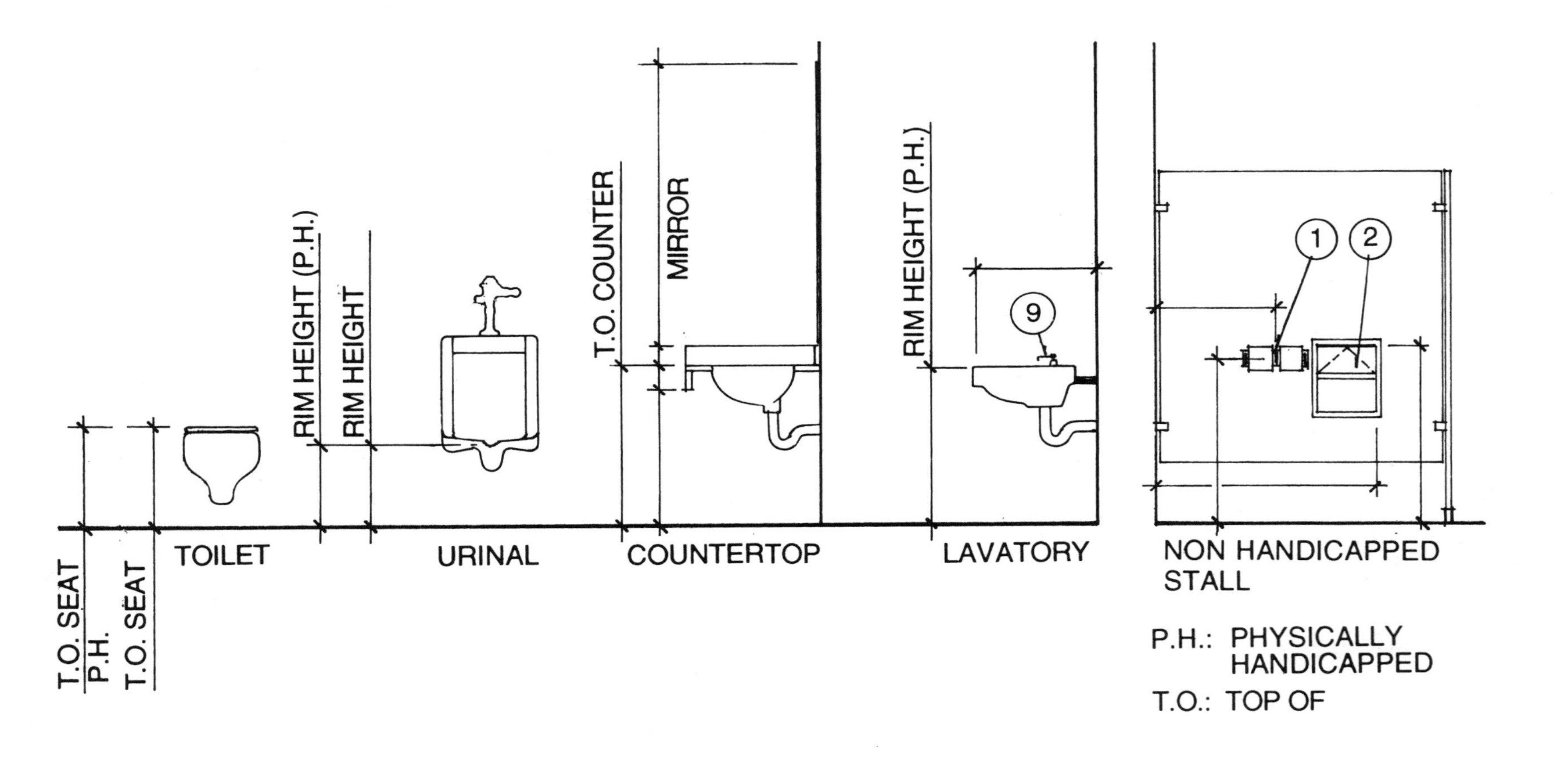

TOILET FIXTURE MOUNTING HEIGHTS

Guidelines:

1. Identify fixtures for the handicapped on the plan.
2. Fill out the missing dimensions.
3. Add fixtures if necessary. Write "NOT USED" on the ones not included in the project.
4. This is a generic detail to be used on all projects. Place it adjacent to the accessories mounting heights detail.

FIGURE 4-58. Toilet Fixture Mounting Heights.

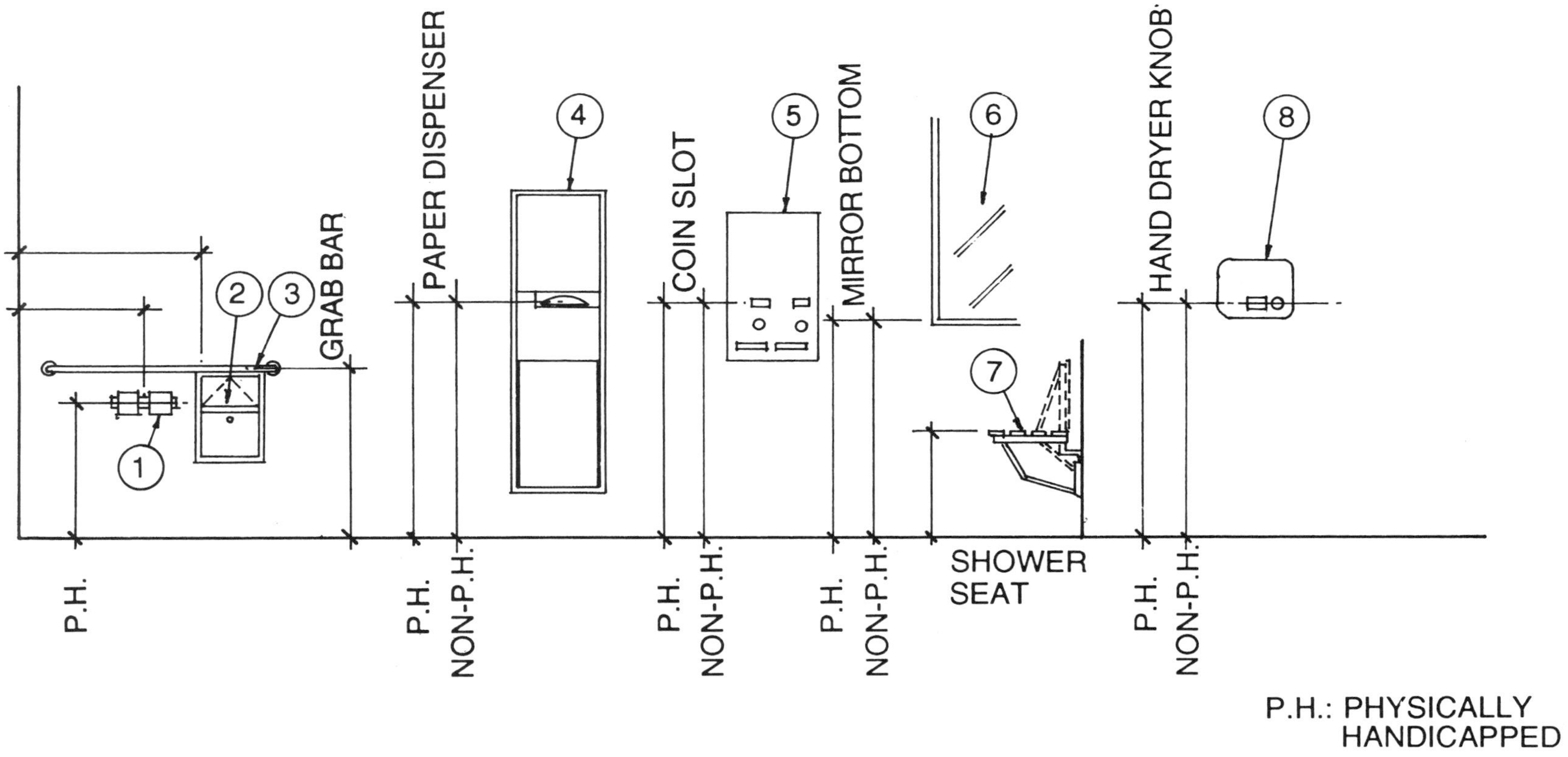

TOILET ACCESSORIES MOUNTING HEIGHTS

3/8"=1'-0"

Guidelines:

1. Fill in dimensions according to code. Buildings occupied predominantly by children should have lower mounting heights.
2. Hair dryers are similar to hand dryers except that they are mounted higher.
3. Condom dispensers are similar to feminine napkin dispensers.
4. Use recessed accessories whenever possible except in sound-rated partitions.
5. Use this as a generic detail on all projects. Add accessories as needed and write "NOT USED" under accessories not required for the project. Place a legend adjacent to this detail.

FIGURE 4-59. Toilet Accessory Mounting Heights.

1. TOILET PAPER DISPENSER
2. NAPKIN DISPOSAL
3. GRAB BAR
4. PAPER TOWEL DISPENSER & DISPOSAL
5. TAMPON/NAPKIN VENDOR
6. MIRROR
7. SHOWER SEAT
8. HAND DRYER
9. HAIR DRYER
10. TOILET SEAT COVER DISPENSER
11. SOAP DISPENSER
12. SHELF
13. MOP & BROOM HOLDER
14. WALL URN
15. PAIL HOOK
16. PAPER CUP DISPENSER
17.
18.
19.
20.

NOTES:
1. ONLY CIRCLED NUMBERS ARE INCLUDED IN THIS PROJECT.
2. ALL ACCESSORIES ARE FLUSH-MOUNTED UNLESS AN ASTERISK IS ADDED TO THE NUMBER TO INDICATE THAT THE **ACCESSORY** IS SURFACE-MOUNTED.

ACCESSORY LEGEND

Guidelines:
1. Add accessories as needed.
2. Use this legend on all projects.

FIGURE 4-60. Accessory Legend.

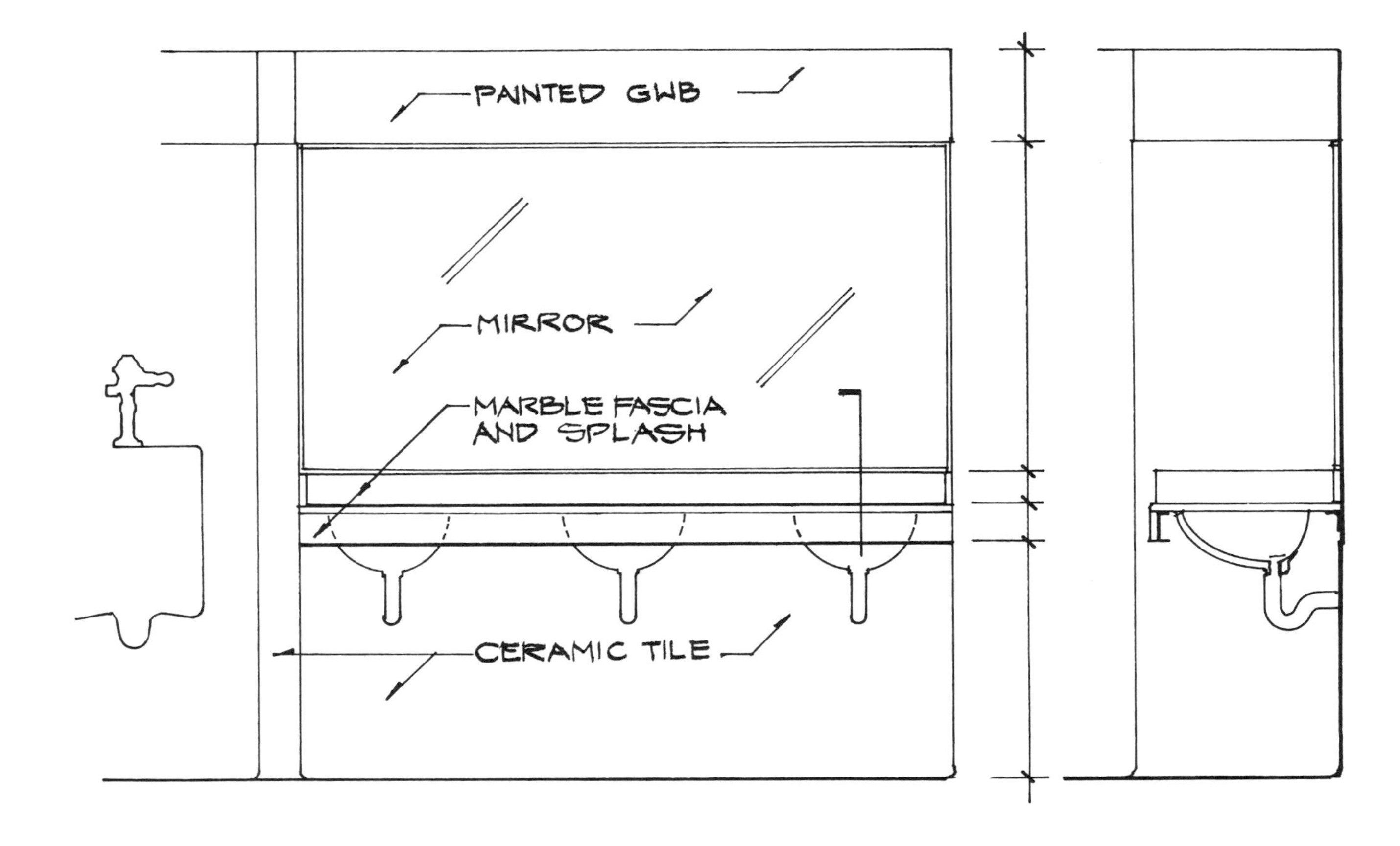

LAVATORY ELEVATION AND SECTION

1/2"=1'-0"

Guidelines:

1. Clearance under the fascia must conform to accessibility code for the physically handicapped.
2. If a light trough is located at the junction of the wall and ceiling, draw the outline.
3. In most cases, this is the only interior elevation required for the toilet room. The room finish schedule describes the finishes and the fixture and accessory mounting height detail provides the rest of the information. Be sure to include the height of the tile in the schedule.

FIGURE 4-61. Lavatory Elevation and Section.

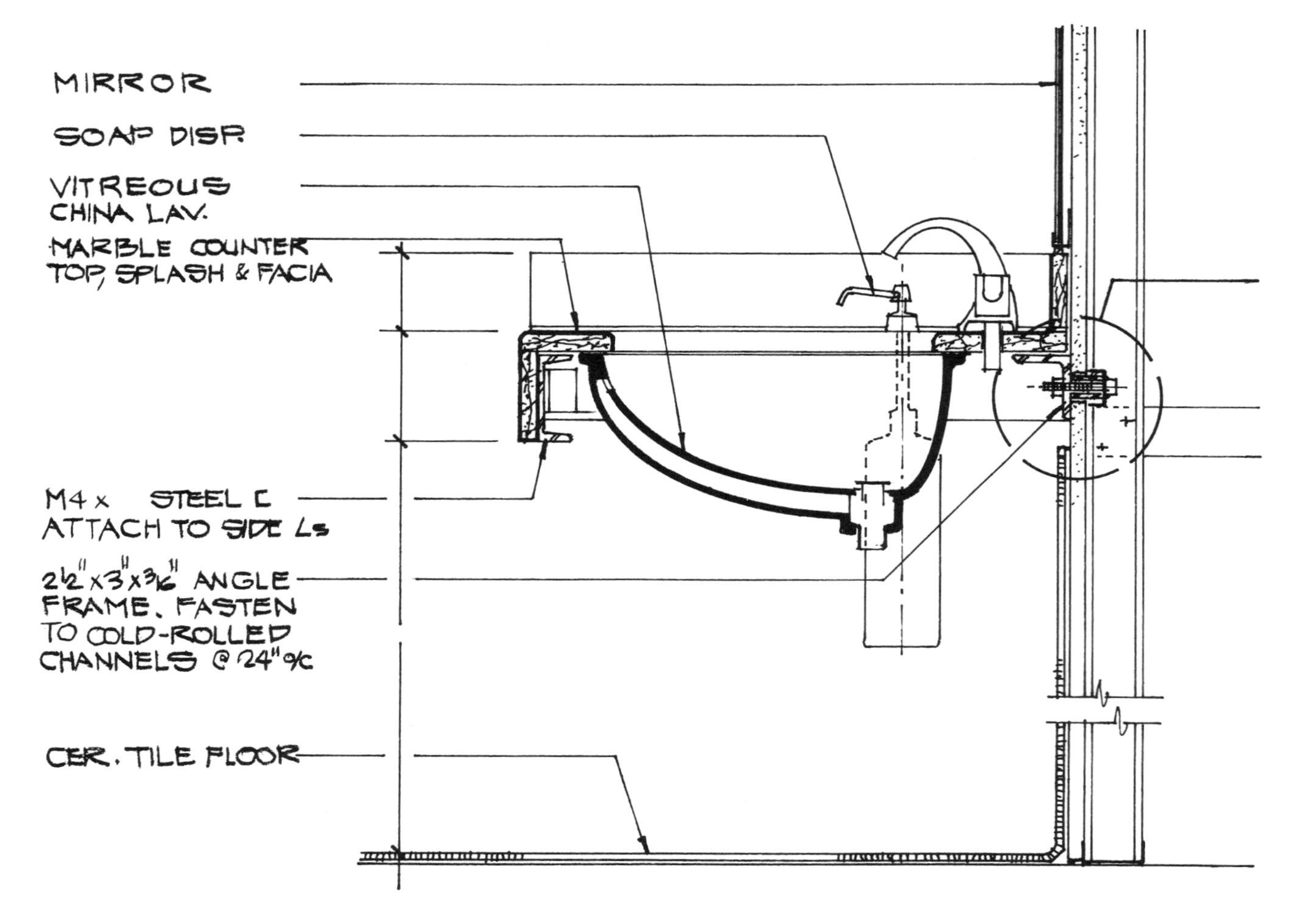

SECTION THROUGH LAVATORY COUNTER

Guidelines:

1. Fill out dimensions. Make sure clearance under counter conforms to the minimum clearance for the physically handicapped.
2. Size steel channel according to span.

FIGURE 4-62. Section Through a Lavatory Counter.

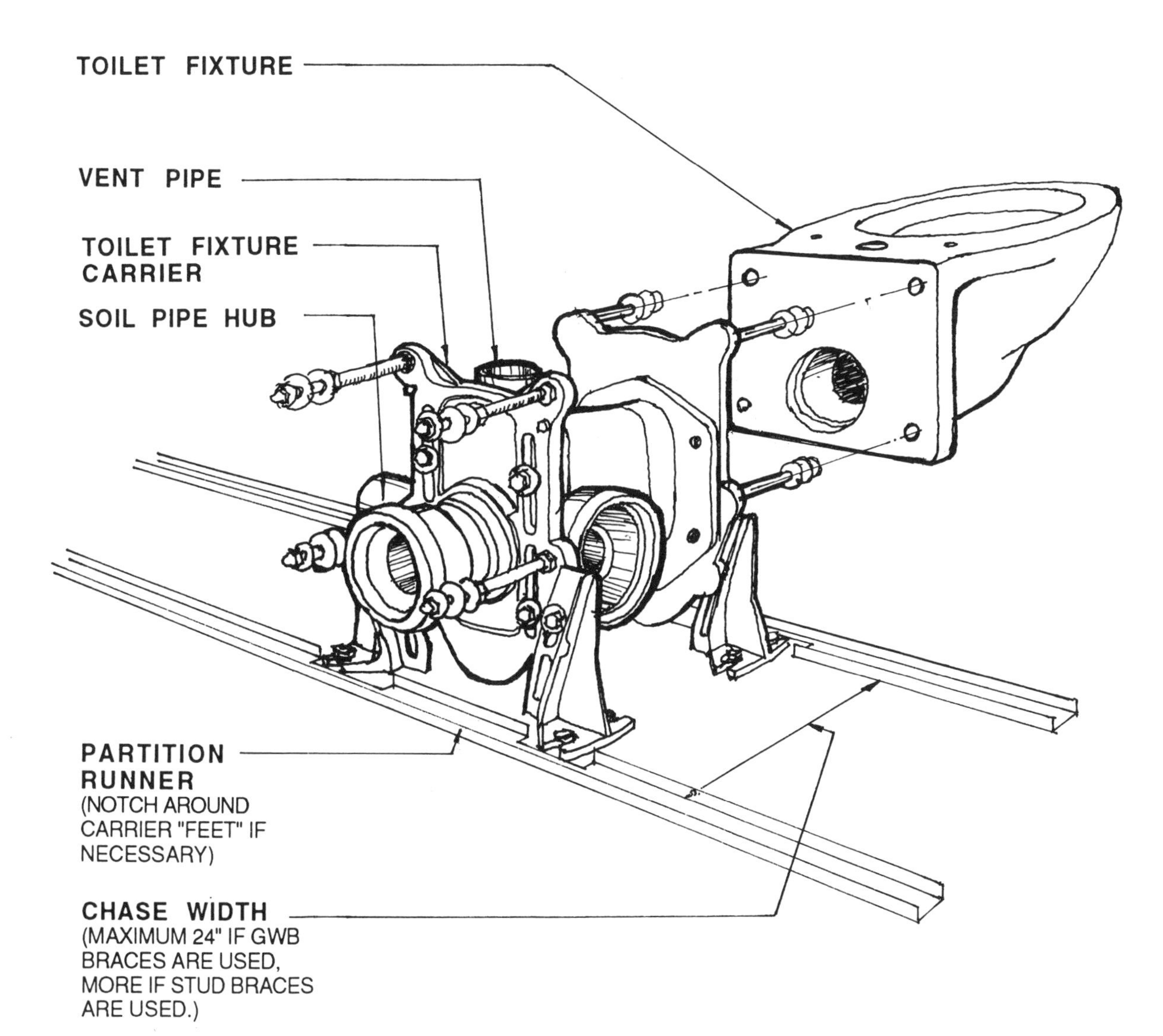

FIGURE 4-63. Toilet Fixture Carrier.

Sometimes, especially in multistory projects occupied by a single tenant, it is advantageous to compute the required number based on the total occupant load for the building rather than the occupant load for each floor. The following example demonstrates this advantage.

Example

An office building contains ten floors used by an insurance company. The gross area for each floor is 11,340 sq. ft. and the building code requires 100 sq. ft. gross per person for an office occupancy. Calculate the number of toilets if the plumbing code requires the following:

0–15 persons: 1 toilet
16–35 persons: 2 toilets
36–55 persons: 3 toilets
56–80 persons: 4 toilets
81–110 persons: 5 toilets
Add one toilet for each additional 40 persons.

Solution

The Single Floor Method:

$$\text{Number of occupants per floor} = \frac{\text{gross area}}{\text{area per person}} = \begin{array}{c} 11{,}340 \div 100 = 113.4 \\ \text{(say 113) persons} \end{array}$$

$$\begin{aligned} \text{Number of toilet fixtures} &= 5 \text{ for the first } 110 \\ &\quad + 1 \text{ for the remaining } 3 \\ &= 6 \text{ fixtures} \end{aligned}$$

$$\begin{aligned} \text{Number of toilet fixtures for the building} &= 6 \times 10 \text{ floors} \\ &= \textit{60 fixtures} \end{aligned}$$

The Whole Building Method:

$$\text{Number of building occupants} = \frac{\text{gross area for ten floors}}{\text{area per person}} = \frac{11{,}340 \times 10}{100} = 1{,}134 \text{ persons}$$

Number of toilet fixtures = 5 for the first 110
+ one toilet for each 40 persons of the remaining 1,024

$$= 5 + \frac{1024}{40} = \textit{30.6 (say 31) fixtures}$$

While the second method represents about half the number required by the first method, it must be checked against the actual requirements of the occupants. The ground floor may require no fixtures while a typical floor may require more than the minimum. Check with the authority having jurisdiction to find out if this method is acceptable before applying it.

Plumbing codes stipulate the minimum number of fixtures. Common sense may indicate the need for more fixtures for the convenience of the occupants. For instance, ladies' rooms in some public buildings such as theaters are notoriously inadequate as can be noticed from the length of lines leading to them between performances. If the budget and space allow, these facilities should exceed the code minimum.

4.3.4 Pipe Chases

Chase width varies depending on the number of fixtures, whether they are located on one or both sides of the wall and whether they are wall- or floor-mounted. Most medium and large nonresidential projects include wall-mounted toilets and urinals. The fixtures are bolted to carriers bolted to the floor within the chase (Fig. 4-63). These carriers are one of the factors that determine the minimum width of the chase. The chase may contain other pipe runs or exhaust vent ducts that may require more space. Check with the mechanical engineer to make sure that the width you choose is acceptable. See *Architectural Graphic Standards* for the recommended widths.

4.3.5 Guidelines

The following observations may come in handy during the design of toilet rooms:

1. Toilet room entrances must be designed to block sight lines from penetrating into the interior of the facility. Be sure to avoid direct view into mirrors (Fig. 4-64).
2. Partition types around and between toilet rooms must have a sound-transmission class (STC) rating of at least 47 (52 in executive areas) to prevent the transmission of noises into adjoining areas.
3. If acoustical tile is used in the ceiling, it must be moisture resistant and easy to clean.

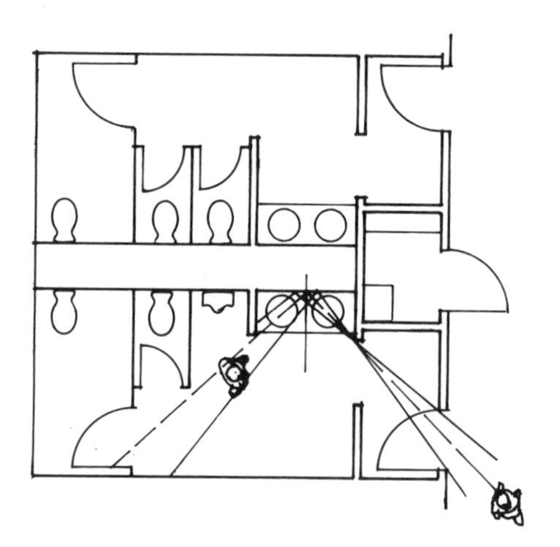

FIGURE 4-64. Sightlines into a Toilet Room Mirror.

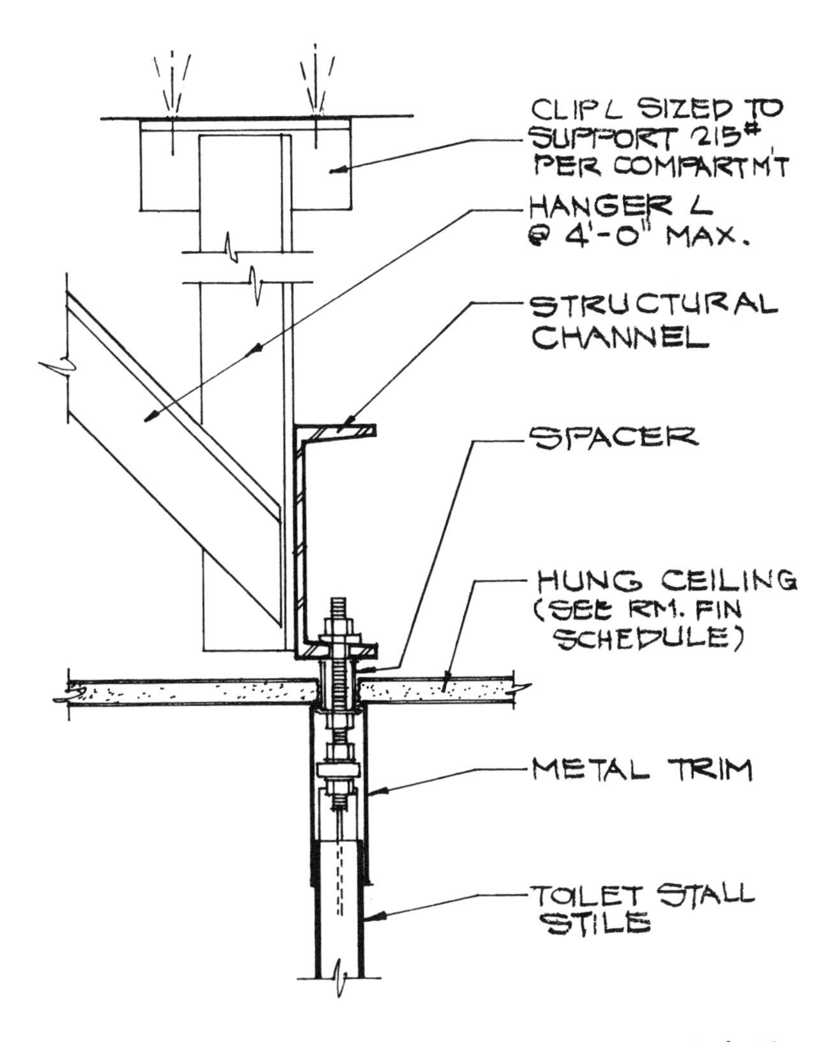

CEILING-HUNG PARTITION

Guidelines:
If space does not allow for frequent hangers and braces, which is the case when wide ducts are located above the ceiling, use a wide flange instead of the channel. Consult with the structural engineer for size.

FIGURE 4-65. Ceiling-Hung Partition Detail.

4. There are two methods to construct toilet room floors. The traditional method requires a depressed slab and a mortar setting bed of 1 to 2 inches. This method has the advantage of providing a positive slope to the drain. The other method is to thin-set the tile directly to the undepressed slab. Drainage is accommodated by creating a dish-shaped depression around the floor drain. This method costs less and is more prevalent, especially in commercial construction.

   If the slab is to be depressed, the structural engineer must be alerted as early as possible to create the necessary framing.
5. In some multistory office buildings, janitorial supplies are stored in the basement or ground floor. The janitor uses the service elevator to transport his cleaning equipment and uses a hose bib located under the lavatory counter to mop the floors. This saves the space and the cost of repetitive service sinks at each floor. A small supply closet is required at each floor, in most cases, to store toilet paper, paper towels, and other supplies.
6. Ceiling-hung partition stiles must be attached to a braced structural beam (Fig. 4-65) capable of supporting 200 lbs. or more. Contact manufacturer for recommendations and inform the structural engineer.
7. An attractive alternative to the conventional 2′ × 4′ light fixtures is to place a special linear light fixture at the chase wall or at both sides of the toilet room (Fig. 4-66). These fixtures are specially designed to reflect light off the walls and are more suitable than framing a GWB trough in the ceiling for that purpose. This fixture does not require a detail. Similar fixtures may also be available from other manufacturers.
8. Urinal screens should not be installed in high-abuse locations such as schools, sports arenas, and similar projects. If installed in moderate-abuse areas, such as office buildings, they should not be wider than 14″ to lessen the likelihood of frequent damage.

Because this section addresses plumbing as well as toilet rooms, water cooler details (Fig. 4-67) and fire hose cabinet furring guidelines (Fig. 4-68) and details (Fig. 4-69 and 4-70) have also been included.

4.4 DOOR DETAILS

4.4.1 General

The door detail sheet or sheets along with Division 8 of the specifications provide all the information needed by the door manufacturer to execute the work. The specifications describe the construction, finish, and hardware for each door type included in the project. Figure 4-71 shows wood doors and hardware options. The following is a description of each item included in the detail sheet.

4.4.2 The Door Schedule

The schedule (Fig. 2-17) identifies each door by its number on the plan. For small projects, it may be sufficient to number

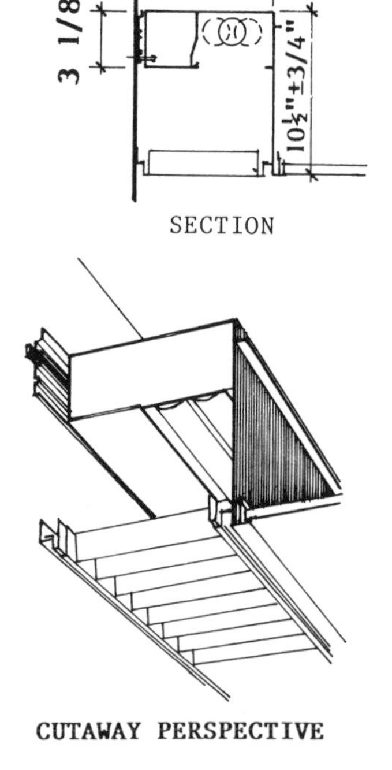

FIGURE 4-66. An Example of a Light Strip Assembly Suitable for a Toilet Room (Wall/Slot-II by Litecontrol).

WATER COOLER DET.

1/2"=1'-0"

Guidelines:

1. Refer to physically handicapped code for rim height, depth, and clearance below the fixtures.
2. If the wall is firerated, be sure to extend the wall at the back of the enclosure all the way to the structure and designate it as a rated partition.

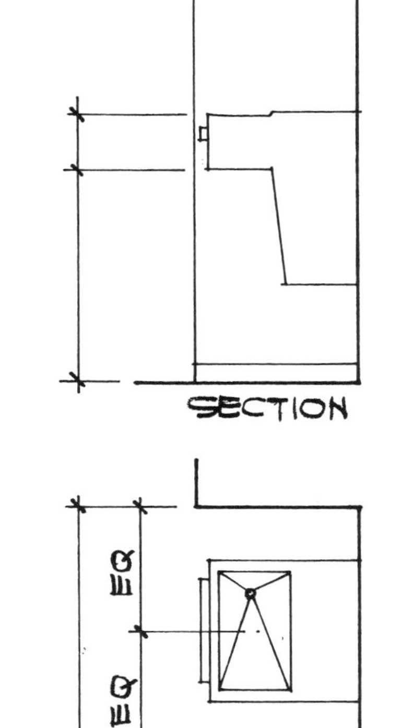

WATER COOLER DET.

1/2"=1'-0"

Guidelines:

This fixture can be mounted without an alcove. Its height may be adjusted to conform to any code.

FIGURE 4-67. Water Cooler Details.

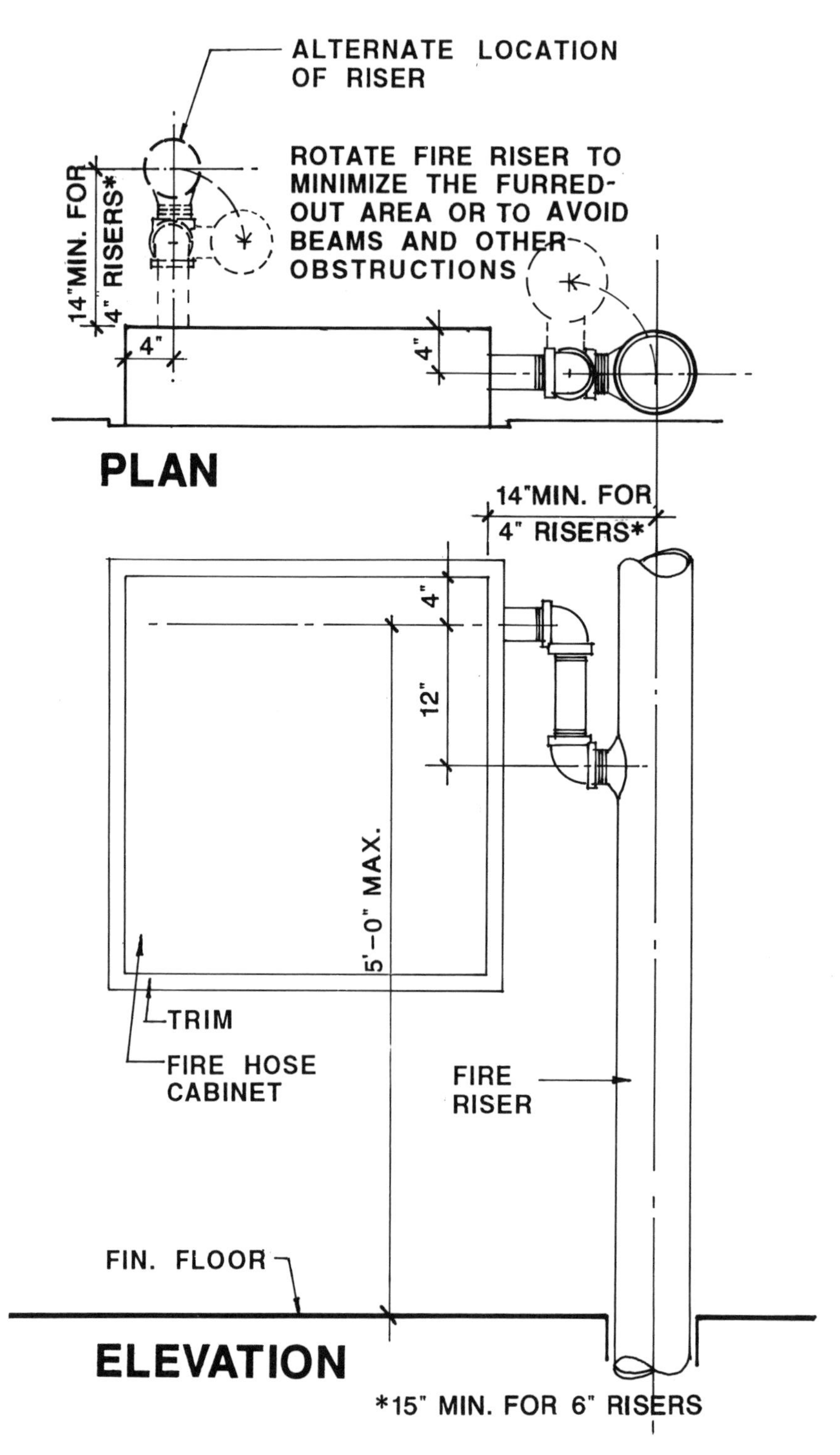

FIGURE 4-68. Fire Hose Cabinet Furring Guidelines.

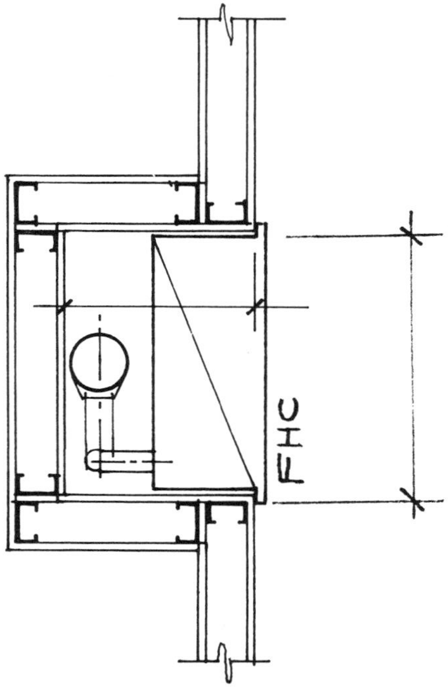

FHC

FIRE HOSE CABINET

1"=1'-0"

Guidelines:
Find maximum fire riser pipe diameter from plumbing riser diagram. Add an allowance for pipe wall thickness, and hubs. Determine width of enclosure accordingly. Fax to plumbing engineer to check before final drawing.

FIGURE 4-69. Fire Hose Cabinet Detail.

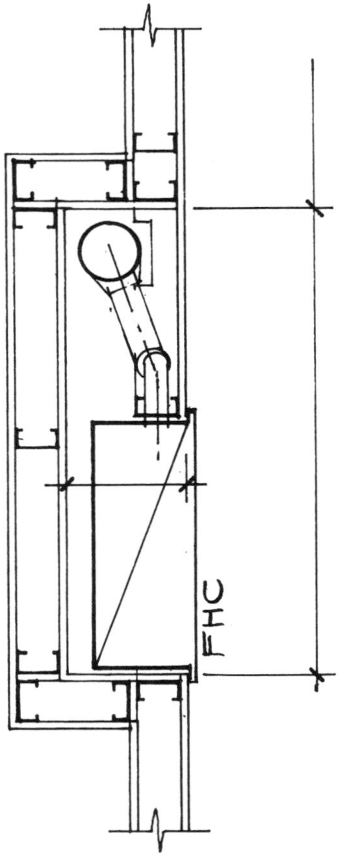

FHC

FIRE HOSE CABINET

1"=1'-0"

Guidelines:
See Fig. 4-69.

FIGURE 4-70. Fire Hose Cabinet Detail.

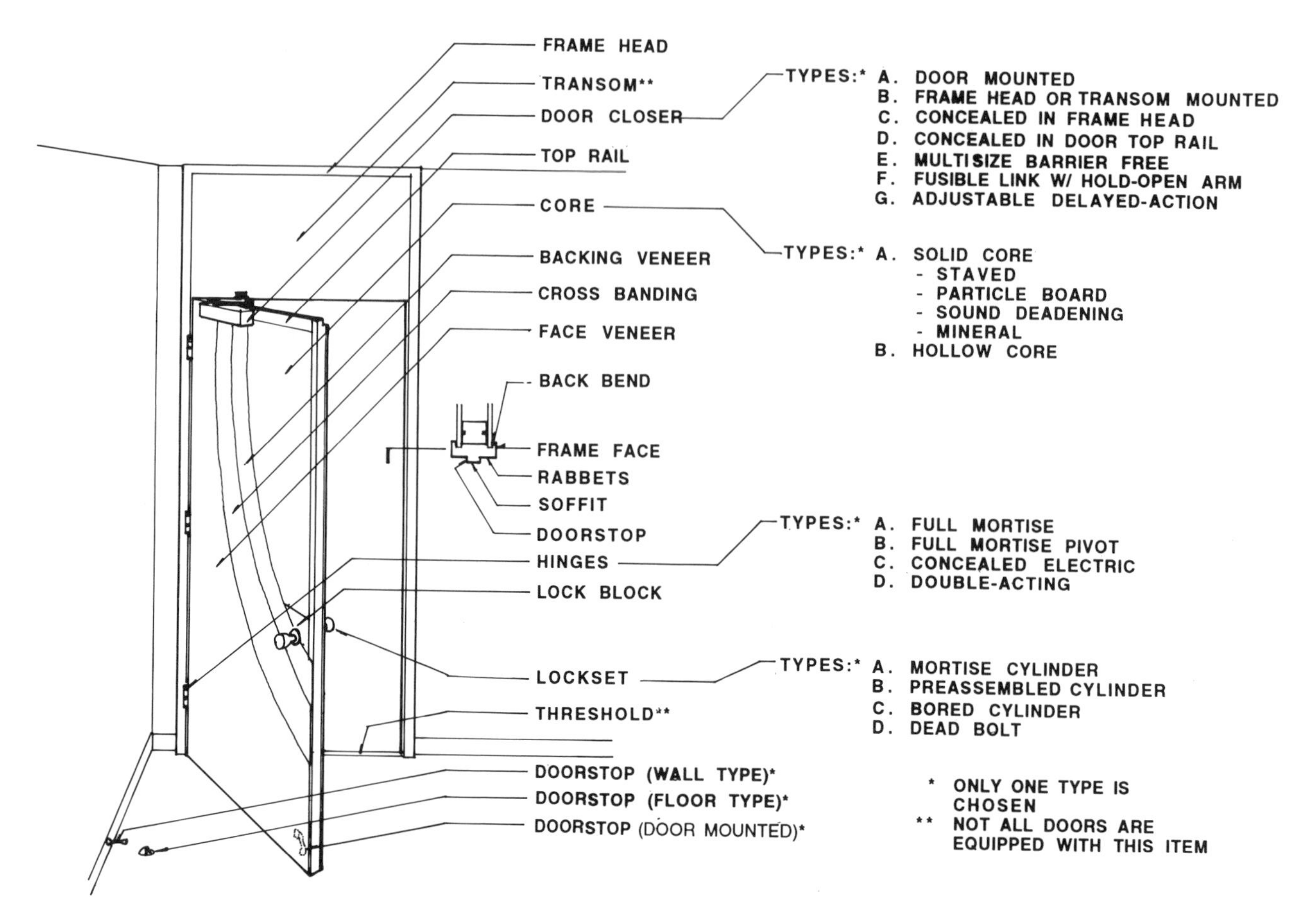

FIGURE 4-71. Wood Doors and Hardware.

the doors consecutively—1, 2, 3. For larger buildings, the numbering should be the same as the room number. The first digit of a room number identifies the floor and the rest of the number identifies the room. If more than one door opens into the room, a letter is added to the number, for instance 2.02A, 2.02B, 2.02C, etc. The schedule also provides information about the nominal door size, ignoring any modifications required for clearance or undercutting. If windows or louvers are to be located on a fire-rated door, the code must be checked to find out whether they are allowed and, if allowed, what size limitation is stipulated. Fire-rated doors bear the U.L. label. The National Fire Protection Association (NFPA) classifies fire-rated doors as follows:

(a) Class A—Openings in fire walls and in walls that divide a single building into fire areas.
(b) Class B—Openings in enclosures of vertical communications through buildings and in 2-hour-rated partitions providing horizontal fire separations.
(c) Class C—Openings in walls or partitions between rooms and corridors having a fire resistance rating of 1 hour or less.
(d) Class D—Openings in exterior walls subject to severe fire exposure from outside of the building.
(e) Class E—Openings in exterior walls subject to moderate or light fire exposure from outside of the buildings.[1]

[1](Reprinted with permission from NFPA 80-1990, *Fire Doors and Windows*, Copyright © 1990, National Fire Protection Association, Quincy, MA 02269. This reprinted material is not the complete and official position of the National Fire Protection Association, on the referenced subject which is represented only by the standard in its entirety.)

4.4.3 Door Frame Details

Frame details form the bulk of the door detail sheet. The door schedule forces the detailer to check each jamb of each door to make sure that it is represented by one of the frame details. The frames are dimensioned to give enough information to the fabricator to manufacture the frames. They must be double-checked to avoid any errors. Please note in the accompanying details that the frame soffit dimension is omitted. This allows the fabricator to set the clearance required between the backbend and the face of the wall. These clearances may vary for different types of walls.

Figures 4-72 and 4-73 are generic door types that can be used on any sizable project. Smaller projects that do not include the majority of these types may be drawn from scratch. These types may also be stored in the computer and recalled selectively to compile the needed door types. The

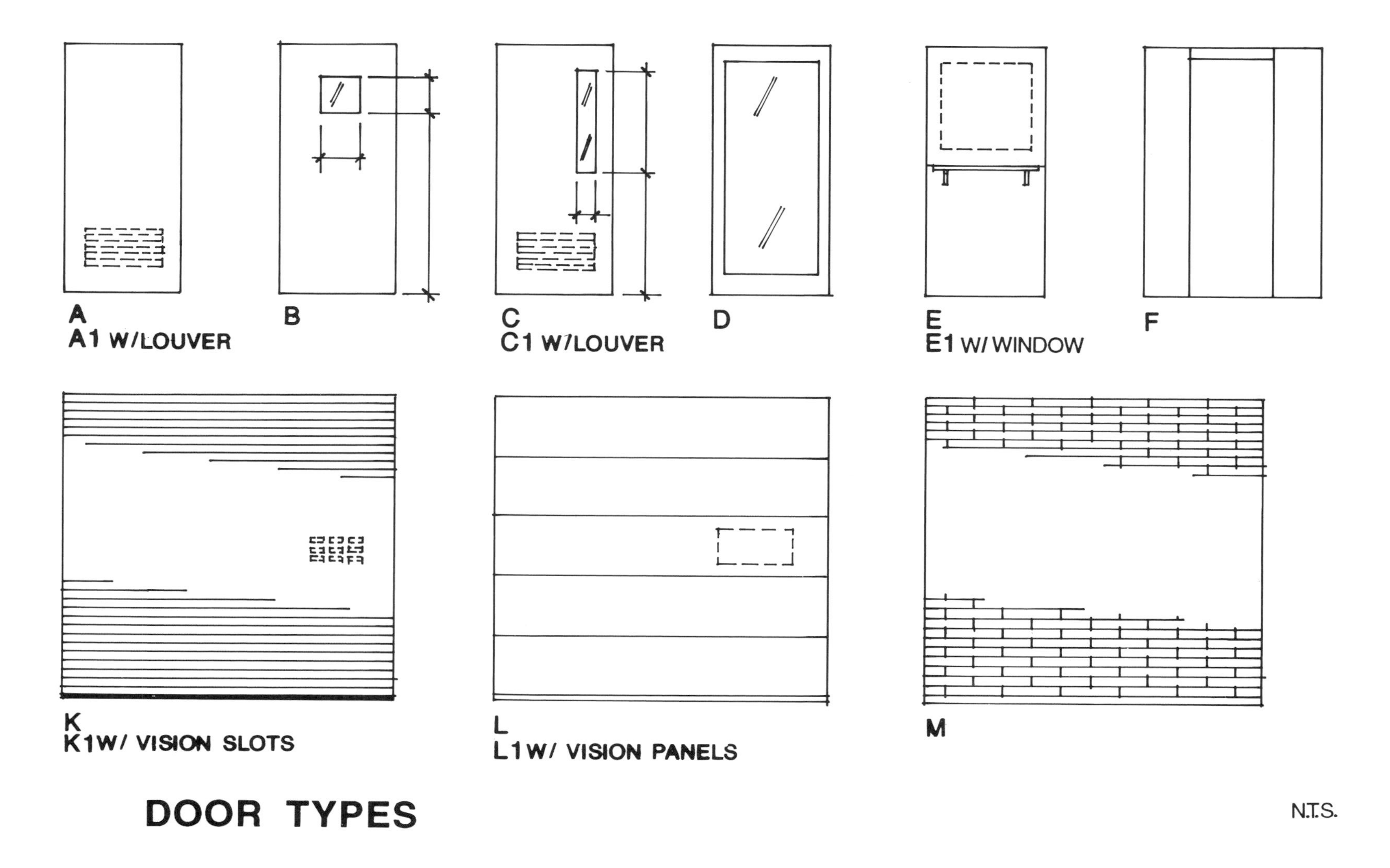

Guidelines:

1. This is a generic detail to be used on all projects. Add types as required.
2. Common usage for types shown are:
   A – Room Access; B – Stair; C – Lab or Classroom; D – Storefront; E – Dutch Door
   F – Dark Room, revolving; K – Roll-up; L – Sectional; M – Roll-up Grille
3. Designate glass types.
4. See Fig. 4-73 for continuation.

FIGURE 4-72. Door Types.

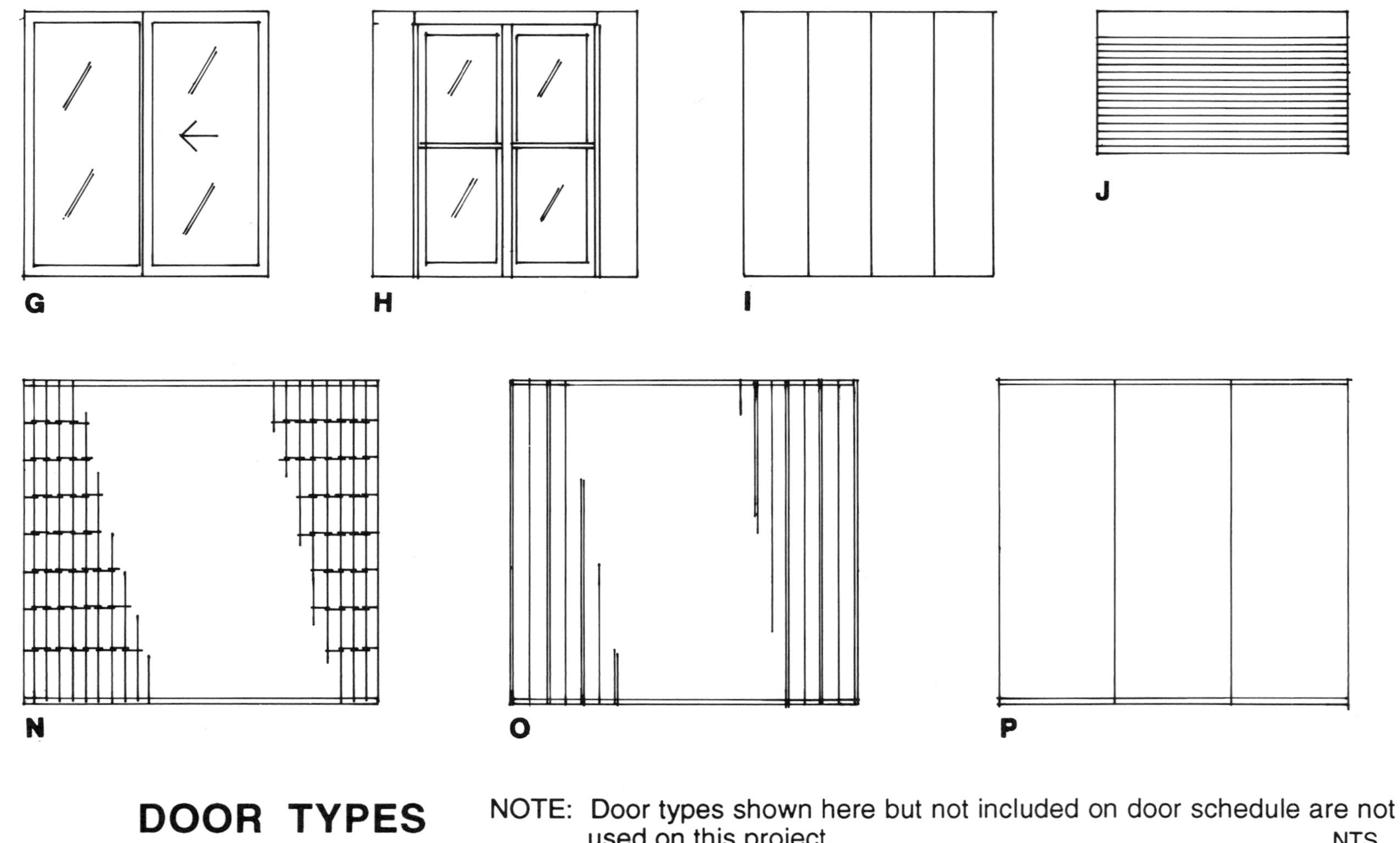

DOOR TYPES

NOTE: Door types shown here but not included on door schedule are not used on this project.

N.T.S.

Guidelines:

1. This is a continuation of door types shown in Fig. 4-72. Add more types as needed.
2. Common usage for types shown:
   G – Sliding Glass Door (mostly residential)
   H – Revolving Door
   I – Folding
   J – Roll-up Counter Shutter
   N – Side-Coiling Grilles (shops & malls)
   O – Accordion
   P – Folding Partitions

FIGURE 4-73. Door Types (cont.)

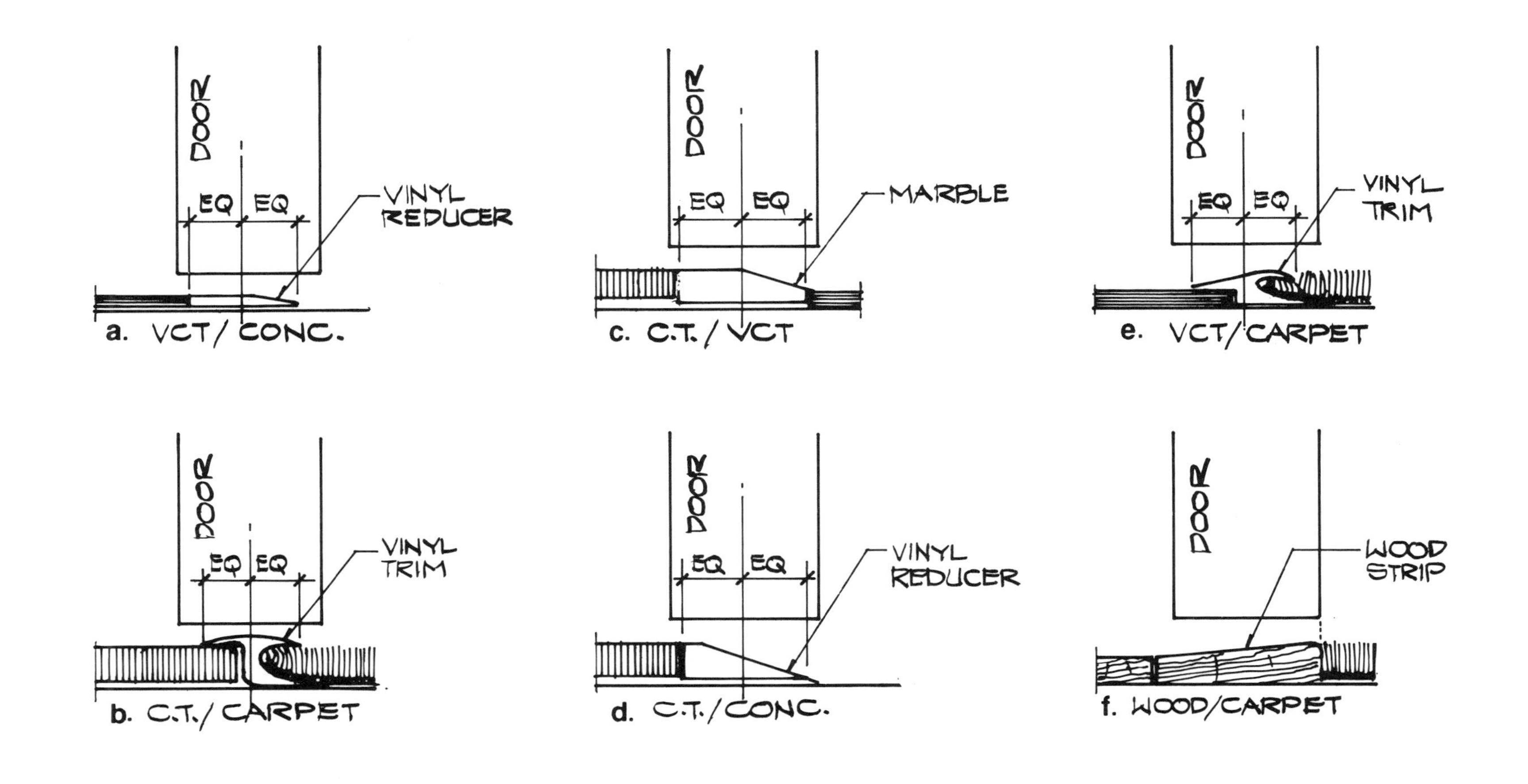

THRESHOLD DETAILS

N.T.S.

Guidelines:

1. This is a generic group of details to be used on all projects. Write "NOT USED" on details not occurring in the project at hand. Add details as required.
2. In the door schedule under the "threshold" column, write the applicable detail number, for example 5b/A_____, where 5 is the generic detail for all thresholds, b is C.T./CPT.
3. If specs describe door clearances, delete dimensions.

FIGURE 4-74. Threshold Details.

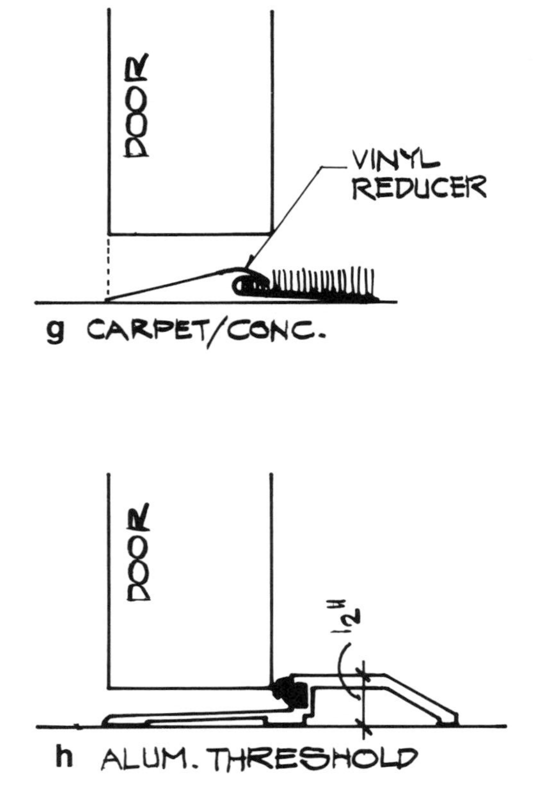

THRESHOLD DETAILS

N.T.S.

Guidelines:

1. This is a continuation of the generic details shown in Fig. 4-74.
2. Add details as needed (see Guidelines under Fig. 4-74.)

FIGURE 4-75. Threshold Details (cont.)

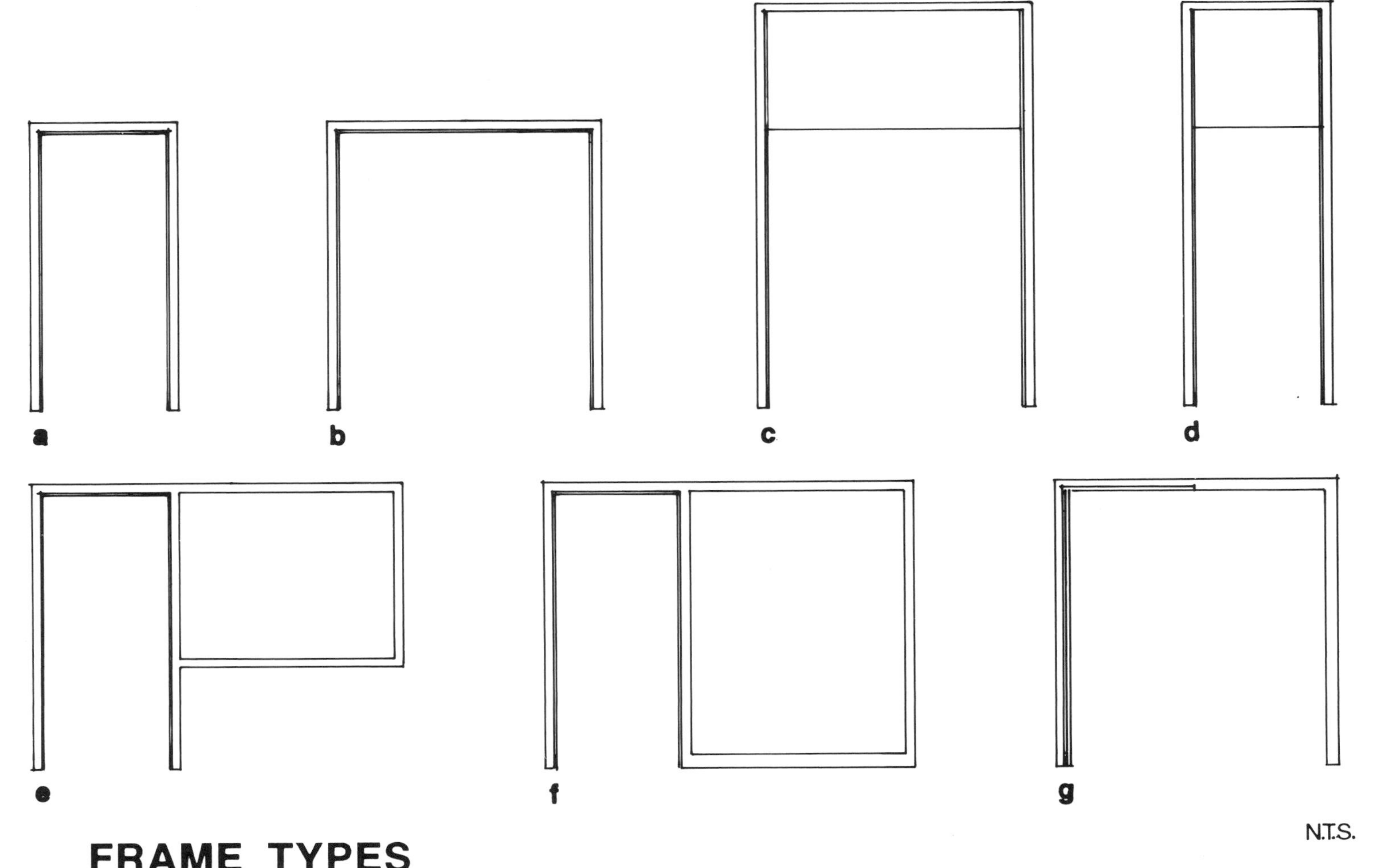

FRAME TYPES

Guidelines:

1. Use this generic detail only if four or more of the frame types shown are used on the project. Add frame types as needed. A removable mullion may be added to frame type "B" for doors that require access for wide equipment occasionally.
2. Type "c" and "d" are available with 1 1/2-hour firerating if transom is wood, 3-hours if hollow metal. Frames "g" and "f" are available with a maximum rating of 20 minutes with a 20-minute rated door. Maximum glass area is 1,296 square feet for clear glass, 5,268 square feet for wire glass. Consult frame and glass manufacturers. A clear fire-rated glass manufactured by Nippon Electric Glass Co. is also available through Firelite. Frame "g" is for double-egress doors.

FIGURE 4-76. Door Frame Types.

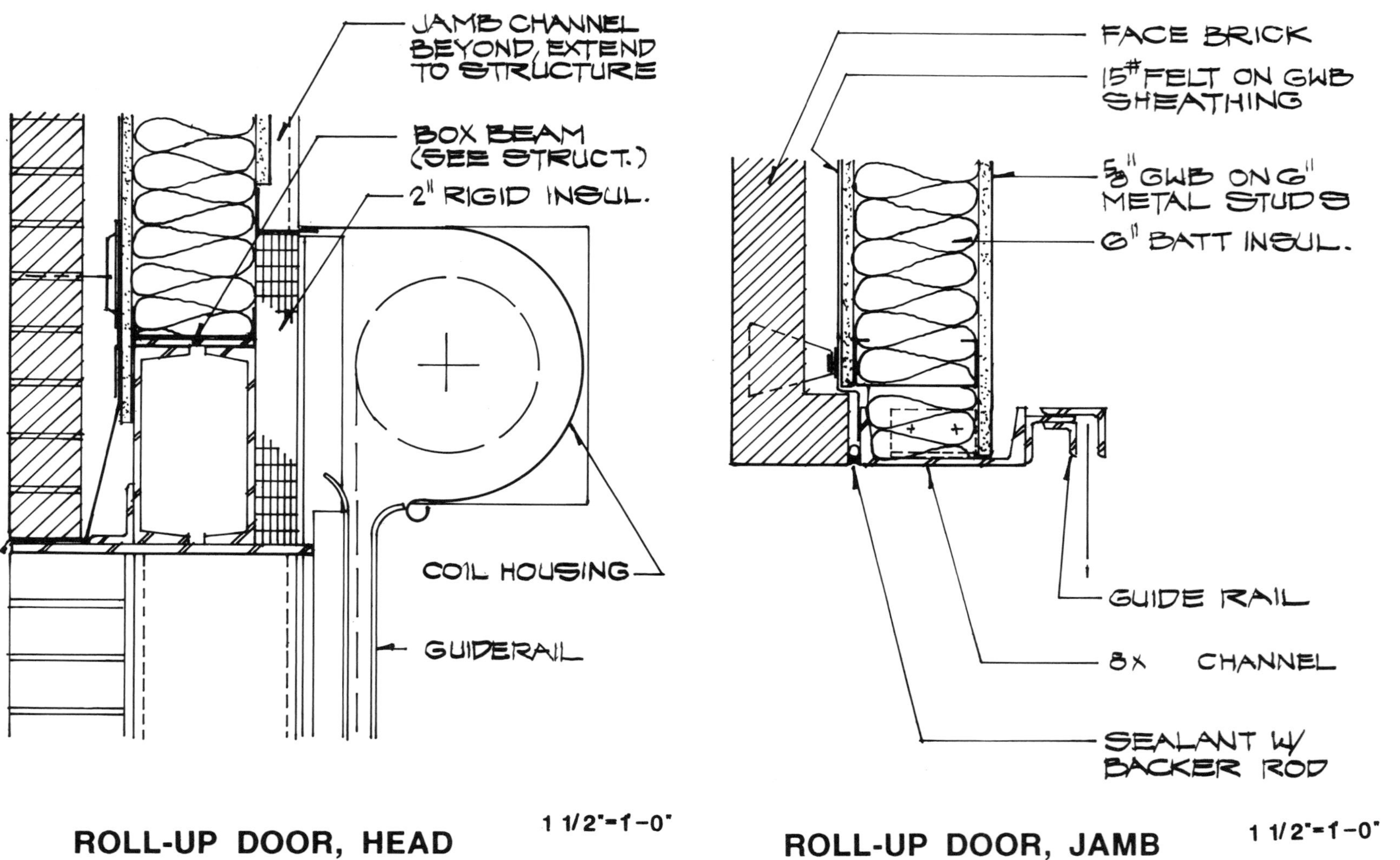

ROLL-UP DOOR, HEAD 1 1/2"=1'-0"

ROLL-UP DOOR, JAMB 1 1/2"=1'-0"

Guidelines:

1. Decide whether the door is insulated or not, has vision panels or is solid, is manual or motor-operated.
2. Choose manufacturer (preferably with local rep.). Compute coil dimension from size chart (see *Sweet's Catalog* or brochure).
3. If motor-operated, make sure there is enough space to accommodate the motor and send motor power requirements to the electrical engineer.
4. Make sure track has a sound structural support at the top.

FIGURE 4-77. Roll-Up Door Head and Jamb Details.

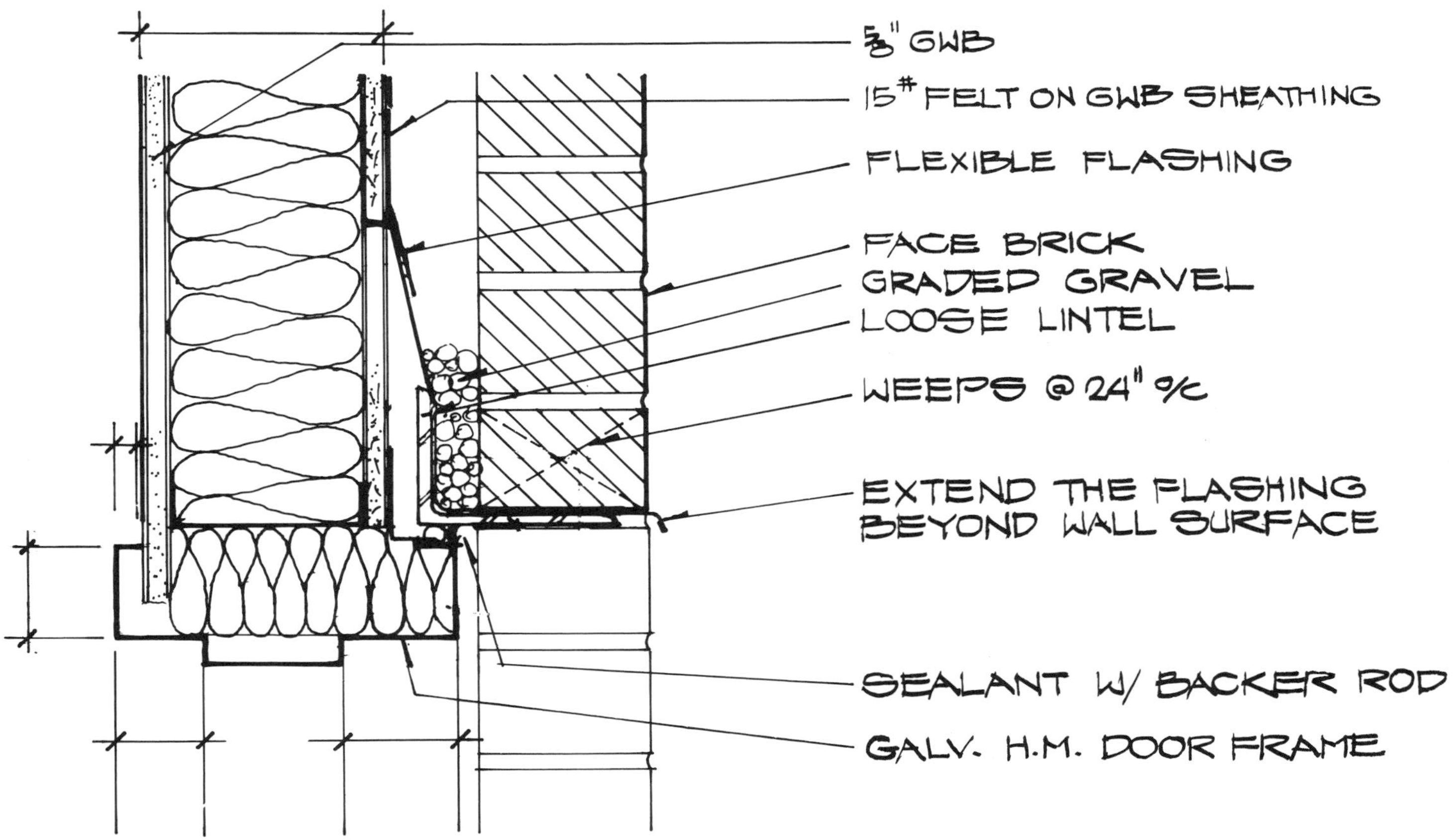

5/8" GWB
15# FELT ON GWB SHEATHING
FLEXIBLE FLASHING
FACE BRICK
GRADED GRAVEL
LOOSE LINTEL
WEEPS @ 24" o/c
EXTEND THE FLASHING BEYOND WALL SURFACE
SEALANT W/ BACKER ROD
GALV. H.M. DOOR FRAME

3"=1'-0"

HEAD DETAIL

Guidelines:

1. If several types of flashing are used on the project and the specs do not describe their location, identify the type on each detail.
2. If there is no lintel schedule, specify the size of the angle and its bearing length.
3. Coordinate door frame dimensions with brick dimensions. See Fig. 4-79 for plan detail.
4. Draw axonometric to detail flashing dammed at the ends.

FIGURE 4-78. Door Head Detail (Brick Veneer Wall).

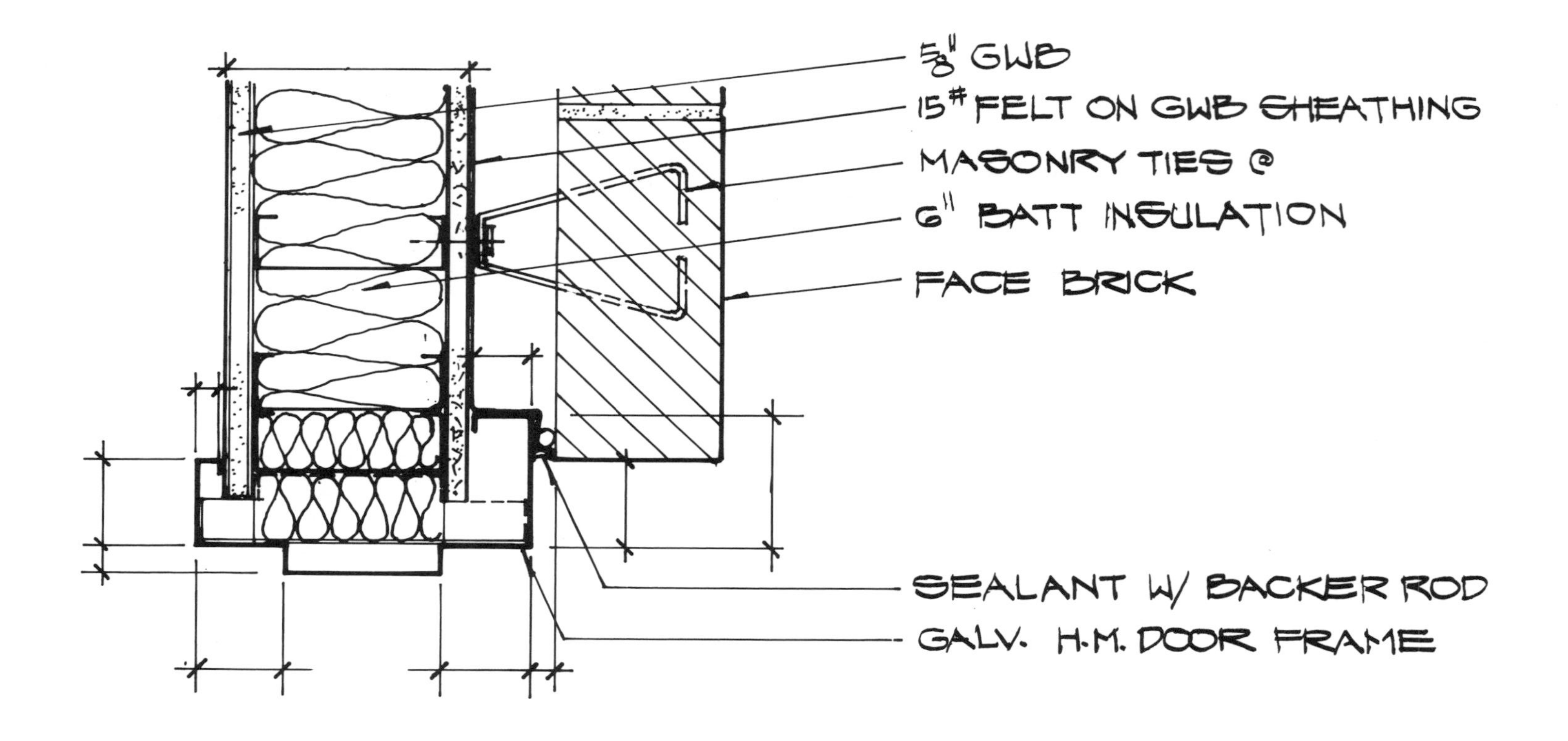

JAMB DETAIL

3"=1'-0"

Guidelines:

1. Refer to notes applicable to Door Head Detail (Fig. 4-78).
2. Fill in thickness of GWB, masonry tie spacing, insulation, and frame dimensions.
3. Although sheathing provides protection against water infiltration, adding the felt is advisable to provide redundancy.

FIGURE 4-79. Doorjamb Detail (Brick Veneer Wall).

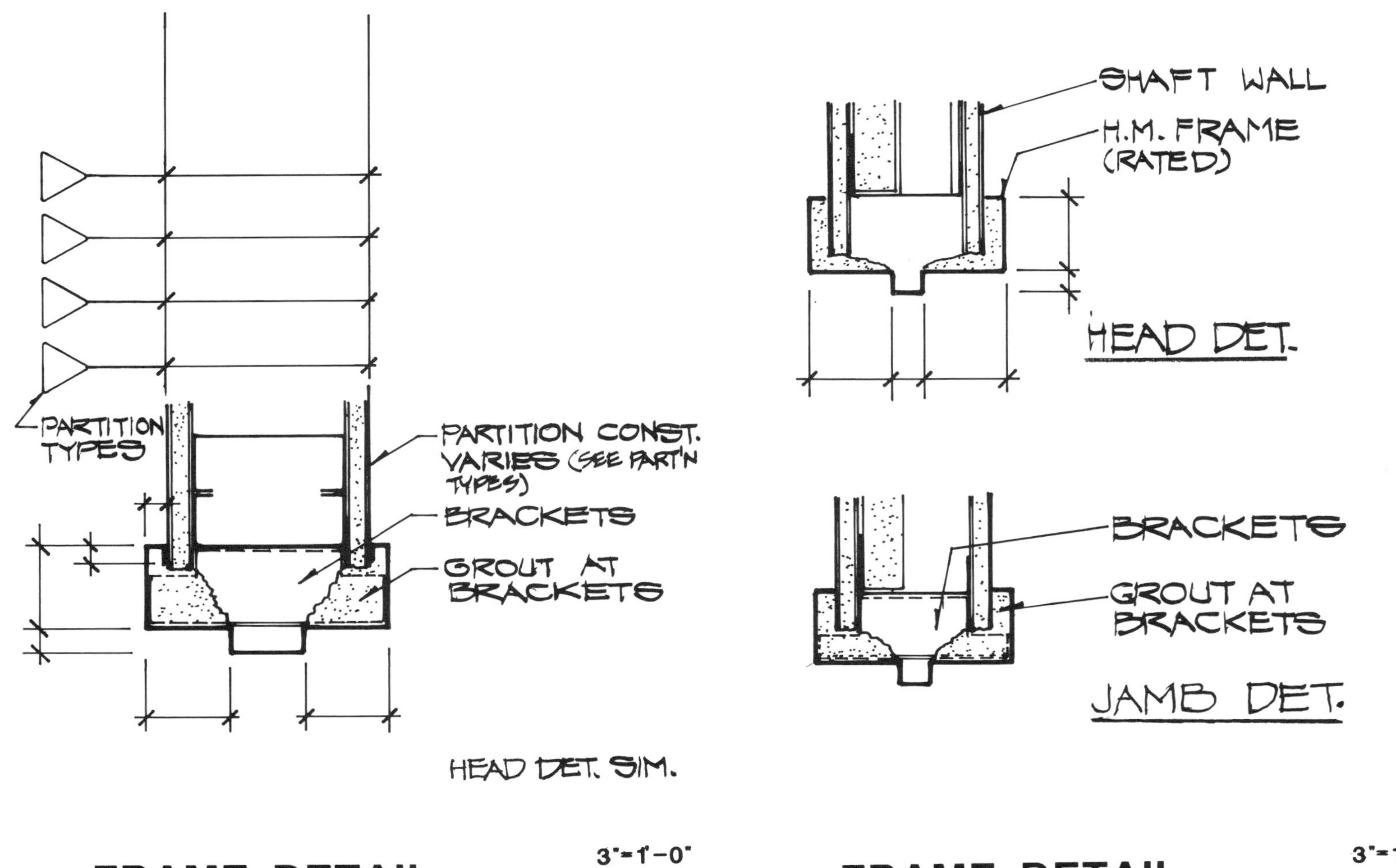

Guidelines:

1. Fill in dimensions. Add dimensions for throat width if more partition types are required.
2. Soffit dimensions must not be less than 1" in width. Face dimension may be reduced to 1 1/2" or 1" if narrow frame is required for aesthetic reasons.
3. Double 20 GA. studs are required for doors weighing 200 lbs. and up.

FIGURE 4-80. Frame Details (GWB Wall).

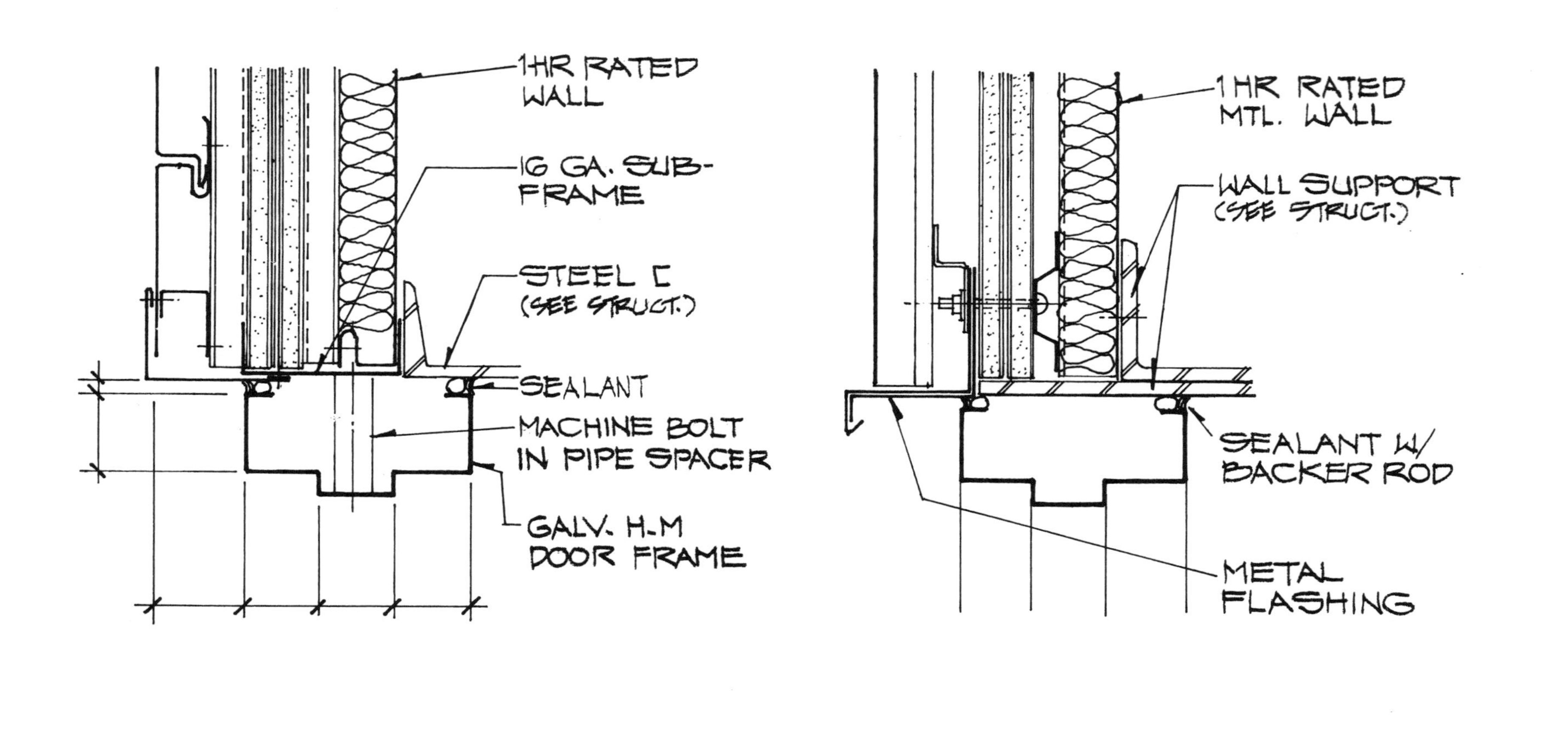

JAMB DETAIL 3"=1'-0"

HEAD DETAIL 3"=1'-0"

Guidelines:

1. This wall construction is common for use in mechanical penthouses. Two- and three-hour rated walls are available.
2. Coordinate wall support information with structural engineer.

FIGURE 4-81. Jamb and Head Details (Metal Wall).

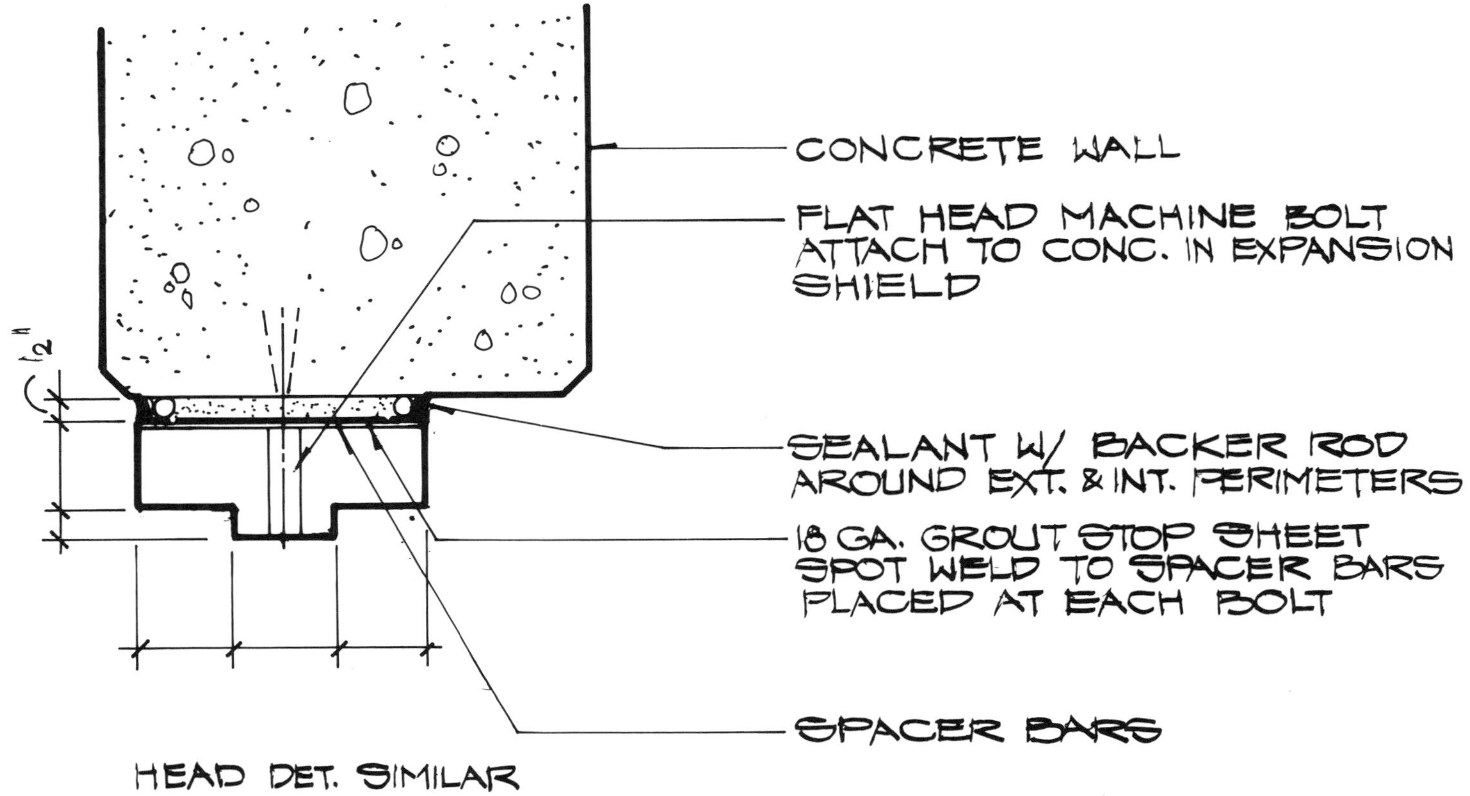

JAMB DETAIL

3"=1'-0"

Guidelines:
1/2" clearance between frame and concrete allows the grout to be placed properly. It also allows for any inaccuracy in concrete forms.

FIGURE 4-82. Jamb and Head Details (Concrete Wall).

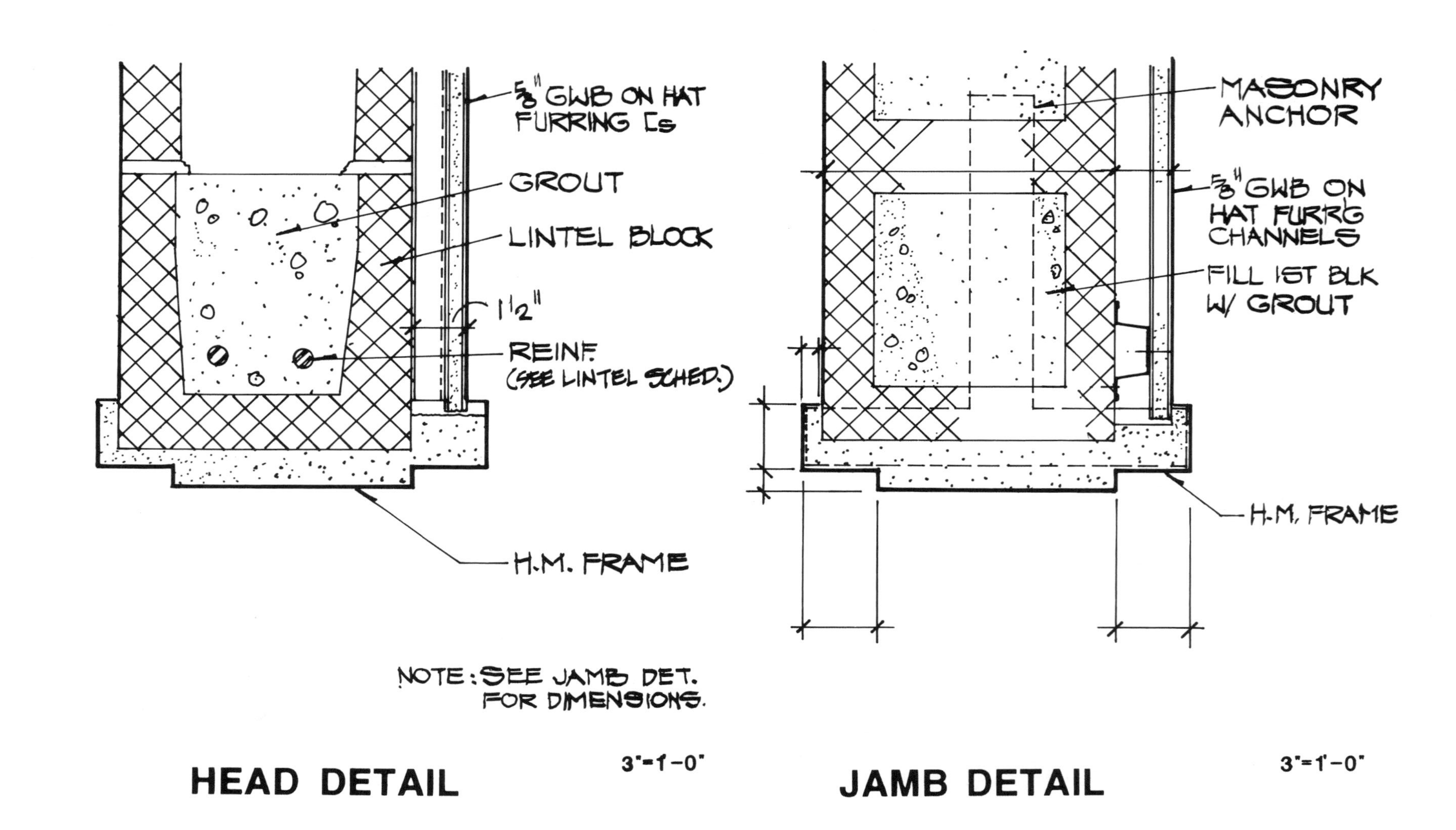

Guidelines:

1. 7 5/8" CMU shown. Modify or redraw for other CMU thicknesses.
2. Use flush frame detail (Fig. 4-84) for 11 5/8" CMU (maximum frame depth should not exceed 12 3/4").
3. For exterior walls, use "Z" furring channels or other systems to accommodate rigid insulation under the GWB. Other methods of insulation may also be used.

FIGURE 4-83. Jamb and Head Details (CMU Wall).

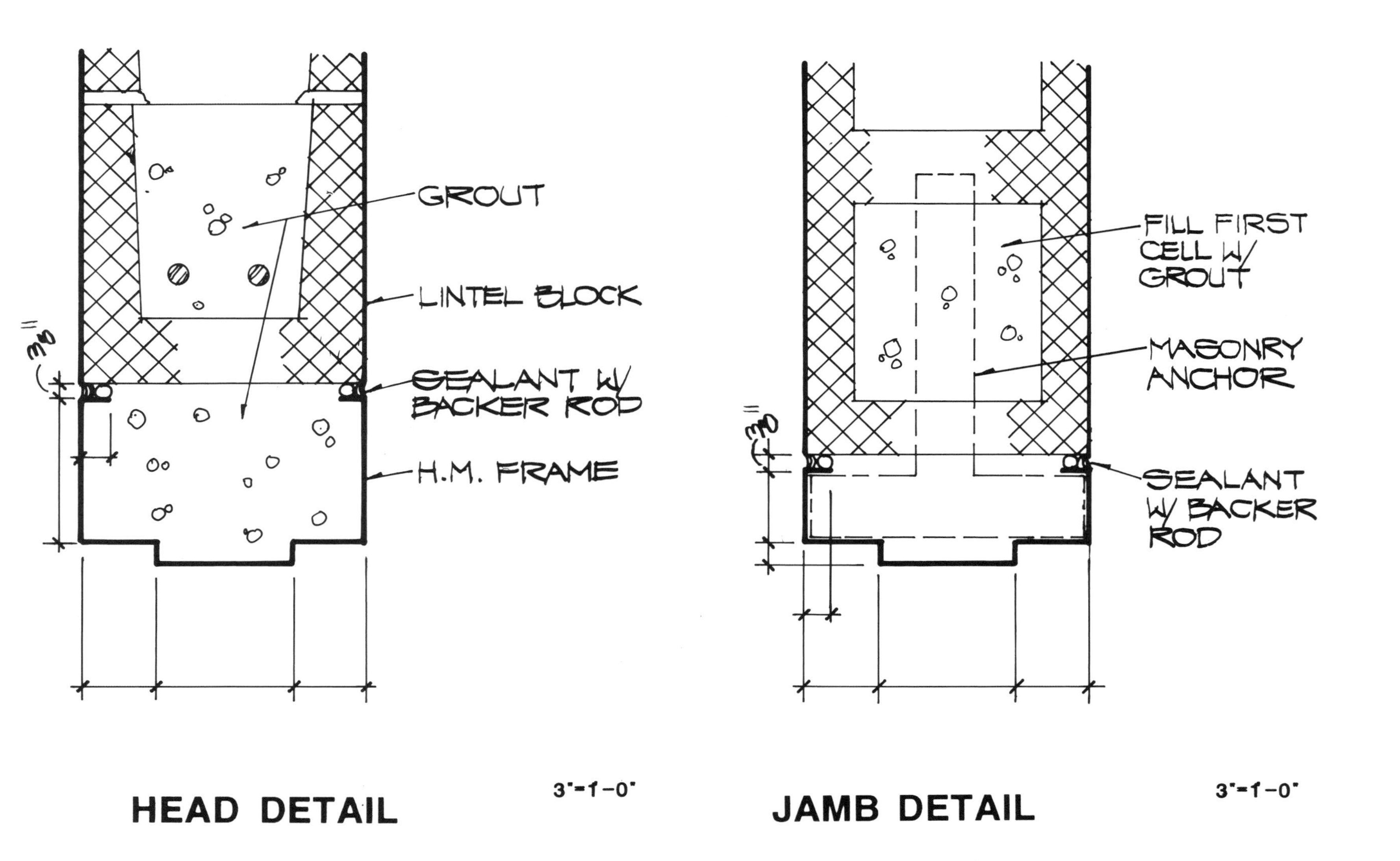

Guidelines:

1. Door head is designed for a 7'-0" nominal height door. If a 7"-2" nominal height door is used, reduce head height to 2".
2. 8" nominal wall thickness shown. If other thickness is used, write "NTS" under the soffit (middle) dimension. Fill out all dimensions.

FIGURE 4-84. Jamb and Head Details (CMU Wall).

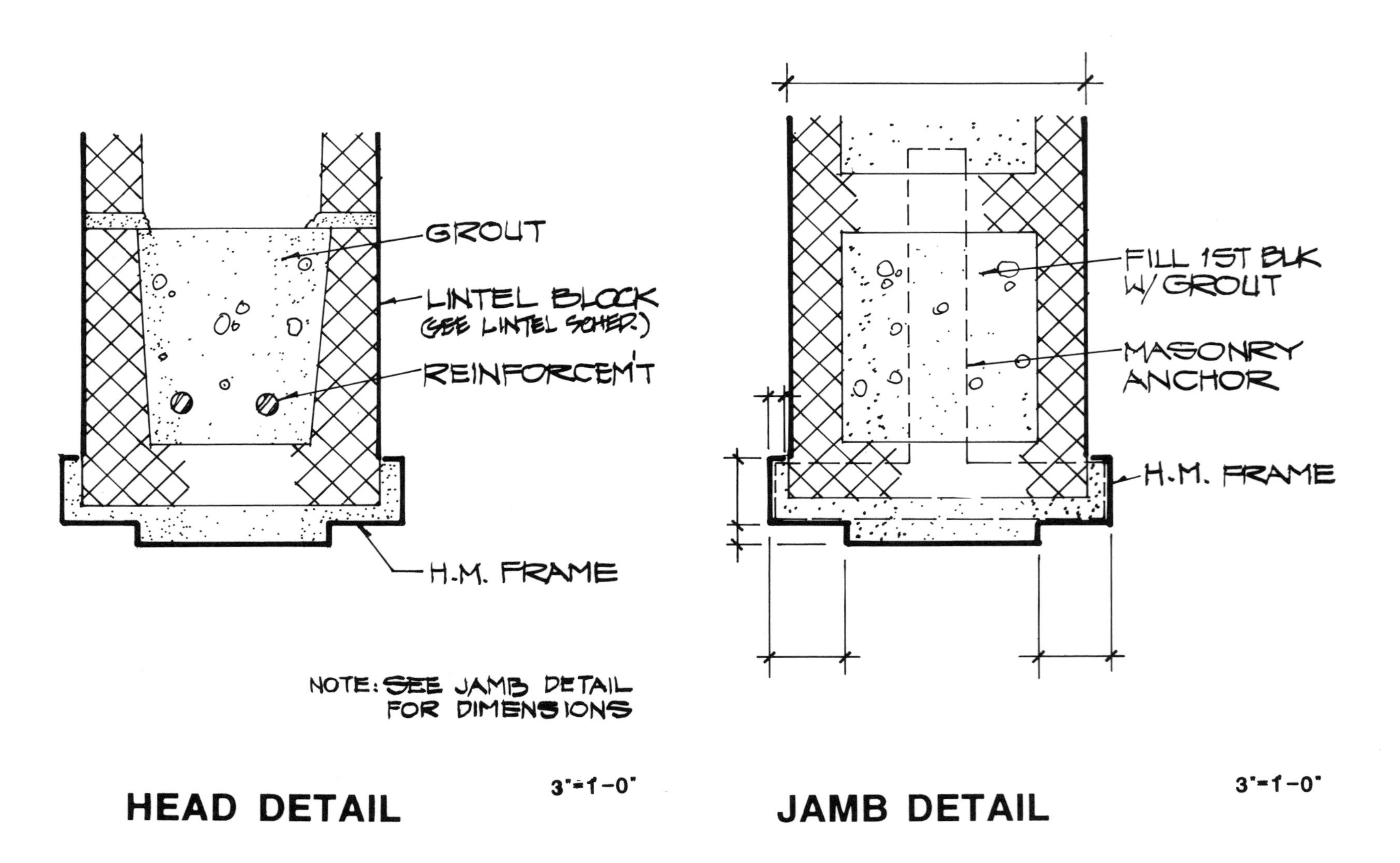

Guidelines:

1. Except for new openings in existing walls, frames are anchored to the floor slab before the wall is built. This allows the mason to place the frame anchor as the wall is being built.
2. Specifications indicate the number of anchors required.
3. See Fig. 4-84 for flush-mounted frames.

FIGURE 4-85. Jamb and Head Details (CMU).

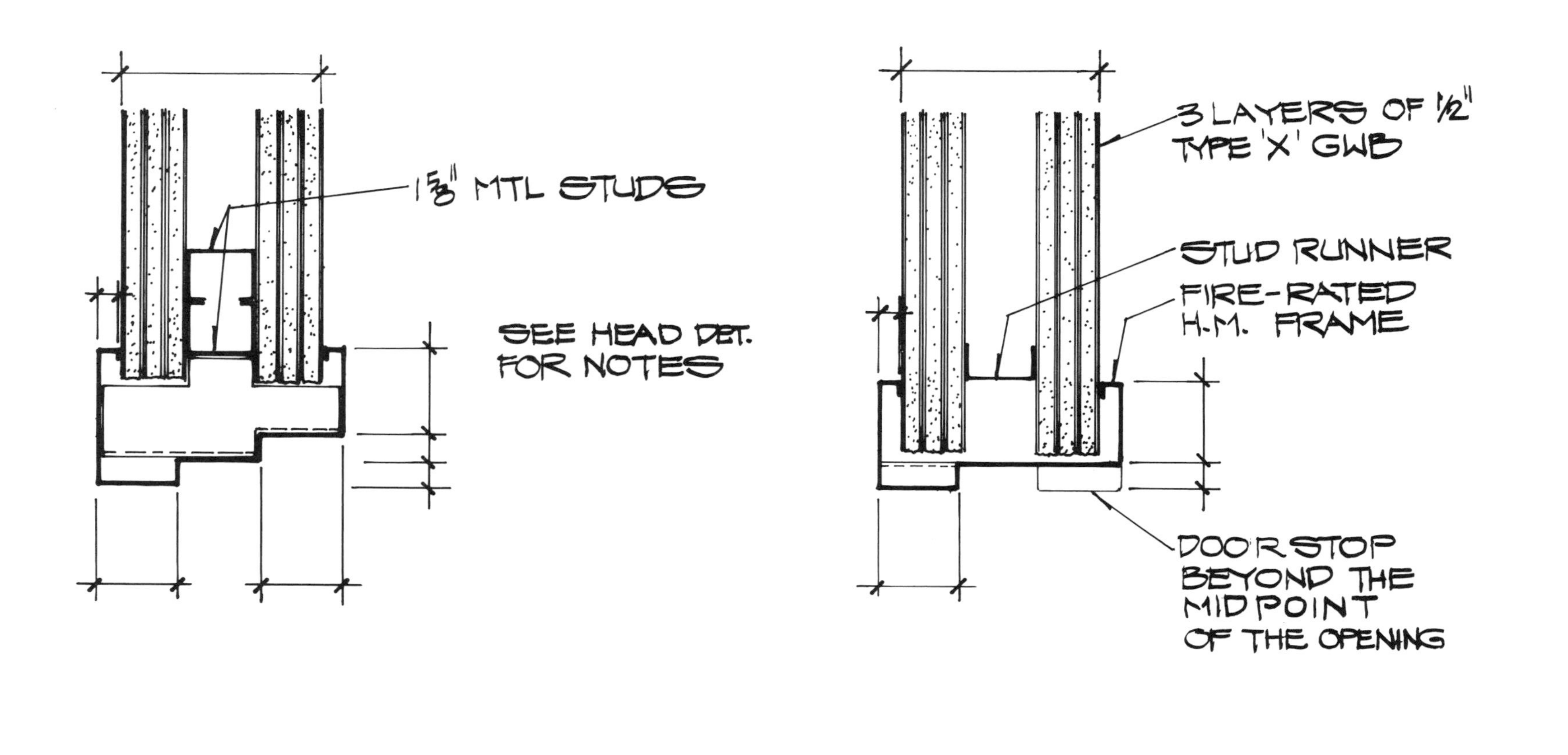

Guidelines:

1. This frame is for a 2-way fire door. The profile of the head detail reverses at the middle of the frame. The jamb profile also reverses to receive the door leaf opening in the opposite direction.
2. This frame type is used in horizontal exits to allow for easy exit from either side of the wall. The door is usually in an open position. A smoke- or fire-activated device automatically closes the door in case of fire.

FIGURE 4-86. Jamb and Head Details (Fire Separation Wall).

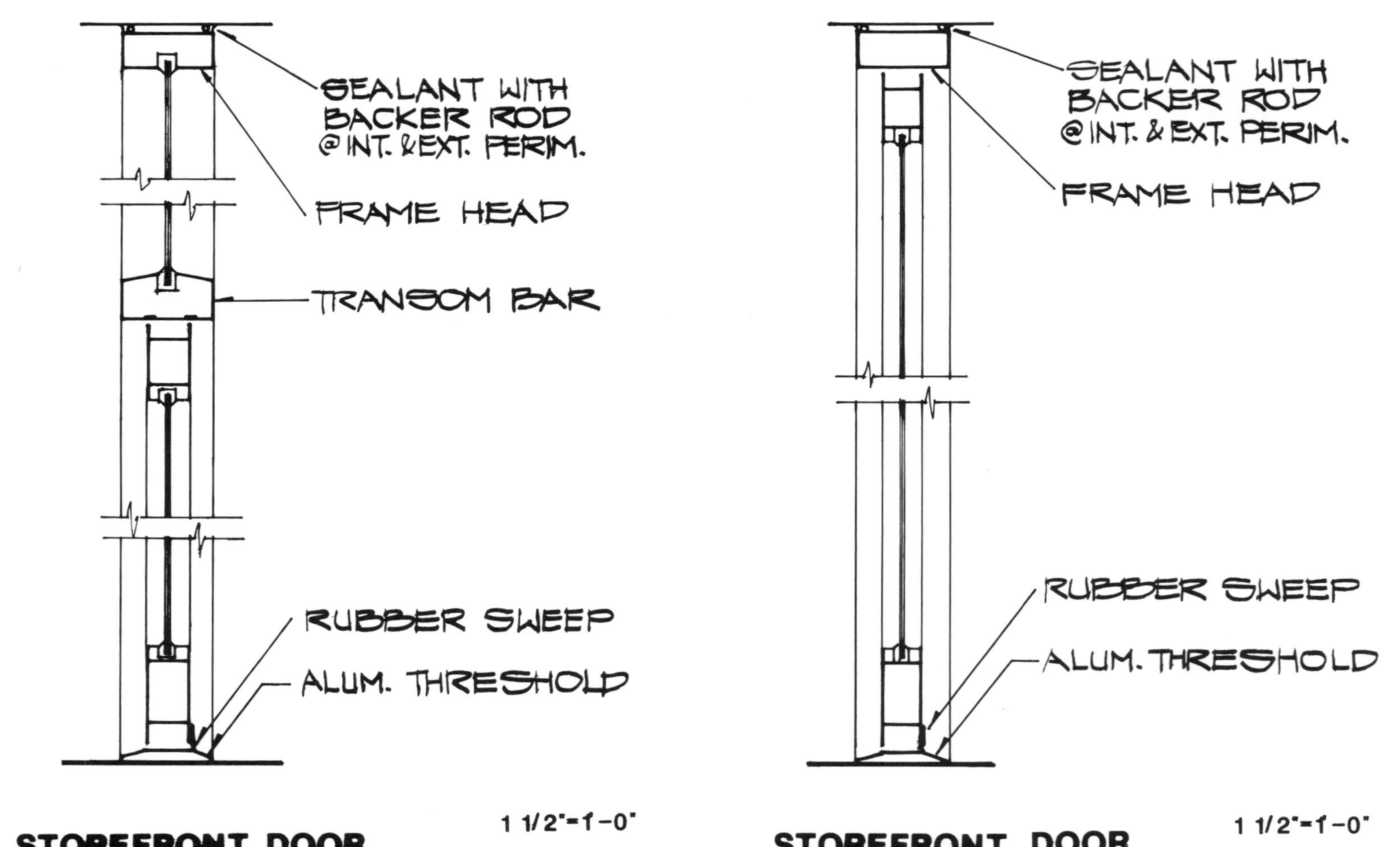

Guidelines:

1. Rubber sweep is optional but it is advisable to include it.
2. Transom bar may contain a concealed closer or closer may be surface-mounted.
3. Modify detail if double-glazing is required for energy conservation.

FIGURE 4-87. Storefront Door Details.

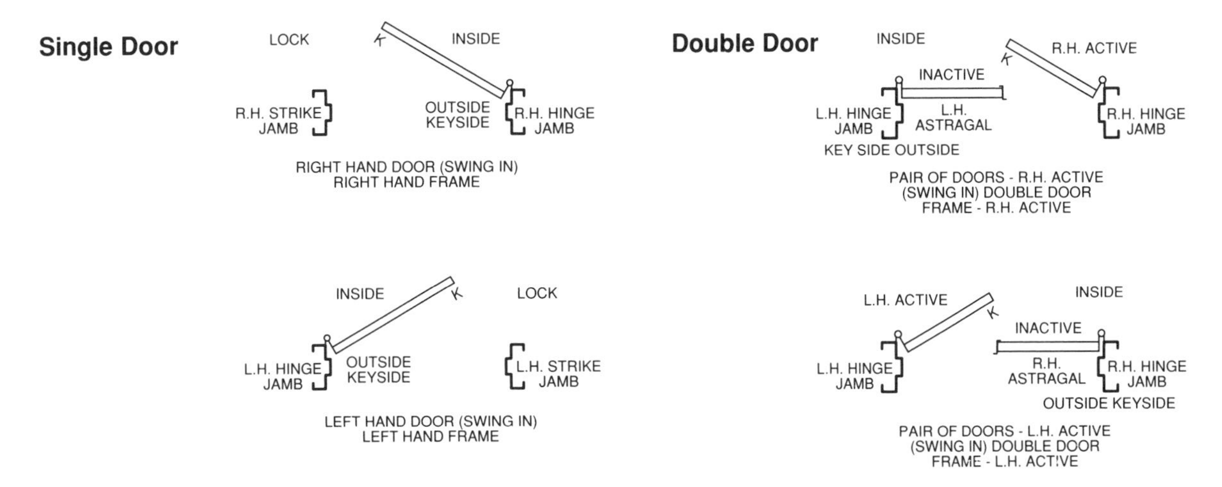

FIGURE 4-88. Door Handing (Reprinted, by permission, from Steelcraft, *Steel Doors and Frames Art Kit*, Catalog No. 505).

same applies to the threshold types shown in figures 4-74 and 4-75, and the frame types shown in figure 4-76. Typical frame types for selective door types are shown in figures 4-77 through 4-87.

4.4.4 Hardware

Hardware supplier representatives are knowledgeable individuals. They can be called upon to provide a list of hardware sets to be included in the specifications and keyed in the door schedule. This service used to be rendered free of charge. Some representatives are discontinuing this practice and asking for compensation as consultants, especially on large projects. The architect should include this fee as part of the reimbursable items during fee negotiations with the client.

Study the sets of hardware to make sure that they satisfy the client's requirements for security and check the code, especially the requirements for panic hardware and door closers.

4.4.5 Door Handing

Door handing (Fig. 4-88) is indicated by the manufacturer in shop drawings. Its function is to indicate which side of the door is the key side. Handing is also used by the fabricator to indicate which side of the door receives the screwed-on glazing trim if the door design includes a lite. The trim must be installed on the side of the door opposite the key side to prevent would-be intruders from gaining access by unscrewing the trim and removing the glass.

The architect must check the handing to ensure that the manufacturer's understanding coincides with the design intent for the security of the building.

References

Elmiger, A. 1976. *Architectural & Engineering Concrete Masonry Details for Building Construction*. Herndon, VA: The National Concrete Masonry Association.

National Association of Architectural Metal Manufacturers. 1992. *Metal Stairs Manual*. 5th ed. Chicago, IL: National Association of Architectural Metal Manufacturers.

National Association of Architectural Metal Manufacturers. 1985. *Pipe Railing Manual*, 2nd ed. Chicago, IL: National Association of Architectural Metal Manufacturers.

National Fire Protection Association. 1990. *Fire Doors and Windows*. NFPA 80-1990. Quincy, MA: National Fire Protection Association.

Steelcraft. 1990. *Steel Doors & Frames Art Kit*. Catalog No. 505. Cincinnati, OH: Steelcraft.

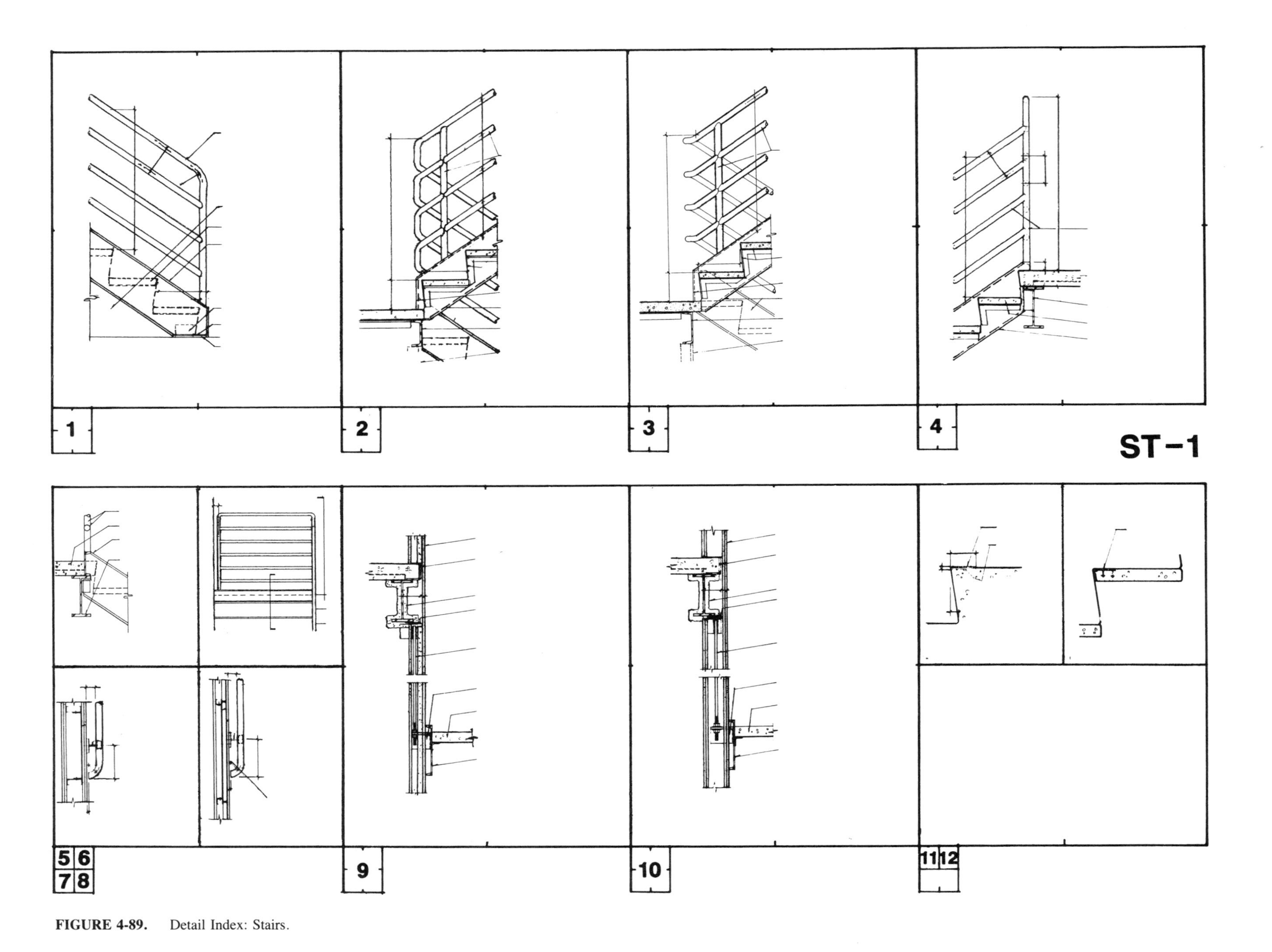

FIGURE 4-89. Detail Index: Stairs.

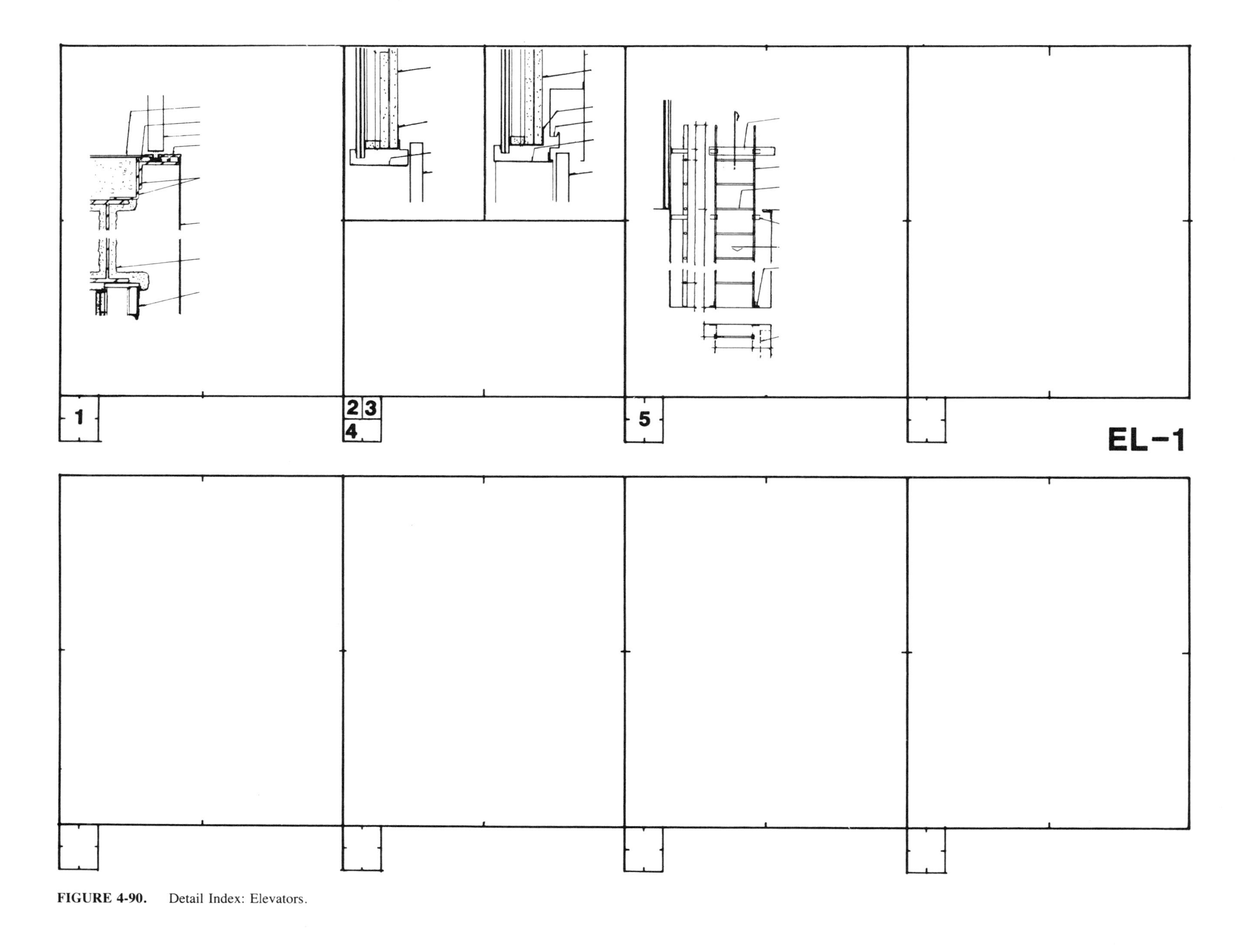

FIGURE 4-90. Detail Index: Elevators.

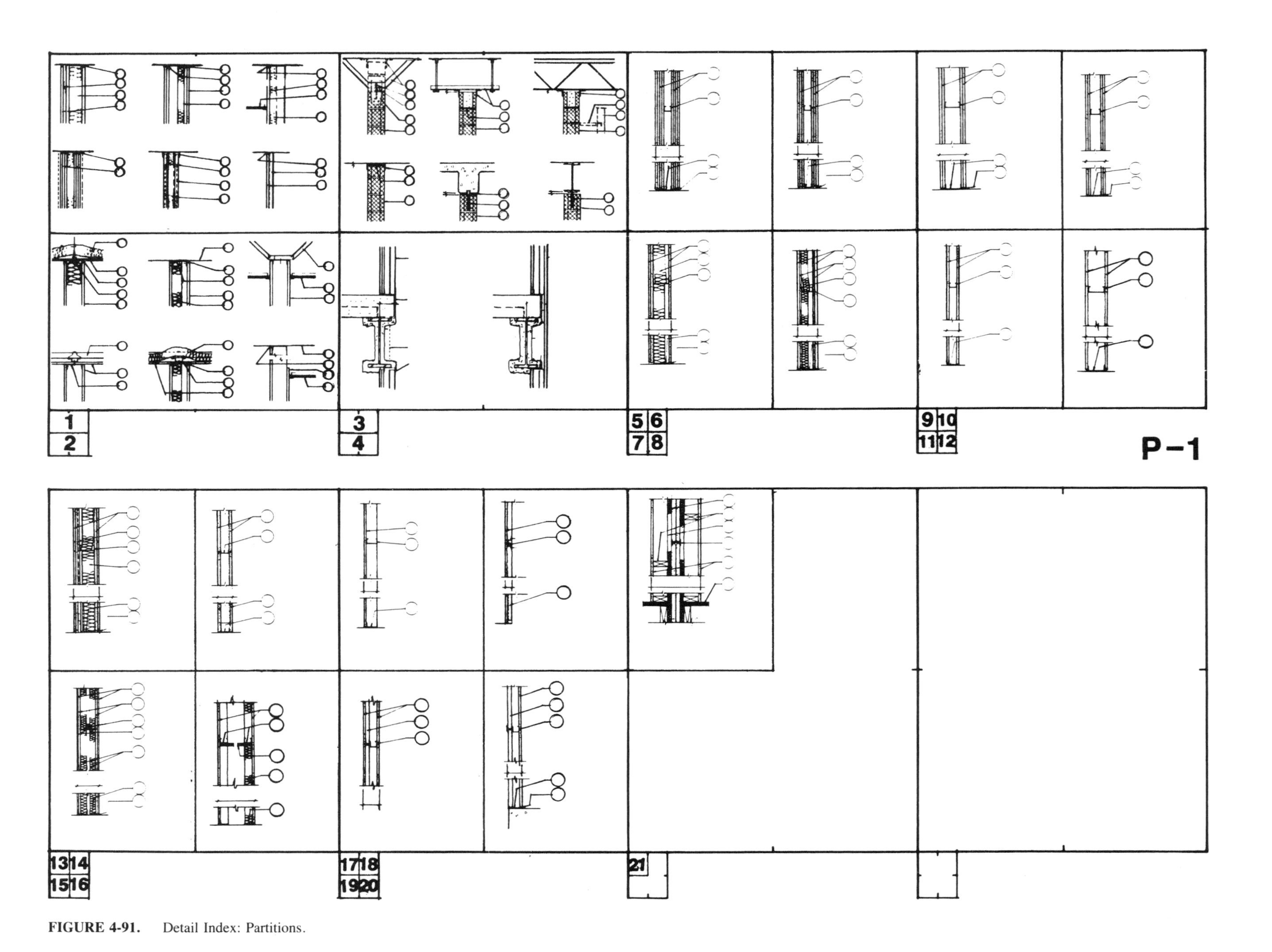

FIGURE 4-91. Detail Index: Partitions.

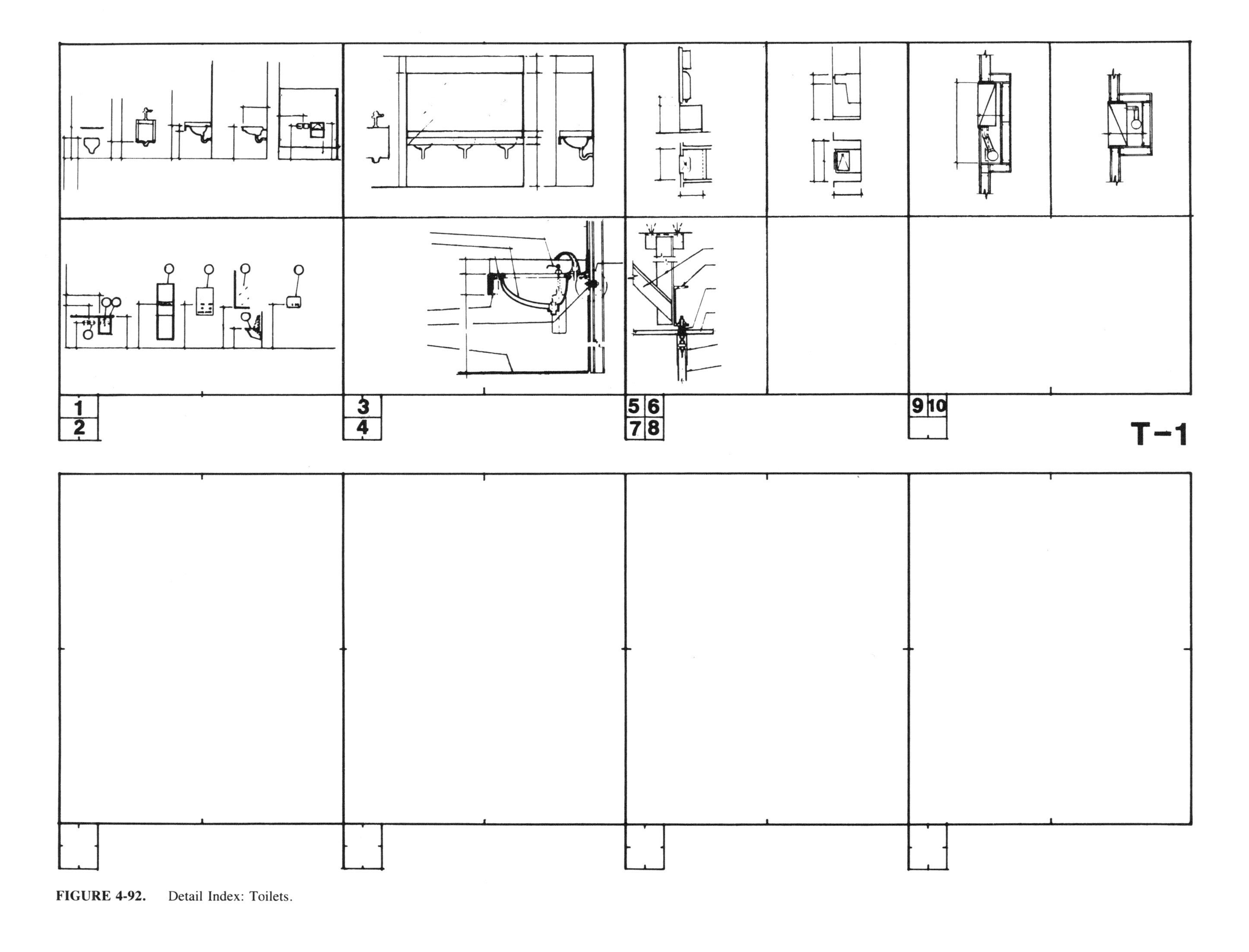

FIGURE 4-92. Detail Index: Toilets.

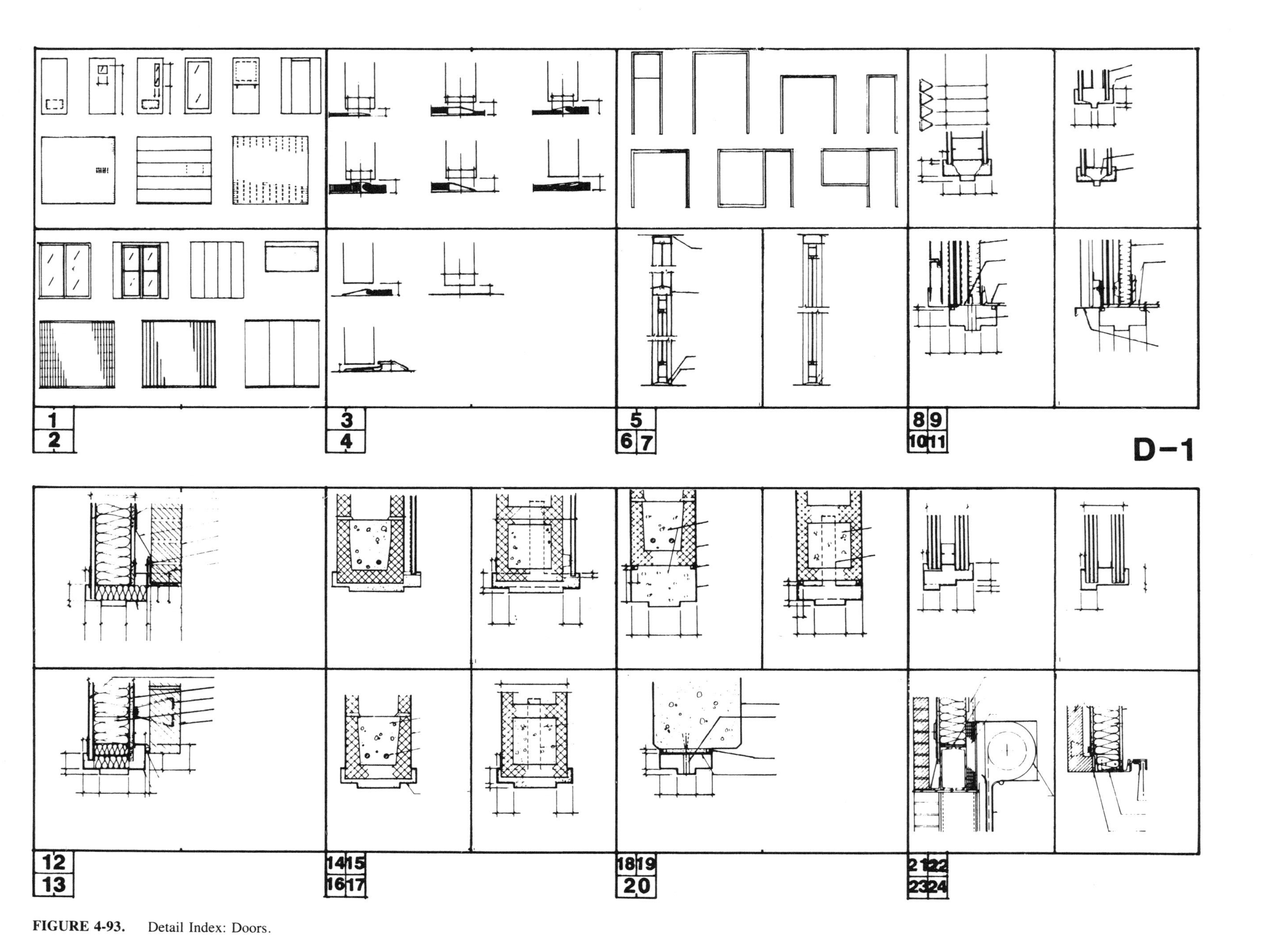

FIGURE 4-93. Detail Index: Doors.

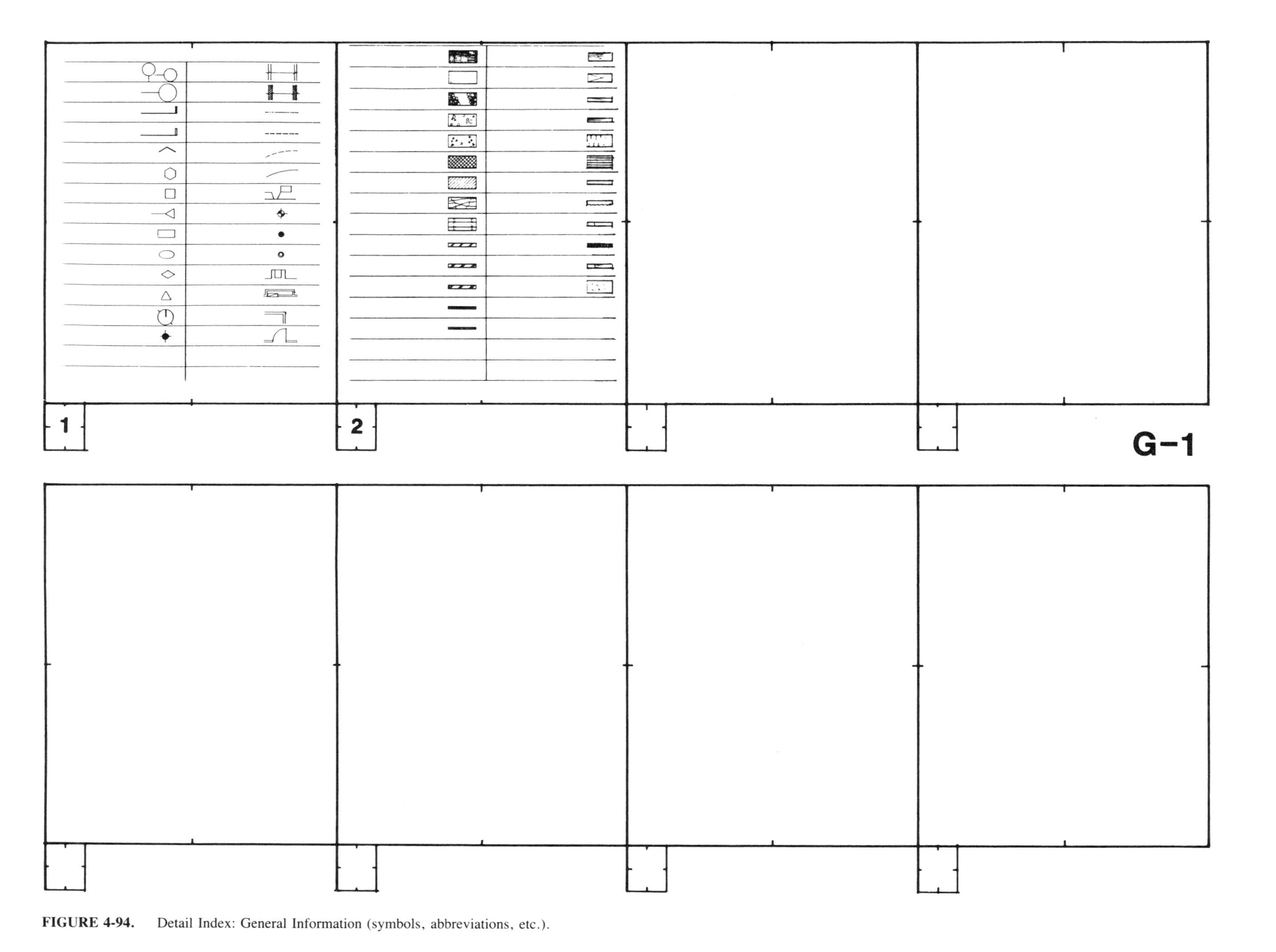

FIGURE 4-94. Detail Index: General Information (symbols, abbreviations, etc.).

DETAIL INDEX KEY

D	**DOOR DETAILS**
D1.1	DOOR TYPES
D1.2	DOOR TYPES
D1.3	THRESHOLD TYPES
D1.4	THRESHOLD TYPES
D1.5	FRAME TYPES
D1.6	STOREFRONT
D1.7	STOREFRONT
D1.8	FRAME TYPES (STANDARD GWB)
D1.9	FRAME TYPES (SHAFT WALL)
D1.10	FRAME JAMB (METAL WALL)
D1.11	FRAME HEAD (METAL WALL)
D1.12	FRAME HEAD (BRICK VENEER WALL)
D1.13	FRAME JAMB (BRICK VENEER WALL)
D1.14	FRAME HEAD (CMU WALL W/ FURRING)
D1.15	FRAME JAMB (CMU WALL W/ FURRING)
D1.16	FRAME HEAD (CMU WALL)
D1.17	FRAME JAMB (CMU WALL)
D1.18	FRAME HEAD (CMU WALL, FLUSH)
D1.19	FRAME JAMB (CMU WALL, FLUSH)
D1.20	FRAME JAMB/HEAD (CONCRETE WALL)
D1.21	FRAME HEAD (FIRE DOOR)
D1.22	FRAME JAMB (FIRE DOOR)
D1.23	FRAME HEAD (ROLL-UP DOOR IN BRICK VENEER WALL
D1.24	FRAME JAMB (ROLL-UP DOOR IN BRICK VENEER WALL
M	**MISCELLANEOUS**
M1.1	GRAPHIC SYMBOLS
M1.2	MATERIALS
ST	**STAIR DETAILS**
ST1.1	DETAIL AT BOTTOM OF STAIR
ST1.2	DETAIL AT INTERMEDIATE LANDING
ST1.3	DETAIL AT INTERMEDIATE LANDING
ST1.4	DETAIL AT TOP LANDING
ST1.5	PARTIAL DETAIL AT TOP LANDING
ST1.6	GUARDRAIL DETAIL
ST1.7	WALL BRACKET
ST1.8	WALL BRACKET
ST1.9	ROD HANGER (IN SHAFT WALL)
ST1.10	FOR HANGER (IN 2-HR. GWB PARTITION)
ST1.11	STAIR NOSING (CONC.)
ST1.12	STAIR NOSING (STEEL)
EL	**ELEVATOR**
EL1.1	SILL DETAIL (STEEL STRUCTURE)
EL1.2	DOOR JAMB DETAIL
EL1.3	DOOR HEAD DETAIL
EL1.4	
EL1.5	PIT LADDER DETAIL

P	**PARTITIONS**
P1.1	TOP OF PARTITION DETAILS
P1.2	TOP OF PARTITION DETAILS
P1.3	TOP OF PARTITION DETAILS
P1.4	TOP OF PARTITION DETAILS AT SHAFTS
P1.5	TOP OF PARTITION DETAILS AT SHAFTS
P1.6	4-HR. GWB PARTITION
P1.7	3-HR. GWB PARTITION
P1.8	ACOUSTIC PARTITION
P1.9	ACOUSTIC PARTITION
P1.10	2-HR. GWB PARTITION
P1.11	2-HR. GWB PARTITION
P1.12	1-HR. GWB PARTITION
P1.13	1-HR. GWB PARTITION
P1.14	ACOUSTIC PARTITION
P1.15	DE-MOUNTABLE PARTITION
P1.16	CHASE WALL
P1.17	CHASE WALL
P1.18	FURRING PARTITION
P1.19	FURRING OVER MASONRY
P1.20	2-HR. SHAFT WALL
P1.21	1-HR. SHAFT WALL
P1.22	PARTY WALL
T	**TOILET DETAILS** (Includes FHC details)
T1.1	FIXTURE MOUNTING HEIGHTS
T1.2	ACCESSORY MOUNTING HEIGHT
T1.3	LAVATORY ELEVATION (GROUP OF 3)
T1.4	LAVATORY DETAIL (MARBLE COUNTER)
T1.5	ELECTRIC WATER COOLER DETAIL
T1.6	ELECTRIC WATER COOLER DETAIL
T1.7	CEILING-HUNG PARTITION
T1.8	
T1.9	FIRE HOSE CABINET
T1.10	FIRE HOSE CABINET

FIGURE 4-95. Detail Index Key.

APPENDIX
BLANK FORMS

TARGET ACTIVITY

TASKS

A B C D VACATION

NOTE: Refer to Fig. 1-8 for more information

A-1. Team Task Assignment Schedule

Product Information Report

Date:
Product Name:
Company Name:
Represented By:

1. Number of years on the market:

2. Cost per unit (sq. ft., unit, etc.)
   Installed:
   Material Only:
   Labor Only:

3. Product characteristics (including limitations):

4. Address of projects that used the product:

5. Availability and delivery lead time:

6. Comparable products to be specified "as equal":

7. Contact for technical assistance:

   Name: Phone:

8. Local architects who used the product:

   Name: Phone:

   Name: Phone:

9. Standard and extended warranty information:

10. Other information:

A-2. Product Information Report Form

PROJECT SHEET STATUS

Project No.:
Phase:
Prepared By:
Date:

Sht. of

Sheet No.	SHEET TITLE	Drawn By	Hrs. to Date	Hrs. this Wk.	Budget Hrs	Percentage of Completion										Progress Hrs.	REMARKS
						10%	20%	30%	40%	50%	60%	70%	80%	90%	100%		
	Subtotal														Subtotal		
	Total Hrs.														Total Progress		____ X ____ = %

NOTE: Refer to Fig. 1-20 for explanations

Standard Details
Completed Parts

A-3. Project Sheet Status Report

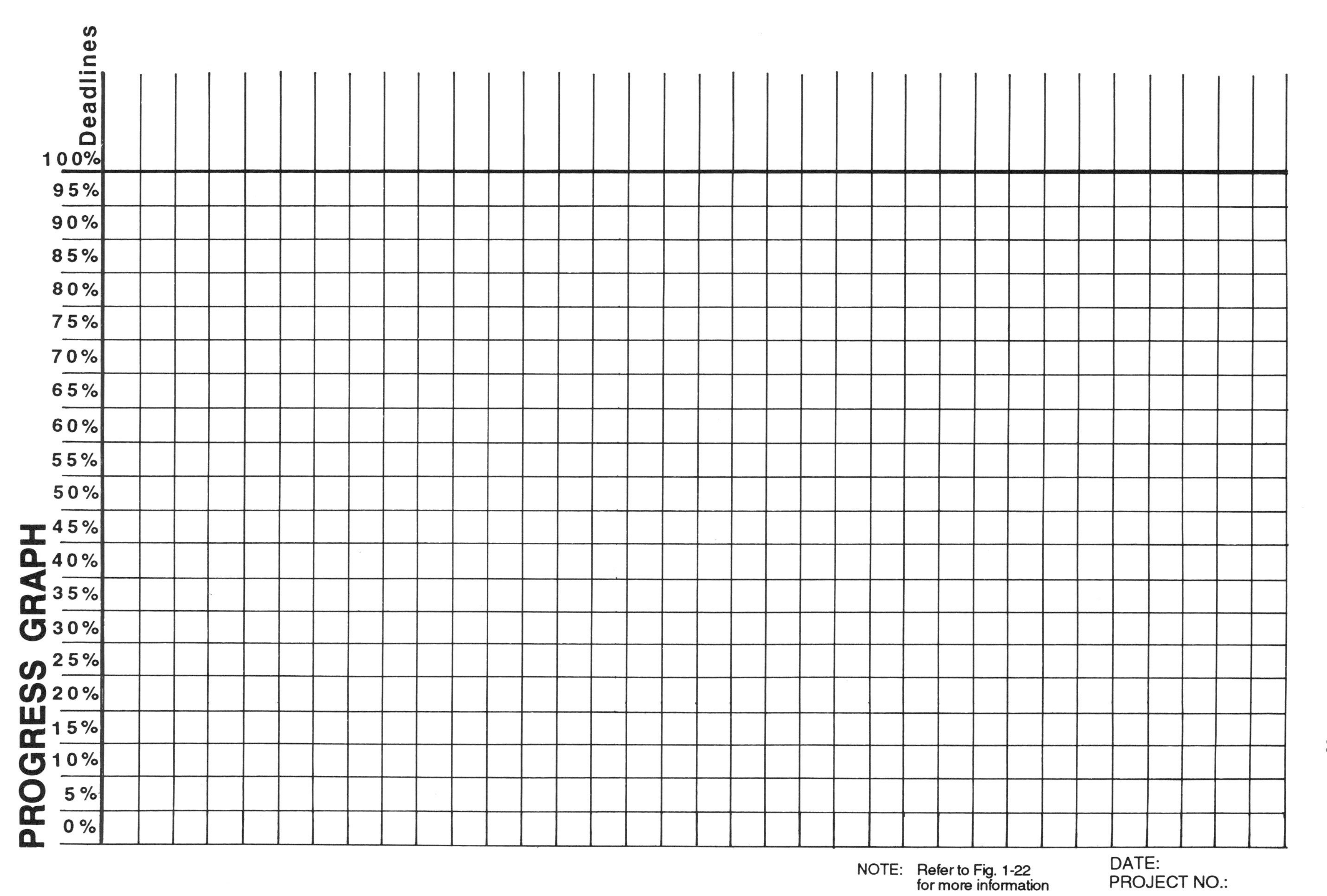

A-4. Progress Graph

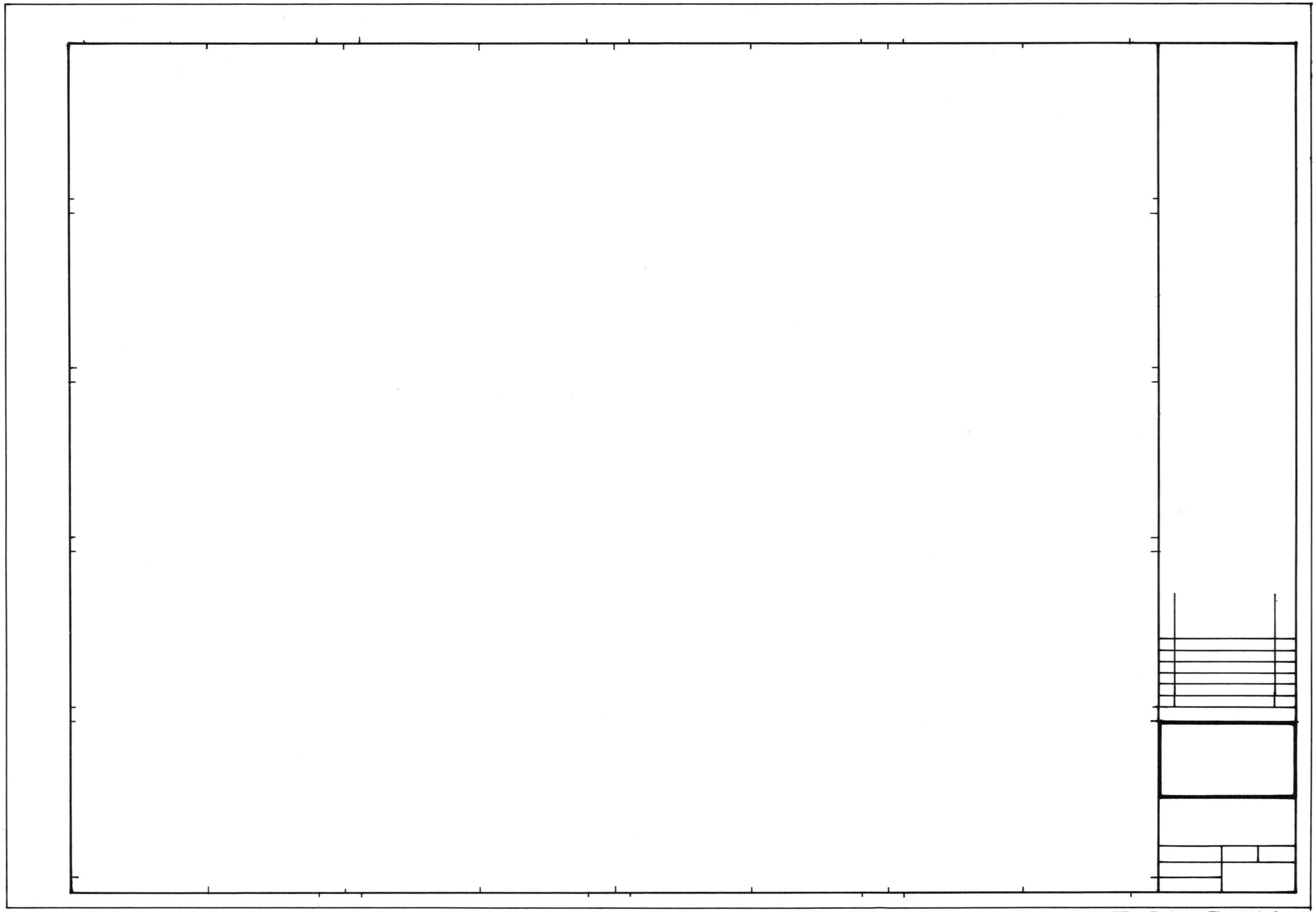

NOTE: Refer to Fig. 2-1b & 2-3 for more information.

A-5. Project Mock-Up Sheet

ROOM FINISH SCHEDULE

ROOM NUMBER	ROOM NAME	FLOOR	BASE	WALLS				CEILING	CEILING HT.	REMARKS
				NORTH	EAST	SOUTH	WEST			

NOTE: Refer to Fig. 2-16 for more information

A-6. Room Finish Schedule

WINDOW SCHEDULE

WINDOW TYPE	ROUGH OPENING (W X H)	OPERATION	GLASS TYPE	DETAILS					REMARKS
				HEAD	JAMB	SILL	RAIL	MULLION	

NOTE: Refer to Fig. 2-18 for more information

A-7. Window Schedule

DOOR SCHEDULE

DOORS						FRAMES				HDW	RATING	REMARKS
NO.	H	W	T	TYPE	MAT'L	HEAD	JAMB	THR'D	TYPE	SET NO.		

NOTE: Refer to Fig. 2-17 for more information

A-8. Door Schedule

COLUMN FIREPROOFING SCHED.

COLUMN				UL DESIGN			
NUMBER	SIZE	W/D RATIO	FLOOR NO.	NUMBER	FIRE RATING	W/D RATIO	DETAIL NO.

NOTE: Refer to Fig. 2-19 for more information

A-9. Column Fireproofing Schedule

PARTITION TYPE SCHEDULE

USED ■ / NOT USED □	TYPE	WALL THICKNESS	STUDS		FURRING			BOARD TYPE (THICKNESS)				CMU THICKNESS	INSULATION				T.O. WALL DET.	FIRE RATING	FIRE TEST	STC RATING	REMARKS
			SIZE	SPACING	HAT	ZEE (DEPTH)	RESILIENT	GWB	GWB, TYPE"X"	GWB, M R	CEMENT BD.		MINERAL FIBER	FIBERGLASS	RIGID	OTHER**					

NOTE: Refer to Fig. 2-20 for more information

A-10. Partition Type Schedule

FIRE RATING:
FIRE TEST:
STC RATING:

TYPE:

FIRE RATING:
FIRE TEST:
STC RATING:

TYPE:

A-11. Partition Type (Draw additional partition types on this blank format, fill out the missing information, and sticky-back on the Project Detail sheet.)

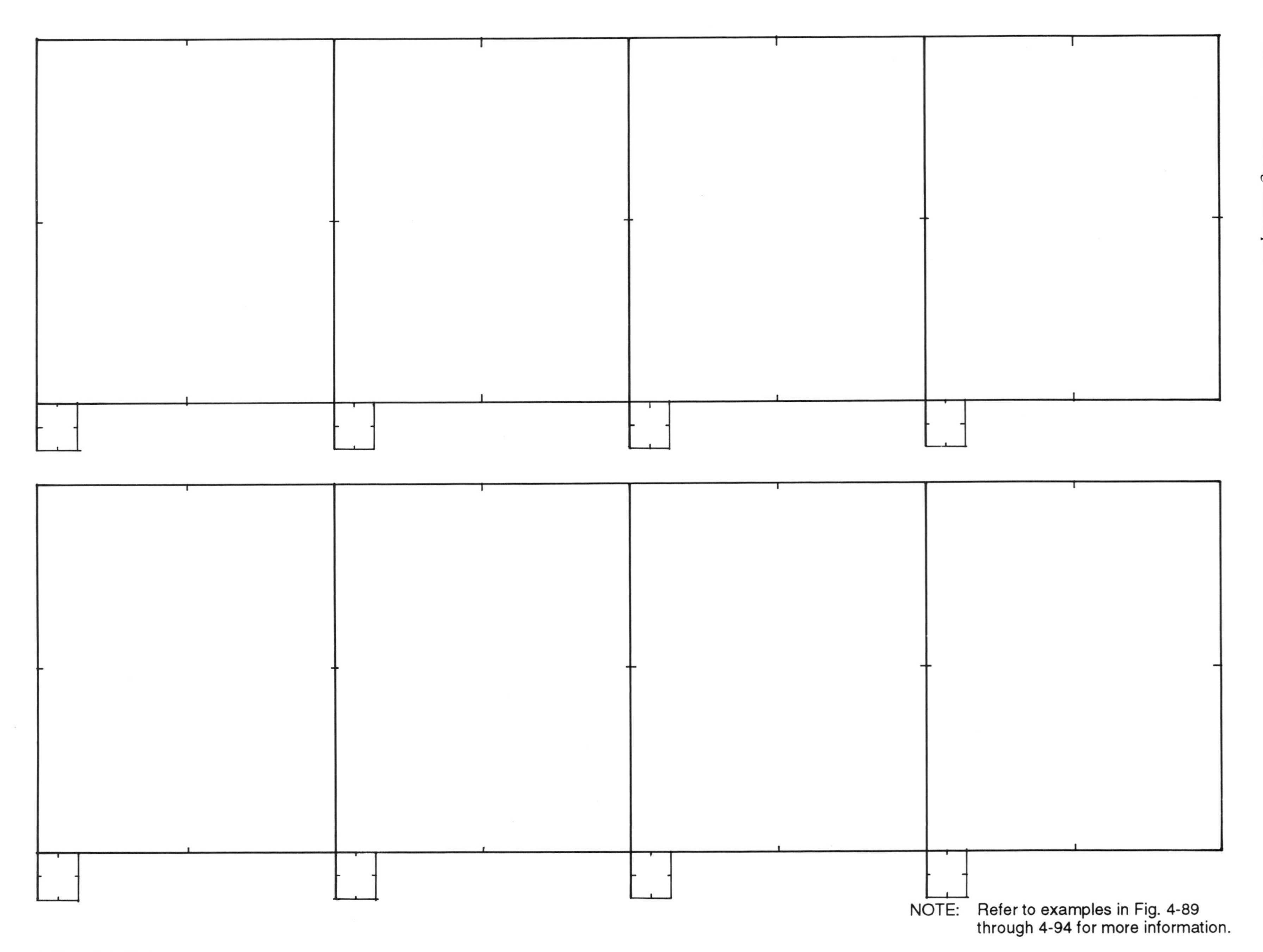

A-12. Detail Index

LABORATORY EQUIPMENT WORK SHEET

PROJECT NAME:
DEPARTMENT:

EQUIPMENT							POWER				UTILITIES							REMARKS
ITEM NO.	DESCRIPTION	N OR E *	CFCI *	OFCI *	OFOI *	CATALOG #	VOLTS	PHASE	AMPS	KILOWATTS	C. WATER	H. WATER	DIST. WATER	CH. WATER	COMP. AIR	N2	DRAIN	

INFO. BY:
DATE:

RM. NAME:
RM. NO.:

REV NO.:
DATE:

* ABBREVIATIONS
N: NEW
E: EXISTING
C: CONTRACTOR
F: FURNISHED
I: INSTALLED
O: OWNER

A-13. Laboratory Equipment Work Sheet (Adapted from a form used at CTJ&D Architects, Houston, TX)

LABORATORY CASEWORK WORK SHEET

PROJECT NAME:
DEPARTMENT:

CASEWORK				POWER					UTILITIES						SINK	
ITEM NO.	ITEM DESIGNATION	MANUFAC.	CATALOG #	OUTLETS	VOLTS	PHASE	AMPS	KILOWATTS	C. WATER	H. WATER	N2	DIST. WATER	DRAIN	COMP. AIR	CATALOG #	COUNTER TOP MATERIAL

INFO. BY:
DATE :

RM. NAME:
RM. NO.:

REV. NO.:
DATE:

A-14. Laboratory Casework Work Sheet (Adapted from a form used at CTJ&D Architects, Houston, TX)

FURNITURE WORK SHEET

RM. NAME: INFO. PROVIDED BY: REV. NO.:
RM. NO.: DATE: DATE:

ITEM NO.	MOVABLE FURNITURE & FURNISHINGS	QTY.	SIZE	REMARKS

SPECIAL FEATURES WORK SHEET

RM. NAME: INFO. PROVIDED BY: REV. NO.:
RM. NO.: DATE: DATE:

ITEM NO.	SPECIAL FEATURES & FINISHES	QTY.	SIZE	REMARKS
	CHALK BOARD			
	TACK BOARD			
	MANUAL SCREEN			
	MOTORIZED SCREEN			
	REAR PROJECTION SCREEN			
	LIGHTPROOF WINDOW			
	LIGHTPROOF DOOR			
	SAFETY SHOWER & EYE WASH			
	FLOOR TRENCH			
	SPECIAL WALL, FLR. & CLG. FINISHES			

A-15. Furniture Work Sheet Special Features Work Sheet

POWER REQUIREMENTS WORK SHEET

PROJECT NAME:　　INFO. PROVIDED BY:　　REV. NO.:
PROJECT NO.:　　DATE:　　DATE:

NO.	ROOM #	DESCRIPTION	LOAD*	VOLTAGE & PHASE	MANUF. BY	CATALOG NO.†

† Cut sheet or data from manufacturer
* AMPS, WATTS, KW, KVA, HP, etc.

Sheet ___ of ___

A-16. Power Requirements Work Sheet

GRAPHIC SYMBOLS

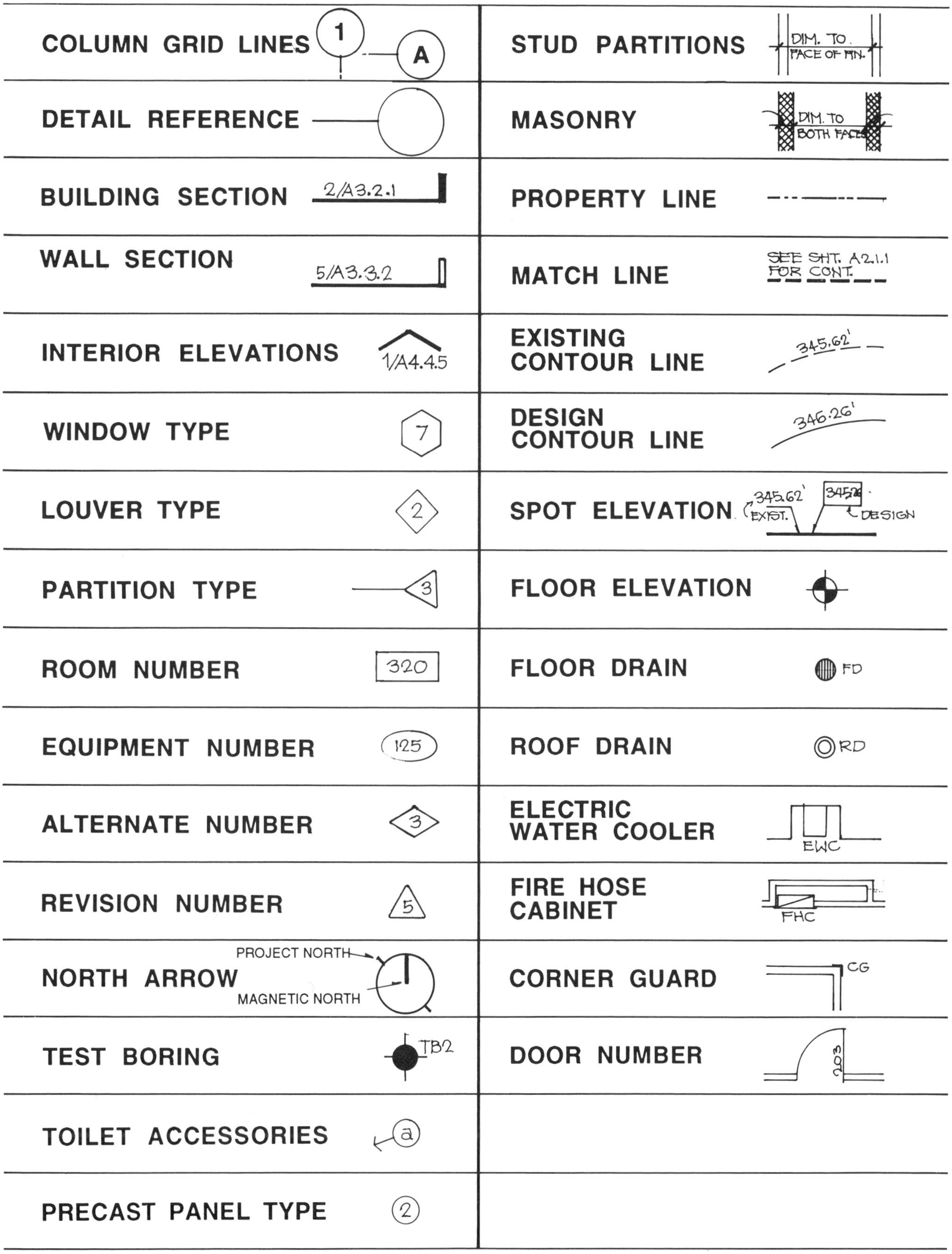

Symbol	Example	Symbol	Example
COLUMN GRID LINES	1 A	STUD PARTITIONS	DIM. TO FACE OF FIN.
DETAIL REFERENCE		MASONRY	DIM. TO BOTH FACES
BUILDING SECTION	2/A3.2.1	PROPERTY LINE	
WALL SECTION	5/A3.3.2	MATCH LINE	SEE SHT. A2.1.1 FOR CONT.
INTERIOR ELEVATIONS	1/A4.4.5	EXISTING CONTOUR LINE	345.62'
WINDOW TYPE	7	DESIGN CONTOUR LINE	346.26'
LOUVER TYPE	2	SPOT ELEVATION	345.62' EXIST. 345.26 DESIGN
PARTITION TYPE	3	FLOOR ELEVATION	
ROOM NUMBER	320	FLOOR DRAIN	FD
EQUIPMENT NUMBER	125	ROOF DRAIN	RD
ALTERNATE NUMBER	3	ELECTRIC WATER COOLER	EWC
REVISION NUMBER	5	FIRE HOSE CABINET	FHC
NORTH ARROW	PROJECT NORTH / MAGNETIC NORTH	CORNER GUARD	CG
TEST BORING	TB2	DOOR NUMBER	203
TOILET ACCESSORIES	a		
PRECAST PANEL TYPE	2		

A-17. Graphic Symbols

MATERIAL SYMBOLS

Material	Material
EARTH	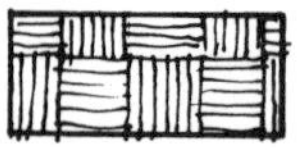 WOOD - FINISH 
SAND OR LIMESTONE	 WOOD - ROUGH 
GRAVEL	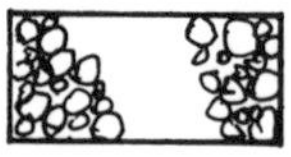 SHIMS 
CONCRETE OR PRECAST	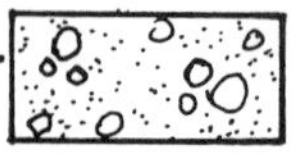 PLYWOOD 
LIGHTWEIGHT CONCRETE	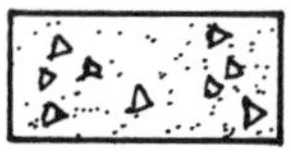 BATT INSULATION 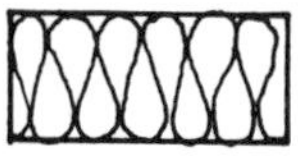
CMU	 RIGID INSULATION 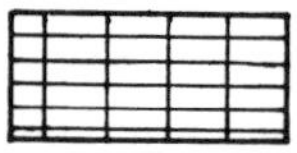
BRICK	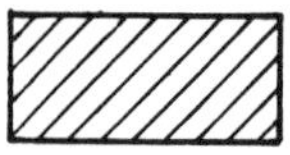 GYPSUM WALL BOARD 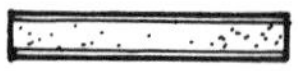
MARBLE OR GRANITE	 METAL LATH AND PLASTER 
GLASS BLOCK	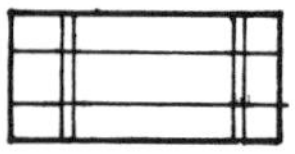 ACOUSTICAL TILE 
STEEL	 CARPET 
ALUMINUM	 CERAMIC TILE 
BRASS OR BRONZE	 TERRAZZO 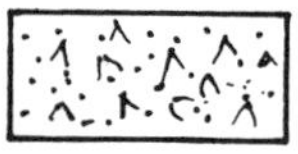
GLASS	
WATERPROOFING	

A-18. Materials Symbols

Index

*Note: page numbers in *italics* refer to illustrations